WIRELESS NETWORKS AND MOBILE COMPUTING

WIRELESS NETWORKS AND MOBILE COMPUTING

Koushik Sinha
Hewlett-Packard Labs, Bangalore, India

Sasthi C. Ghosh
Indian Statistical Institute, Kolkata, India

Bhabani P. Sinha
Indian Statistical Institute, Kolkata, India

CRC Press
Taylor & Francis Group
Boca Raton London New York

CRC Press is an imprint of the
Taylor & Francis Group, an **informa** business
A CHAPMAN & HALL BOOK

CRC Press
Taylor & Francis Group
6000 Broken Sound Parkway NW, Suite 300
Boca Raton, FL 33487-2742

Printed on acid-free paper
Version Date: 20150929

International Standard Book Number-13: 978-1-4822-2793-2 (Hardback)

Visit the Taylor & Francis Web site at
http://www.taylorandfrancis.com

and the CRC Press Web site at
http://www.crcpress.com

Contents

List of Figures

List of Tables

Preface

Wireless communication constitutes one of the fastest growing industry segments in recent years, encompassing a number of application domains such as cellular mobile networks, ad hoc networks, cognitive radio networks, ubiquitous and pervasive computing, sensor networks and so on. All these applications are becoming increasingly popular these days due to the multifaceted benefits they provide to society ranging from personal communication services, growth of business and economy to healthcare, distant education services, disaster management, security services and defense applications. For sustained growth in these fields, we need trained personnel for further development and research in the areas of wireless networking and mobile computing. In a natural way, such a need has triggered the inclusion of formal courses on these topics in the undergraduate- and graduate-level engineering degree programs of many leading institutes and universities all over the world.

Also, there exist various issues and challenging research problems in all these areas which need urgent attention of scientists and engineers. In the area of cellular mobile networks, the major research areas include bandwidth management, mobility management, network security and energy-efficient communication, amongst others. Over the last decade, the number of such cellular mobile users has been increasing almost exponentially. Further, with the advances in technology, there has been an increasing demand from mobile users for providing multimedia services like voice, text, still-image and video through wireless networks. On the other hand, the available bandwidth required for such multimedia data communication for a large number of mobile users is very limited. Such limited availability of the radio spectrum, signal distortion and the interference caused by the environment and other mobile users imposes an inherent bound on the capacity of such wireless networks. Developing methods to utilize the scarce radio spectrum efficiently is more critical than ever. Along with this, secure communication through wireless and energy-efficient communication through the cellular mobile environment constitute other important problems which need immediate attention of scientists for an effective solution. In the same way, in the area of ad hoc networks, the challenging research problems include network security as well as energy-efficient communication requirements.

Similar to these cellular and ad hoc mobile networks, sensor networks also play a vital role in many real-life applications ranging from agriculture, health care, habitat monitoring, environmental monitoring, defense applications, and so on. Due to the limited battery life as well as limited computing and communication abilities of the constituent sensor nodes, efficient design and management of these sensor networks, handling user queries in various application domains over streaming sensor data and also network security appear to be challenging problems to researchers. Summarily, there are various research problems in all these application domains of wireless communication, solutions of which are urgently demanded for the benefit of society.

The purpose for writing this book is two-fold. The primary objective is to introduce the basic concepts of wireless networks and mobile computing which will be enough for an engineering student at the undergraduate or graduate level to build a solid background in this area to enter into a professional life or to take up a research career. We have presented the materials in various chapters in a cohesive manner so that there will be a natural flow of concepts from one chapter to the next, which will be easy for a student to follow. This is in contrast to books whose chapters are written by different sets of authors, each with a different style and notations which often become difficult to follow during course work, as experienced by the authors while teaching this course to the students using available books in the market.

The second but no less important objective is to include recent research results along with the challenging problems in the areas mentioned above with extensive references to the existing literature at the end of each chapter. Recent results on various topics including those by the authors from their research endeavors over one and a half decades, are presented for comprehensive knowledge of the topics. This will certainly help researchers in these areas to get state-of-the-art pictures in the relevant fields and motivate them to innovate novel ideas for solutions to the unsolved challenging research problems for future generation wireless networks and mobile computing systems.

Taking into account the previously mentioned ideas this book has been surprisingly written containing only nine chapters. Chapter 1 gives an overall introduction to wireless networks and mobile computing. Chapter 2 deals with mobility management issues, discussing location management, handoff management and mobility models. Chapter 3 is devoted to bandwidth management in cellular mobile networks, starting from the basic concepts to various existing techniques, critical comparisons of their performances and future research challenges in the context of next-generation mobile communication systems. Chapter 4 discusses the localization issue in detail. In Chapter 5, we discuss the various existing techniques and protocols for message communication in ad hoc networks and future directions of research in this area. In Chapter 6, we introduce wireless local area networks (WLANs) along with recent research results and future directions of research in this area. Chapter 7 is devoted to the placement of base stations in a cellular network as well as deployment of sensor nodes in a wireless sensor network where we first model the problem from a geometrical point of view and then discuss existing techniques for an efficient solution to this problem. In Chapter 8, we address energy efficiency in wireless communication, with special emphasis on physical, MAC and network layers. Chapter 9 covers various aspects related to secure wireless communication including authentication, jamming, intrusion detection and stream cipher generation.

Exercises are given at the end of each chapter which will help the students get a better insight of the topics while solving those problems. Star (*) marked problems in the exercises indicate a higher level of difficulties and are, in fact, research problems that may be taken up by the researchers in the respective areas.

The authors are grateful to Ansuman Bhattacharya, Dibakar Saha, Pratham Majumder and Sayan Sen Sarma for their generous help in drawing a few figures and formatting some tables and algorithms while preparing the LaTeX source file for the book.

We will consider their efforts worthwhile if this book is found useful to the academic community, for both teaching and research, in the discipline of wireless networks and mobile computing.

Koushik Sinha

Sasthi C. Ghosh

Bhabani P Sinha

Chapter 1

Introduction

During recent years, wireless communication has emerged as the most widely used technology in various application domains involving almost all spheres of our daily activities. Most notable of these applications is, of course, the mobile telephone system which has proved extremely useful and popular worldwide. Starting with about 200 million registered mobile users in 1997 [1], the number of users of the mobile telephone system increased to more than five billion by early 2014.

However, prior to the introduction of mobile telephone systems, wireless communication technology was also used extensively in other household and commercial devices like cordless telephones, remote controls for TV sets, cordless microphones, and so on. Over the past two decades, the world has experienced a phenomenal growth in communication science and technology, particularly in optical communication and wireless communication. Many types of communication networks like ZigBee networks, Wi-Fi networks, cellular mobile networks, ad hoc networks, sensor networks and many other such networks based on wireless communication technology have been introduced for real-life applications ranging from personal communication, entertainment, to rural and urban healthcare, smart home building, inventory control, surveillance, police and security control, military and defense systems, and so on. A brief review of various application areas of wireless communication technology is given in the following section.

1.1 Applications Involving Wireless Communication

1.1.1 Cordless Telephones

A cordless telephone is a replacement of the handset cord of a landline telephone receiver set by means of a radio link so that the handset can be physically detached from the remaining part of the receiver set (may be considered the *base station* of the cordless telephone) [2]. This modified handset communicates with the base station which, in turn, is connected to a fixed telephone line. The communication range of the handset is usually very small, just sufficient for covering short distances within a room or building. Unlike a corded telephone, a cordless telephone needs supply of power from the mains for its base station to work. The cordless handset is powered by a rechargeable battery, which is charged when the handset is placed on its cradle.

In 1994, digital cordless telephones operating around 900 MHz frequency range were introduced. Use of digital signals enabled the phones to be more secure (decreased eavesdropping) than those using analog signals. In 1995, the digital spread spectrum (DSS) technique was introduced for cordless phones. This technology enabled the digital information between the receiver and the base station of the cordless telephone to be spread in pieces over several frequencies , thereby making it very difficult for anyone to eavesdrop on the cordless conversations.

1.1.2 Cordless Microphones

A cordless microphone has no physical cable connecting it directly to the sound recording or amplifying equipment with which it is associated [3]. This is also alternatively known as a radio microphone or a wireless microphone. Its body contains a small, battery-powered radio transmitter which transmits the audio signal from the microphone by modulated radio waves to a nearby radio receiver unit, which recovers the audio from the received radio signal by appropriate demodulation technique. The rest of the audio equipment is connected to the receiver unit by a cable. Cordless microphones are widely used in the entertainment industry, television broadcasting, seminars and interviews to allow the audience to speak and also the interviewers or performers or entertainers to move about freely while using a microphone to amplify their voices.

There are many different standards, frequencies and transmission technologies used to replace the microphone's cable connection and make it into a cordless or wireless device. Users may transmit, for example, radio waves over the UHF (300 MHz to 3 GHz) or VHF (30 MHz to 300 MHz) frequency bands and use FM, AM, or various digital modulation schemes. Some low cost models may, however, use infrared light. Infrared microphones require a direct line of sight communication between the microphone and the receiver, while the costlier radio frequency models do not have such constraints. Some models operate on a single fixed frequency, but the more advanced models operate on a user selectable frequency to avoid interference and allow the use of several microphones in nearby places at the same time.

1.1.3 Wi-Fi Networks

Wi-Fi, also spelled Wifi or WiFi [4], is a technology that allows electronic devices to exchange data or connect to the Internet through wireless communication using 2.4 GHz signals (in the UHF band) or 5 GHz signals (in the super high frequency or SHF band of 3 GHz to 30 GHz). Wi-Fi is also used synonymously to indicate a wireless local area network (WLAN) based on the IEEE 802.11 protocol.

Many devices can use Wi-Fi, e.g., personal computers, laptops, video-game consoles, smartphones, some digital cameras, tablet computers and digital audio players. These can connect to a network resource such as the Internet via a wireless network access point. Such an access point (or hotspot) has a range of about 20 meters (66 feet) indoors and a greater range outdoors. Hotspot coverage can cover an area as small as a single room with walls that block radio waves or as large as many square kilometres achieved by using multiple overlapping access points.

Wi-Fi can be less secure than wired connections (e.g., Ethernet) because an intruder does not need a physical connection to such a network. Web pages that use the secure sockets layer (SSL) - a protocol used above the transport layer for encrypting information over the Internet, are secure but unencrypted Internet access can easily be detected by intruders. Because of this, Wi-Fi has adopted various encryption technologies. The early WEP (wired equivalent privacy) encryption, introduced in 1999 for IEEE 802.11 wireless networks has been proven easy to break. Higher quality protocols such as WPA (Wi-Fi protected access) and WPA2 were introduced in 2003 and 2004, respectively.

1.1.4 ZigBee Networks

ZigBee [5] is a specification for a suite of high level communication protocols used to create personal area networks built from small, low-power digital radios. It was conceived in 1998, standardized in 2003, and revised in 2006.

ZigBee is used in applications that require a low data rate, long battery life, and secure networking. It has a defined rate of 250 Kbit/s, best suited for periodic or intermittent data or a single signal transmission from a sensor or input device. Typical applications include wireless light switches, electrical meters with in-home displays, traffic management systems, and other consumer and industrial devices that require short-range wireless transfer of data at relatively low rates.

ZigBee is based on the IEEE 802.15 standard. Though low-powered, ZigBee devices can transmit data over long distances by passing data through intermediate nodes (devices) to reach more distant ones in the network. The ZigBee network layer supports star, tree and generic mesh networks. Every network has a coordinator node. Within star networks, the coordinator must be the central node. Trees and meshes allow the use of ZigBee routers to extend communication at the network layer.

ZigBee networks are secured by 128-bit symmetric encryption keys. In home automation applications, transmission distances range from 10 to 100 meters line-of-sight (LOS), depending on power output and environmental characteristics.

The technology defined by the ZigBee specification is intended to be simpler and less expensive than other wireless personal area networks (WPANs) using Wi-Fi or Bluetooth standards (Bluetooth is a wireless technology standard for exchanging data over short distances using short wavelength UHF radio waves in the ISM (industrial, scientific and medical) band from 2.4 to 2.485 GHz introduced by telecom vendor Ericsson in 1994 as wireless alternative to RS-232 data cables).

ZigBee operates in the ISM radio bands: 868 MHz in Europe, 915 MH in the USA and Australia and 2.4 GHz in other countries. Data transmission rates vary from 20 Kbps in the 868 MHz frequency band to 250 Kbps in the 2.4 GHz frequency band.

ZigBee builds upon the physical layer and media access control defined in IEEE Standard 802.15.4 (2003 version) for low data rate WPANs . The specification goes on to complete the standard by adding four main components: network layer, application layer, ZigBee device objects (ZDOs) and manufacturer-defined application objects that allow for customization and favor total integration. The ZDOs are responsible for a number of tasks, which include executing the device roles, management of requests to join a network, device discovery and security.

1.1.5 Cellular Mobile Networks

As already mentioned above, mobile telephone systems have emerged as the most widely used and popular technology over the last few years. This is mainly based on the cellular mobile networks in which the geographical region under consideration is spatially divided into a number of cells that may be either overlapping or non-overlapping. A *base station (BS)* (also known as a *base transceiver station (BTS)*) is established in each cell (Figure 1.1) which coordinates the activities of all mobile phones (also commonly termed as mobile terminals (MTs)) residing in that cell. Each mobile phone in the cell communicates through this base station via a wireless channel assigned by the base station. In Figure 1.1, two MTs X and Y are shown in two cells and both communicate to the BSs in their respective cells. Several such base stations may be concentrated at a *base station controller (BSC)* and several such BSCs may be connected to a *mobile switching center (MSC)* to constitute a hierarchical network structure. Such a structure makes the design and development of the network easy, and makes it easy to manage the mobility of the MTs. A hierarchical structure also makes it easy to maintain the network in case of any failure, update or modification of the network components. The connection links from the BSs to BSCs and from the BSCs to the MSCs are usually through wireline links.

Several techniques such as frequency-division multiplexing (FDM), time-division multiplexing (TDM) and code-division multiplexing (CDM) can be used to distribute the available radio spectrum into appropriate communication channels. More efficient techniques based on combining the above can also be designed to allocate the available radio spectrum among several users. A channel can simultaneously be used by multiple base stations if their separation is sufficient to satisfy the inter-

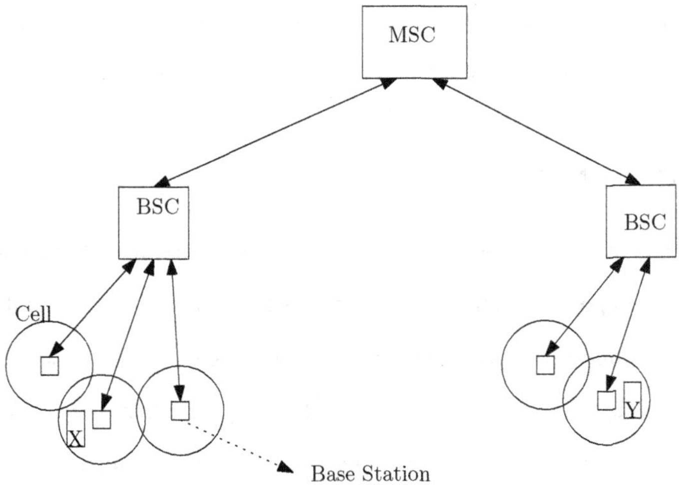

FIGURE 1.1: Cellular mobile network architecture

ference constraints. With the advances in technology, there has been an increasing demand from mobile users for multimedia services like voice, text, still image and video over wireless networks. On the other hand, the available bandwidth required for such multimedia data communication for a large number of mobile users is very limited. Such a limited availability of the radio spectrum, signal distortion and the interferences caused by the environment and other mobile users impose an inherent restriction on the capacity of such wireless networks. Therefore, developing methods to utilize the scarce radio spectrum efficiently is more critical than ever.

Interference is a function primarily of transmitter power, receiver sensitivity, antenna gains, and channel loss. Channel loss is a function of distance, frequency and weather, and is quantified by the minimum acceptable signal-to-noise ratio. Assuming that most of these factors are already determined or beyond our ability to influence, the interference can be defined directly as a function of frequency and distance. Based on that, three types of interference are generally taken into consideration which can be avoided by satisfying the following constraints: i) *co-channel constraint*, due to which the same channel cannot be assigned to certain pairs of cells simultaneously, ii) *adjacent channel constraint* , for which adjacent channels cannot be assigned to certain pairs of cells simultaneously, and iii) *co-site constraint*, which implies that any pair of channels assigned to the same cell must be separated by a certain number of channels [7].

The task of assigning frequency channels to the cells satisfying the interference constraints and using as little bandwidth as possible is known as the *channel assignment problem* (CAP). In its most general form, the CAP is equivalent to the generalized graph coloring which is a well known NP-complete problem [10, 11, 12]. In a subsequent chapter, we provide detailed discussion about this channel allocation problem in cellular mobile networks.

1.1.6 Ad Hoc Networks

Based on the application scenario, the common ways to design wireless networks are: i) infrastructure-based networks, ii) infrastructure-less networks and iii) hybrid networks which uses a mix of both infrastructure and infrastructure-less network components. The cellular mobile network discussed above is an example of an infrastructure-based network where base stations are the existing infrastructures in the network. Absence of such base stations in a network leads to an infrastructure-less network of which the ad hoc network is an example. In such an ad hoc network, the component telephone devices themselves create a communicating network of their own, without the support of any base station. Such ad hoc networks may be static (where the component units may not change their geographical positions) or dynamic (the component telephone units may be mobile, changing their geographical positions with time). An ad hoc network of mobile telephone devices or handsets is commonly known as a *mobile ad hoc network (MANET)*. Nodes in an ad hoc network usually communicate through a single channel by means of broadcasting or multicasting of data over the network.

Ad hoc networks are particularly useful where an infrastructure, (e.g., a base station) is difficult, if not impossible, to set up due to unfriendly terrain conditions, or in a disaster situation where the infrastructure has been destroyed or damaged. Accordingly, these networks have proven very effective in disaster management, military and surveillance operations, communication in underground mines and communication under seas.

Due to the absence of any infrastructure, ad hoc networks may be considered as distributed systems with no arbiters. However, the computing facilities with such portable or mobile devices are also limited in terms of available memory, processing capability and speed of computation. Moreover, each component node is operated by a battery to allow mobility of the nodes. This makes the network highly energy constrained.

Due to the energy constraint, a component node of an ad hoc network may go out of the network if it runs out of battery power. That is, this node will no longer be a part of the network and this situation may effectively be abstracted by a node failure in the network. Similarly, in a MANET where the nodes have mobility, if a node moves out of the communication range from all of the remaining nodes of the network, that node will be disconnected from the rest of the network. This situation may be abstracted by a link failure in the network. On the other hand, new nodes are allowed to join an ad hoc network (static or mobile) at any time, establishing connection links to some node(s) already existing in the network. All these facts make the topology of an ad hoc network and a mobile ad hoc network highly dynamic in nature. The communication protocols developed on such networks must consider this dynamic network topology arising from energy-constraints, mobility and new arrivals in designing a robust communication network.

1.1.7 Cognitive Radio Networks

Similar to the mobile computing environment that typically operates in licensed spectrum bands, the explosive growth in the number of wireless devices operating in the unlicensed radio spectrum bands created a severe shortage of available, unlicensed radio spectrum in recent times. The multitude of wireless networks and protocols (e.g., Wi-Fi, Bluetooth, Zigbee, etc.) operating in the unlicensed bands and vying for their share of the spectrum to meet their respective operational parameters leads to interference and performance degradation for all. Interestingly though, while the unlicensed bands are heavily loaded, in contrast, certain portions of the spectrum, particularly the licensed bands, are relatively underutilized. As an example, recent studies by the Federal Communication Commission (FCC) [6] in the US have shown that at any given time and in any given geographic locality, less than 10% of the available spectrum in the TV band (from 470 MHz to 698 MHz) is utilized. Similar spectrum utilization statistics were observed in the UK by the OFCOM regulatory body, in other parts of Europe, and in other parts of the world [13].

To exploit these underutilized parts of the spectrum (also referred to as *white spaces* or *spectrum holes*), the FCC advocates the development of a new generation of programmable, smart radios that can dynamically access various parts of the spectrum including the licensed bands. Such radios would operate as *secondary users* in the licensed bands and are required to possess the capabilities of spectrum usage sensing, environment learning and interference avoidance with the *primary users* of the licensed spectrum bands while simultaneously ensuring the *quality of service* (QoS) requirements of both the primary and secondary users.

Software defined radios (SDRs) with such capabilities of cognition of primary and secondary channels and allocating channels to different users on demand are referred to as *cognitive radios* (CRs). A schematic view of a cognitive radio network is shown in Figure 1.2 depicting different functional requirements of such a network.

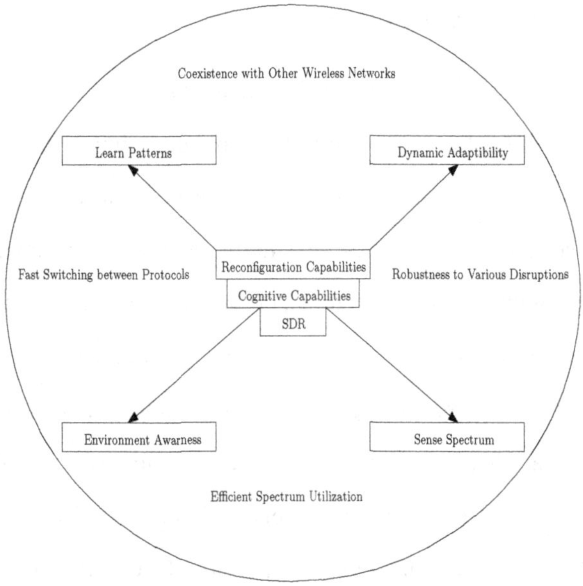

FIGURE 1.2: Schematic view of a cognitive radio network

Cognitive radios are widely accepted as the wireless technology of the future and it is envisioned that most radios on portable devices (e.g., cellphones (telephones used in cellular mobile networks) and Wi-Fi enabled notebook PCs) will possess some amount of cognitive capability [14] by the time this book is published.

In the context of networks operating in licensed bands, the ability to exploit unused parts of the spectrum can provide a multitude of benefits. Taking the case of cellular networks as an example, such frequency agility can dramatically reduce operational, deployment and licensing costs, increase the number of users that can be supported under a single base station and improve the operator's ability to provide guaranteed service through the intelligent exploitation of the large additional unused spectrum pool. Moreover, such cognitive radio-based cellular networks can offer benefits in the events of attacks or natural disasters. While a conventional cellular system is centralized and can only operate in licensed bands, a cognitive system is capable of establishing communications even if some network elements are out of order or the total requested bandwidth by the users exceeds the operator's licensed bandwidth. It is thus apparent that cognitive radios will play a pivotal role in the coming generations of wireless networks [18, 19, 8, 9, 20, 13, 14, 17, 16, 15].

1.1.8 Wireless Mesh Networks

As an alternative to the cellular and ad hoc network architectures, there has been a great deal of interest in recent times in wireless mesh networks (WMNs). The operational characteristics of WMNs may be considered to be somewhere in between infrastructure-based networks such as wireless local area networks (WLANs) and ad hoc networks with no pre-installed infrastructure support. In WMNs, the network consists of a set of *fixed* gateway nodes and a set of non-gateway nodes (either stationary or mobile). The non-gateway nodes must access a gateway node (similar to WLANs) to establish communication to the outside world. These non-gateway nodes can act as hosts or as wireless routers by forwarding packets from other users (similar to ad hoc networks), enabling other non-gateway nodes to establish links with the gateway nodes in a multi-hop fashion, if required.

This fusion of ad hoc network and infrastructure-based network properties allows significant extension of the coverage area of the network and enables nodes in areas with poor signal propagation properties to contact the outside world with higher probability of success. The overall effect of such a deployment is thus a substantial reduction in the number of deployed gateway nodes needed to cover a zone and consequently the overhead for installation, operational, and maintenance expenses.

Furthermore, such a network structure also allows for scalable operations simply by adding new gateway nodes when and where required. The reliability also improves as multiple paths to gateway nodes are more common and re-routing of packets in case of node failures is handled as in ad hoc networks.

Because of the above mentioned advantages, even traditional infrastructure-based networks such as cellular networks are beginning to look at possibilities of migrating to either WMN-based or WMN-like topologies [21, 22, 24, 28]. The promises are several: reducing infrastructure setup and maintenance cost, improving fault tolerance and robustness in the face of unnatural events and natural calamities and enhancing network coverage, among others. However, with the use of a mesh architecture, the importance of the channel assignment problem becomes even more accentuated due to the increased probability of interference between the links of a WMN, as most applications of WMNs (e.g., dense sensor networks) tend to contain more links than traditional cellular and ad hoc networks in an attempt to improve reliability through path diversity [28, 27, 26, 25]. As an example, Lan and Trang [23] proposed a technique for channel assignment in WMNs with multiple channels and multiple radios at each node of the network for increasing the network throughput.

1.1.9 Vehicular Ad Hoc Networks (VANETs)

A vehicular ad hoc network (VANET) is a mobile ad hoc network (MANET) consisting of moving cars. Each participating car acts as a wireless router or node of the network, with a range of about 100 to 300 metres. If cars move out of the signal range, they will drop out of the network, while other cars may also join the network. Design of such a network was first intended to enable police and fire vehicles to communicate with each other for security and safety purposes on highways.

1.1.10 Wireless Sensor Networks

A wireless sensor network (WSN) or simply a sensor network , is a network of spatially distributed autonomous sensors to monitor physical or environmental conditions, such as temperature, sound, pressure, etc. and cooperatively pass their data through the network to a sink node at some location, as shown in Figure 1.3. The more modern networks are bi-directional so as to enable control of sensor activity.

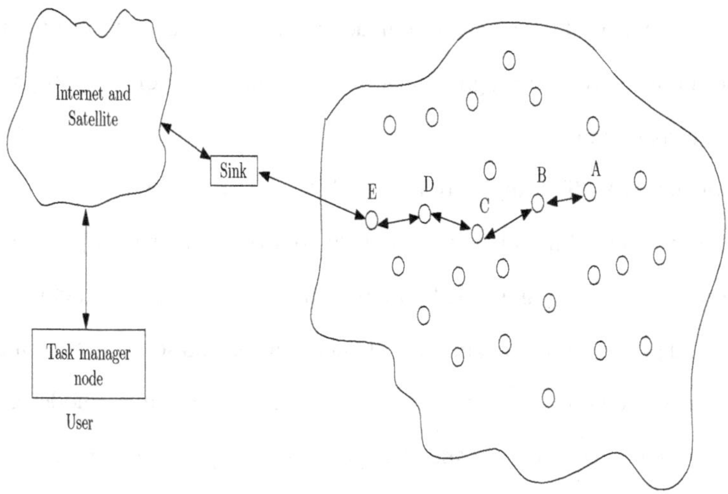

FIGURE 1.3: Wireless sensor network

Each such sensor network node has several parts: a radio transceiver with an internal antenna or connection to an external antenna, a microcontroller, an electronic circuit for interfacing with the sensors and an energy source, usually a battery or embedded form of energy harvesting. A sensor node may vary in size from that of a shoe box down to the size of a grain of dust, although functioning *motes* of genuine microscopic dimensions have yet to be created. The cost of sensor nodes is similarly variable, ranging from a few to hundreds of dollars, depending on the complexity of the individual nodes. Size and cost constraints on sensor nodes result in corresponding constraints on resources such as energy, memory, computational speed and communications bandwidth.

A sensor network may be thought to be an ad hoc network. Just like the ad hoc networks, a sensor network may also be static or mobile. For example, a surveillance system with large static camera sensors deployed in a mall or supermarket is a static sensor network, while very light and small-sized sensors deployed over a terrain where the sensors can float around or may change their positions over time constitute a mobile sensor network. Communication protocols in wireless sensor networks may be developed in a similar way as those in ad hoc networks. Data propagation among the sensor nodes may be through point-to-point communication or through flooding using broadcast or multicast.

Sensor networks find applications in a large number of real-life situations including environmental monitoring, inventory control in industries, process control, machine health monitoring, smart home building, monitoring of structures of large buildings and bridges, military and surveillance control, and so on. Processing of data streams received continuously from the sensor nodes by the sensor nodes even with low computational resources for specific application purposes constitutes a challenging research problem.

1.1.11 Body Area Networks

A body area network (BAN) (also referred to as a wireless body area network (WBAN) or a body sensor network (BSN)) is a wireless network consisting of a set of wearable computing devices. In particular, the network consists of several miniaturized body sensor units (BSUs) together with a single body central unit (BCU). The technology for developing body area networks started around 1995 based on the idea of using wireless personal area network (WPAN) technologies for implementing communications on, near, and around the human body.

1.2 Effects of Mobility of Devices

A wireless mobile network, whether cellular, ad hoc or sensor, consists of mobile devices (telephones, sensors or other types of devices) which are invariably operated by batteries of limited energy because of the mobility of such devices. When the battery of a device is exhausted, the device leaves the communication network. This phenomenon may be effectively abstracted by a node/link failure in the network.

Apart from this energy constraint, such mobile devices, because of their small size for portability, usually have small computing resources in the form of memory, computational power and speed of computation. They are also usually constrained by low communication bandwidth.

Mobility of component nodes in a network also gives rise to additional problems in network connectivity, e.g., when a node moves out of the range of communication from all other nodes in the network, it is dropped from the network. On the other hand, new nodes are also usually allowed to join a network at any time and may request additional communication channels for sending their messages. Such an allocation of new channels to newly arrived nodes must avoid all types of channel interferences with the existing communicating nodes in the nearby region. The communication protocols of the network must take proper account of such a dynamic network topology arising from all these situations.

Further, when a mobile communicating node changes its geographical position, it is possible that the channels assigned to this node may interfere with the channels allocated to other communicating nodes in the vicinity of its new location. Accordingly, some new non-interfering channel needs to be assigned to this node very quickly so as not to disrupt communication which is already in progress.

Quality of Services (QoS) is an important consideration for developing the communication protocols in any network. The quality of service is usually expressed or demanded by the users in several different forms like delay (time required for communicating packets of message), delay jitter (i.e., variation in delay for communicating different packets of the same message), reliability (e.g., bit error rate (BER)), availability of channels, and the bandwidth of communication under different application environments. For example, for voice communication over telephone, packet delay is important and should be very small – not perceptible by human audio recognition systems; otherwise it may give rise to unwanted repetition of a message by the sender, often overlapping with the reply message from the receiver. Also, delay jitter in sending voice signals (e.g., using voice-over IP (VoIP)) is equally annoying to a human audio recognition system if different parts of the same sentence arrive with different lengths of intermittent delay. But it may be possible to compromise with the reliability of communication of voice signals to some extent. For example, if one or two bits of a voice packet are received erroneously, the human ear can possibly tolerate that and if not, the person receiving the message can ask the sender to repeat the message once again. This may also apply to communicating video signals. On the other hand, for text data communication, delay and delay jitter are relatively unimportant, but reliability is extremely important. Quick allocation of channels to a requesting user for a specific purpose needs to be done by the network. Also, usage of minimum bandwidth for effecting the desired communication over the network is a direct concern for the designers of the network system; the network must provide uninterrupted service to the maximum possible number of users with the available bandwidth resource.

Due considerations need to be given for developing robust communication protocols for maintaining the desired QoS under all dynamic network conditions as discussed above that may arise from mobility of the component nodes.

1.3 Issues in Cellular Mobile Networks

As evident from the requirements of the cellular mobile networks described above, we essentially need to i) assign a channel to a new call, avoiding all sorts of channel interferences from other calls and ii) maintain the communication between any two nodes even when one or both of them move from one place to another. Apart from these two essential factors, the next important concerns are the energy-efficiency in implementing the communication protocols and security of messages to be communicated. Thus, broadly speaking, we need the following four issues to be resolved when implementing a cellular mobile network.

- Mobility management
- Bandwidth management
- Energy management
- Security

1.3.1 Mobility Management

In order to achieve the major goal of mobile computing in a cellular mobile network, i.e., the ability of computing at any place at any instant of time even when people are on the move, the important concern is, of course, to allow a user to make a call from any place (cell) to any other person in the network and to continue the call even when one or both of them are moving. The following four activities are involved in implementing this requirement.

- Call setup
- Handoff management
- Internetworking and interoperability

We give a brief introduction to these activities as follows:

1.3.1.1 Call Setup

In order to set up a call from a mobile terminal (MT) to another MT, two distinct activities are involved: i) location registration and ii) call delivery. These two activities are known as location management functions. Each mobile terminal is required to update its location information periodically or at least when it changes its location. For this purpose, usually a number of cells are grouped together to form a *location area (LA)* and a mobile switching center (MSC) controls several such LAs. Whenever an MT moves from one LA to another LA, the MT initiates a request to its current MSC for updating the location databases. On the other hand, call delivery involves querying the location databases to find the current location of the called MT. We discuss these two activities in detail in Chapter 2.

1.3.1.2 Handoff Management

When an MT changes its position from one cell to another cell, the strength of the signal received by it from the base station (BS) of the old cell becomes less than that from the base station of the new cell. Accordingly, the control of this MT is to be handed over from its old BS to the new BS. A smooth handoff of control of the MT from the old BS to the new one without hampering the ongoing communication with this MT, if any, is known as handoff management. In other words, handoff or handover management enables the network to maintain a user's connection when the user (MT) continues to move and changes its access point from one BS to another BS. There are three distinct stages for handoff: i) handoff initiation, ii) new connection generation and iii) data flow control. In the first stage, handoff may be initiated by a user or MT when it receives the signal from its BS at a very low strength (lower than some threshold value) or it may be initiated by network software (either due to weak signals received from an MT or due to some changing network conditions, e.g., the communication channel used by the MT may be changed by executing a channel allocation algorithm at the base station to satisfy the channel demands of other MTs in the network). The second stage of handoff management, new connection generation, may be performed in three different ways. In case of network-controlled handoff (NCHO) or mobile-assisted handoff (MAHO), the network generates a new connection and finds the resources and the additional routing operations. In mobile controlled handoff (MCHO), the MT finds the new resources and the network approves them. The third stage, i.e., data flow control, involves the delivery of data from the old connection path to the new connection path according to the agreed-upon services to be provided to the MT. In Chapter 2, we address handoff management in somewhat more detail.

1.3.1.3 Internetworking and Interoperability

There are different backbone networks which can support cellular mobile communication. Examples of such backbone networks include public land mobile networks (PLMNs) , mobile IP networks, wireless asynchronous transfer mode (WATM) networks, low earth orbit (LEO) satellite networks , and so on. Each such backbone network has its own kind of signal propagation techniques and communication protocols to provide services to cellular mobile users. While roaming, it is quite possible that an MT moves from one place to another where the supporting backbone network is changed. The future generation mobile communication system aims at global roaming capability for mobile users. In order to implement this idea, the associated problem of internetworking must be properly addressed. The mobile devices need to be equipped with the appropriate hardware (physical layer and MAC layer issues) and software (network layer issues) to operate in different backbone networks during roaming. Also, this switching from one backbone network to another should be smooth and transparent to the users without disrupting the ongoing calls of the moving MTs. The location registration, call delivery and handoff management techniques must also be appropriately handled for a smooth switching from one backbone network to another.

Likewise, an MT may also move from one area to another where the backbone network remains unaltered, but the service provider is changed causing corresponding changes in communication techniques, protocols and service level agreements (e.g., tariffs) between the user and the service provider. For example, different service providers may operate over different frequency bands for message communication, and use different channel access mechanisms. One may use frequency division multiple access (FDMA) to access the channel, while the other may use code division multiple access (CDMA). The issues of compatibility of the mobile devices and the associated communication protocols, and designing the appropriate location registration, call delivery, handoff management techniques for such interoperability of the mobile services need also to be addressed to provide roaming facilities across different service providers. All these issues related to internetworking and interoperability constitute an important area of future research and development for providing the effective global roaming facilities to the mobile terminals.

1.3.2 Bandwidth Management

Because of the extreme popularity of mobile telephone systems, the number of cellular mobile users is increasing day by day almost exponentially. With about 200 million listed mobile users in 1997, the number has gone up to more than 6.8 billion in early 2015 and continues to increase at a high rate. Moreover, the users of mobile telephone systems no longer confine their use to voice messages and texts; multimedia signals including audio, still image and video are also communicated through cellular mobile networks. Quite naturally, the bandwidth requirement of the communication channel increases for multimedia signals. Thus, for audio signals, 64 Kbps is considered enough, while for video signals, 384 Kbps is the bare minimum bit rate used in second generation mobile systems. In third and fourth generation mobile communication systems, for better quality video communication, the channel speeds are intended to be 2 Mbps and 100 Mbps, respectively.

On the other hand, the total bandwidth for signal communication in cellular mobile networks is very limited. Apart from the 100 MHz band from 860 MHz to 960 MHz used by the GSM (Global System for Mobile Communications) system, the 1992 World Administrative Radio Conference in Geneva allocated the 1885 to 2025 MHz and 2100 to 2200 MHz bands – a total of only 230 MHz – for terrestrial and satellite communications. Within these two bands, the satellite band covers 1980 to 2010 MHz and 2170 to 2200 MHz for a total of 60 MHz, while the remaining 170 MHz is for terrestrial mobile communication. The frequency gaps between 2025 to 2110 MHz and beyond 2200 MHz are reserved for other services such as remote sensing, cable TV relay service, space research, etc. In 1995, the World Radio Conference changed the above frequency allocation a bit. The satellite band allocation for Region 2 (the Americas and the Caribbean) was modified to 1990 to 2025 MHz and 2160 to 2200 MHz for a total of 75 MHz, leaving only the other 155 MHz for terrestrial services. This has made the situation for cellular mobile communication in the countries of this region more difficult. Even with some later minor modifications, the end result is that the total bandwidth available for terrestrial uses is very small compared to the huge ever-increasing demand due to a large number of mobile users along with the increasing demand for multimedia communication.

One possible way to overcome this problem of *bandwidth crisis* is by intelligent re-use of the frequencies at sufficiently distant cells of the cellular network so that these two frequencies do not interfere with each other. Over the past two decades, research results on this topic have appeared in the literature, and yet a number of problems need to be solved to develop an effective algorithm for channel allocation in cellular mobile networks. We will discuss these issues later in this text.

1.3.3 Energy Management

Energy is a current buzz word. However, there are two distinct groups of scientists and technologists who are interested in energy. One group focuses on the generation of energy from unconventional sources, and the other group looks for efficient utilization of energy in various real-life applications. Our concern in connection with cellular mobile networks is the same as that of the latter group.

In particular, due to the mobility of the terminals, the mobile devices are operated with batteries having limited energy. Also, often the devices are small and hence the batteries used have small capacities of total energy. This has motivated the researchers to develop energy-efficient communication techniques which may be easily adopted in cellular mobile networks. Energy efficiency has been studied over the past few years for minimizing energy consumption in different layers of the communication protocols, e.g., in the physical layer, MAC layer, network layer, and so on. In the physical layer, for example, one may think of i) designing circuits with minimum number of gates and other electronic components to reduce the total power consumption, ii) selecting appropriate modulation and demodulation techniques for designing low-power transceivers. In the MAC layer, we may try to avoid collisions, or at least keep the collisions as few as possible, because a collision requires re-transmission of the collided message leading to waste of power. We may also try to keep the channel in sleep mode to use less power than in the normal mode and tweak the sleep, listen and awake periods for optimal performance in terms of energy consumption. In the network layer, the messages may be routed through a path requiring minimum hops and hence minimum energy consumption.

We will devote Chapter 8 to energy-efficient communication in various low-power wireless networks.

1.3.4 Security

Security in communication is one of the very important aspects to be considered in most real-life applications involving message communication. In particular, the wireless networks are more vulnerable to attacks by intruders than the wired networks. This has given rise to security issues like user authentication, encryption and decryption of messages and jamming. However, often the algorithms designed for these purposes consume a lot of time and energy (through the use of large programs or complex hardware) which may not suitable for online wireless communication. Research to find suitable techniques for online low-energy secure communication in cellular mobile networks is ongoing.

1.4 Issues in Ad Hoc Networks

The important issues that need to be considered by the researchers and developers of ad hoc networks include the following:

- Initialization
- Leader election
- Communication protocol
- Location identification
- Energy
- Security

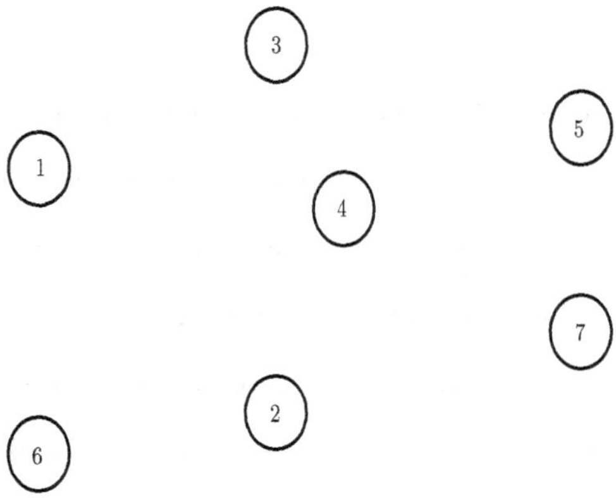

FIGURE 1.4: Initialization in ad hoc networks

1.4.1 Initialization and Leader Election

The first two issues on the above list pertain only to ad hoc networks due to the absence of any infrastructure in the network as in the cellular mobile systems. In order to establish a communication between two nodes in the network, every node of the network must have a unique identifier or Id which should also be known to all other nodes in the network. This assignment of unique Ids to the nodes is known as *initialization*. Referring to Figure 1.4, there are seven nodes of the ad hoc network which are given unique Ids $1, 2, \cdots, 7$ by this initialization process. While assigning the unique Ids, due consideration must be given to the possibility that any node may drop out (be switched off or the battery may be out of power) any time, or go out of communication range from others; and at any time, any new node may join anywhere in the network. The topology of the network should be determined after every such arrival or departure of a node. In other words, the initialization algorithm must consider such dynamic behavior of the network.

The second issue of leader election as given above is also a specific issue of ad hoc networks in absence of an infrastructure, e.g., a base station in a cellular mobile network. It is often easy to control communication in a network for optimizing other network parameters, e.g., network congestion due to flooding of messages, to-

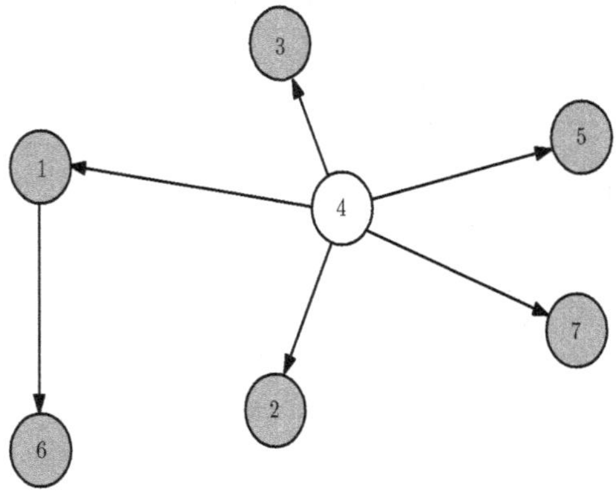

FIGURE 1.5: Leader election in ad hoc networks

tal energy expenditure on the nodes, and average message communication time, if the network has a so-called *leader* that will take control of message routing with a view to optimizing those parameters. Again, referring to Figure 1.5 with seven nodes in the ad hoc network, the node 4 is elected as a leader and after this election, node 4 broadcasts a message to all other nodes informing them of its election as the leader.

Often it is advisable to cluster the network so that the nodes within a cluster are geographically near each other, requiring very little communication power of the transceivers; and a cluster head or leader of a cluster is chosen based on some criterion (for example, the smallest average distance from all the nodes in the cluster, maximum remaining energy, etc.). This is illustrated by the example in Figure 1.6. Communication between a node u in, say, cluster 1 and a node v in, say, cluster 2, may then be effected through the leaders of clusters 1 and 2.

In Chapter 5 of the book, we will discuss the different communication protocols used in ad hoc networks.

1.4.2 Location Identification

Most location estimation systems in cellular mobile and ad hoc networks utilize the fundamental method of using *trilateration* or *triangulation* of received sig-

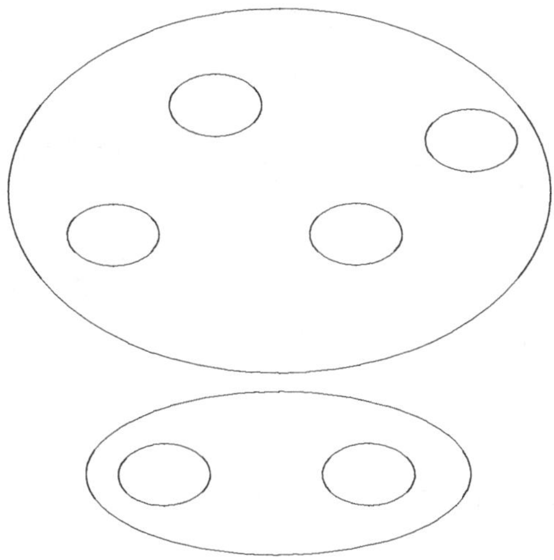

FIGURE 1.6: Clustering in ad hoc networks

nals to obtain an estimate of the receiver's position [29]. In this method, in a two-dimensional geometry, the distances of the mobile terminal in question are measured from three reference points which are not collinear. By drawing three circles centered at these three reference points with these measured distances of the reference points from the terminal as the radii of these circles, we can estimate the coordinates of the mobile terminal. The distance of the mobile terminal from a reference point can be determined by sending a signal (at ultrasonic or radio frequency) from the reference point to the mobile terminal and then measuring the time of flight of the signal. This is also called the time of arrival (TOA) technique for finding the location of the mobile terminal. Such an estimate of the distance is possible if there is a direct path or a line of sight (LOS) communication between the transmitter and the receiver. Non-line of sight or NLOS signals occur under multi-path conditions when the received signals come from reflected, diffracted or scattered paths, thus introducing excess path lengths into the Euclidian distance between the transmitter and the receiver. The NLOS error is defined as the excess distance traversed compared to the direct path and is always positive. The corruption of LOS signals by NLOS signals and Gaussian measurement noise are the major sources of error in all location estimation systems; the former is the dominant factor [2].

Another possible way of locating a mobile terminal is by measuring the angle of arrival (AOA) in which the transmitter sends a signal which is received by the receiver with the help of a highly directional antenna so that the receiver can estimate the angle of arrival of the received signal with respect to a reference direction. By noting the directions of receipt of the LOS signals by the nodes at two reference points, the position of the mobile terminal can be found from the point of intersection of two straight lines.

The first and arguably archetypal indoor location discovery system was the Active Badge Location System [31]. The Active Bat system used time difference of arrival technique [32]. Complementing the Active Bat was the Cricket system [5] which implemented both lateration and proximity techniques. It had a less precision than the Active Bat, but had the advantages of privacy support and a decentralized scalability. RADAR [4] was a building-wide tracking system developed by Microsoft. It used a building signal characteristic model for location estimation. Unfortunately, movements of people or objects may necessitate reconstruction of the pre-defined signal-strength database or creation of an entirely new database.

The Global Positioning System (GPS) [3, 36] is perhaps the most widely publicized location-sensing system. Unfortunately, GPS does not scale well in dense urban areas or in indoor locations. Modeling of the radio propagation environment provides a more accurate location estimate by mitigating the effect of NLOS errors. While reasonably accurate radio propagation models exist for outdoor conditions [37, 38, 39], unfortunately there are no such unanimously accepted models for indoor environments.

Authors in [6] modeled distance errors for a sample office environment, measured using TOA techniques. Their simulation results show that LOS signals can be modeled as a Gaussian distribution while the obstructed LOS signals may be treated as a combination of Gaussian and exponential distributions for the environment considered. Authors in [6, 41] demonstrated that the bandwidth of the channel plays an important role in the magnitude of errors introduced and that a minimum bandwidth of 100 MHz is required for effective location estimation.

Several authors have attempted to mitigate the effect of NLOS errors [6, 42, 43, 1, 45, 4, 46, 47, 48, 49]. The least squares (LS) algorithm [50, 51], uses the Davidon algorithm [52] to find a location estimate that would minimize the *residual* of the estimate. The residual weighting algorithm (Rwgh) was proposed in [42] as another way to mitigate the effects of errors in distance measurements caused by NLOS propagations. Authors in [50] compare different geolocation algorithms for indoor environments, assuming the channel characteristics proposed in [6]. Authors in [43] proposed two algorithms for identification and correction of NLOS errors, based on *a priori* knowledge of the NLOS errors. An NLOS error mitigation algorithm based on maximum likelihood detection has been presented in [53]. Another approach to location estimation was presented in [1], where the authors attempted to distinguish between LOS and NLOS measurements by assuming a statistical distribution of thermal receiver noise and signal characteristics.

Authors in [54] considered the problem of position estimation in large area networks, particularly sensor networks. They based their algorithm on measurement of signal strengths or time difference of arrival of signals and then the use of statistical methods to characterize the separation distances between two nodes. APS [55] is another location estimation algorithm that does not use GPS. However, the use of a hop-by-hop propagation process, such as DV hop or DV distance to propagate location estimates makes it unsuitable for mobile and random graphs. Authors in [29] presented a location estimation algorithm based on triangulation technique of signals in a cellular mobile network that is suitable for outdoor environments. Authors in [56] presented another GPS-free location estimation technique for mobile ad hoc networks. A common characteristic of all of the above algorithms is that they compute a single location point as an estimate. However, any estimation is a probabilistic measure and hence may be considerably different from the true scenario in some cases, especially for dense urban or indoor environments where NLOS errors are functions of location and time.

1.4.3 Energy and Security Issues

The energy and security issues in ad hoc networks are very much the same as those in cellular mobile networks. Some detailed discussions on these two issues in low-power wireless networks, in general, will be given in Chapters 8 and 9.

1.5 Issues in Cognitive Radio Networks

The important issues to be addressed in cognitive radio networks include:

- Channel sensing
- Channel allocation

Detailed discussions on these two issues in the context of both single signals and multimedia signals in the network can be found in [57, 58, 59].

1.6 Issues in Sensor Networks

The important issues to be considered in sensor networks include the following:

- Placement of sensors
- Communication protocols
- Computation with streaming data
- Energy
- Security

The first issue of deploying the sensors at strategic locations is important for several reasons depending on the application requirements. For example, for a video surveillance system using cameras as sensors in a shopping mall or a library, the visibility of all points of the mall or library by the set of camera sensors is essential. In another application of weather sensors, the placement should be such that all the sensors are able to communicate their sensed data to the sink location through other sensors using minimum energy for transmission and reception of all messages. We consider this aspect of placement in more details in Chapter 7.

The issues of communication protocols, energy and security are similar to those in ad hoc networks. However, depending on the typical application requirement of the sensed data, some specific energy-efficient algorithms can be developed for the application layer of the communication protocol stack in some special types of sensor networks by exploiting the impact of the computation process on the sensed data.

1.7 Exercises

1. What are the frequency bands in which Wi-Fi networks usually operate and what is the communication protocol followed by these networks? What are the encryption mechanisms usually followed by the Wi-Fi networks?

2. Name the IEEE standard on which the ZigBee protocol is based. What are the typical application areas of ZigBee networks? What are the operating frequency bands of ZigBee networks?

3. Distinguish between a cellular mobile network and an ad hoc mobile network.

4. Does an ad hoc network suffer any disadvantage for not having infrastructure when compared to a cellular network? Discuss this issue.

5. How often will an ad hoc network need to call its initialization routine? How often must an ad hoc network call its leader election routine? Do the energy considerations of individual nodes have any bearing on the initialization or leader election activities of the network?

6. Can an ad hoc network consist of all static communicating devices or must at least one of these devices be mobile?

7. What are the functional characteristics of a cognitive radio network?

8. Distinguish between a primary user and a secondary user in a cognitive radio network. Should every communicating node in a cognitive radio network necessarily have the channel sensing and channel allocation capabilities?

9. What are body area networks and what are the typical applications for which such networks may be useful?

10. What are internetworking and interoperability as applied to mobile communication environments?

Bibliography

[1] I. F. Akyildiz, J. McNair, J. S. M. Ho, H. Uzunalioglu and W. Wang, "Mobility management in next generation wireless systems," *Proc. IEEE*, Vol. 87, pp. 1345-1382, August 1999.

[2] http://en.wikipedia.org/wiki/Cordless_telephone.

[3] http://en.wikipedia.org/wiki/Wireless_microphone.

[4] http://en.wikipedia.org/wiki/Wi-Fi.

[5] http://en.wikipedia.org/wiki/ZigBee.

[6] Federal Communications Commission, Spectrum Policy Task Force, *Report ET Docket No.* 02-135, Nov. 2002.

[7] C. W. Sung, and W. S. Wong, "Sequential packing algorithm for channel assignment under co-channel and adjacent-channel interference constraint," *IEEE Trans. on Vehicular Technology,* Vol. 46, pp. 676-686, August 1997.

[8] M. D. Silvius et. al., "Smart radio: spectrum access for first responders," *Tech. Report*, Center for Wireless Telecommunications (CWT), Virginia Tech. `http://www.cognitiveradio.wireless.vt.edu`, 2009.

[9] Cognitive Radio Project, `http://kom.aau.dk/project/cognitive/cognitive_radio_project.htm`, 2009.

[10] W. K. Hale, "Frequency assignment: theory and application," *Proc. IEEE,* Vol. 68, pp. 1497-1514, 1980.

[11] Y. Yuan, P. Bahl, R. Chandra, T. Moscibroda and Y. Wu, "Allocating dynamic time-spectrum blocks in cognitive radio networks," *Proc. ACM MobiHoc*, pp. 130–139, 2007.

[12] Y. Yuan, P. Bahl, R. Chandra, P. A. Chou, J. I. Ferrell, T. Moscibroda, S. Narlanka, and Y. Wu, "KNOWS: kognitive networking over white spaces," *Proc. IEEE Dyspan* 2007.

[13] M. Sherman, A. N. Mody, R. Martinez and C. Rodriguez, "IEEE standards supporting cognitive radio and networks, dynamic spectrum access, and coexistence," *IEEE Comm. Magazine*, pp. 72–79, 2008.

[14] P. Koch and R. Prasad, "The universal handset," *IEEE Spectrum Magazine*, vol. 46(4), pp. 32–37, April 2009.

[15] E. G. Villegas, R. V. Ferre and J. Paradells, "Frequency assignments in IEEE 802.11 WLANS with efficient spectrum sharing," *Wireless Communications and Mobile Computing*, vol. 9(8), pp. 1125–1140, 2009.

[16] L. Song and D. Hatzinakos, "Cognitive networking of large scale wireless systems," *Intl. J. of Communication Networks and Distributed Systems*, vol. 2(4), pp. 452–475, 2009.

[17] E. Baccarelli, M. Biagi, C. Pelizzoni and N. Cordeschi, "Cognitive MIMO radio: An emerging paradigm for enhancing wireless access capability," *Intl. J. of Communication Networks and Distributed Systems*, vol. 2(2/3), pp. 302–330, 2009.

[18] Cognitive Radio Projects, Dept. of Electronics, University of York, `http://www.elec.york.ac.uk/research/comms/wireless/cognitive.html`, 2009.

[19] Cognitive Radio Networks, *Broadband Wireless Networking Labaratory*, Georgia Tech. University, `http://www.ece.gatech.edu/research/labs/bwn/CR`, 2009.

[20] M.M. Buddhikot, P. Kolody, S. Miller, K. Ryan and J. Evans, "DIMSUMNet: new directions in wireless networking using coordinated dynamic spectrum access," *Proc. IEEE WoWMoM 2005*, pp. 78-85, June 2005.

[21] Ö. Oyman and S. Sandhu, "End-to-end design principles for broadband cellular mesh networks," *Wireless Mesh Networks*, Springer US, pp. 57–76, 2007.

[22] Y.-D. Lin and Y.-C. Hsu, "Multihop cellular: A new architecture for wireless communication," *Proc. INFOCOM*, 2000.

[23] N. H. Lan and N. U. Trang, "Channel assignment for multicast in multi-channel multi-radio wireless mesh networks," *Wireless Communications and Mobile Computing, Special issue: Next Generation Wireless Communications and Mobile Computing/Networking Technologies*, vol. 9(4), pp. 557–571, 2009.

[24] B. S. Manoj, R. Ananthapadmanabha and C. S. R. Murthy, "Multi-hop cellular networks: architecture and protocols for best-effort and real-time communication," *J. of Parallel and Distributed Comp.* (JPDC), vol. 65(6), pp. 767–791, 2005.

[25] J. M. Rabaey et. al., "Short distance wireless, dense networks, and their opportunities," *Proc. 10th Euromicro Conf. on Digital System Design Architectures, Methods and Tools*, 2007.

[26] R. C. Pereira, R. D. Souza and M. E. Pellenz, "Overlay cognitive radio in wireless mesh networks," `http://arxiv.org/abs/0805.3643`, 2008.

[27] H. E. Gamal, "On the scaling laws of dense wireless sensor networks," *Proc. SenSys*, 2003.

[28] B. Kasiri, M. Naderi and B. Abolhassani, "Diversity-based relay selection for multihop cellular networks," *Proc. Intl. Conf. on Advanced Comp. Theory and Engg.*, 2008.

[29] K. Sinha and N. Das, "Exact location identification in a mobile computing network," *Proc. Intl. Conf. Par. Proc. Workshop* (ICPP), pp. 551–558, 2000.

[30] J. J. Caffery and G. L. Stüber, "Subscriber location in CDMA cellular networks," *IEEE Trans. Vehi. Tech.*, vol 47, pp. 406–416, 1998.

[31] R. Want, A. Hopper, V. Falcao and J. Gibbons, "The active badge location system," *ACM Transactions on Information Systems*, pp. 91–102, Jan 1992.

[32] A. Harter et. al., "The anatomy of a context-aware application," *Proc. of the 5th Intl. Conf. Mobile Computing and Networking* (MOBICOM 99), *ACM Press*, New York, pp. 59 - 68, 1999.

[33] N. B. Priyantha, A. Chakraborty and H. Balakrishnan, "The cricket location-support system," *Proc. of MOBICOM 2000, ACM Press*, pp. 32–43,Boston, USA, 2000.

[34] P. Bahl and V. Padmanabhan, "RADAR: an in-building RF-based user location and tracking system," *Proc. IEEE Infocom*, pp. 775–784, 2000.

[35] J. Hightower and G. Borriello, "Location systems for ubiquitous computing," *IEEE Computer*, vol. 34, No. 8, pp. 57–65, 2001.

[36] B. Parkinson et. al, "Global positioning system: theory and application," *Progress in Astronautics and Aeronautics*, vol. 163, 1996.

[37] A. Falsafi, K. Pahlavan and G. Yang, "Tranmission techniques for radio LAN's - a comparative performance evaluation using ray tracing," *IEEE J. Sel. Areas Commun.*, vol. 14, no. 33, pp. 477–491, 1996.

[38] L. J. Greenstein et. al., "A new path-gain/delay-spread propagation model for digital cellular channels," *IEEE Trans. on Vehi. Tech.*, vol. 46, no. 2, pp. 477–485, May 1997.

[39] S. Fischer et. al, "Time of arrival estimation of narrowband TDMA signals for mobile positioning," *IEEE Intl. Sym. Pers., Indoor and Mobile Radio Comm.*, vol. 1, pp. 451–455, 1998.

[40] B. Alavi and K. Pahlavan, "Modeling of the distance error for indoor geolocation," *Proc. IEEE Wireless Commun. Networking Conf.*, vol. 1, pp. 668–772, 2003.

[41] B. Alavi and K. Pahlavan, "Bandwidth effect on distance error modeling for indoor geolocation," *Proc. of the IEEE Personal Indoor Mobile Communications Conference* (PIMRC), 2003.

[42] P-C. Chen, "A non-line-of-sight error mitigation algorithm in location estimation," *Proc. IEEE Wireless Commun. Networking Conf.*, pp. 316–320, 1999.

[43] L. Cong and W. Zhuang, "Non-line-of-sight error mitigation in mobile location," *Proc. IEEE INFOCOM*, 2004.

[44] M. P. Wylie-Green and S. S. (Peter) Wang, "Robust range estimation in the presence of the non-line-of-sight error," *Proc. 54th IEEE Vehi. Tech. Conf.*, vol. 1, pp. 101–105, 2001.

[45] K. Pahlavan, X. Li and J. Mäkelä, "Indoor geolocation science and technology," *IEEE Commun. Mag.*, pp. 112–118, Feb. 2002.

[46] S-S Woo, H. R. You and J. S. Koh, "The NLOS mitigation technique for position location using IS-95 CDMA networks," *Proc. IEEE Vehi. Tech. Conf.*, vol. 4, pp. 2556–2560, 2000.

[47] A. M. Ladd, K. E. Bekris, A. Rudys, L. E. Kavraki and D. S. Wallach, "Robotics-based location sensing using wireless ethernet," *Wireless Networks*, vol 11, no. 1/2, pp. 189–204, 2005.

[48] S. Guha, R. N. Murty and E. G. Sirer, "Sextant: a unified node and event localization framework using non-convex constraints," *Proc.* 6th *ACM Intl. Symp. on Mobile Ad Hoc Networking and Computing* (MOBIHOC), pp. 205–216, May 25-28, 2005.

[49] J. Bruck, J. Gao and A. A. Jiang, "Localization and routing in sensor networks by local angle information," *Proc.* 6th *ACM Intl. Symp. on Mobile Ad Hoc Networking and Computing* (MOBIHOC), pp. 181–192, May 25-28, 2005.

[50] M. Kanaan and K. Pahlavan, "A comparison of wireless geolocation algorithms in the indoor environment," *Proc. IEEE Wireless Commun. Networking Conf.*, pp. 177–182, 2004.

[51] A. Tarighat, N. Khajehnouri and A. H. Sayed, "Improved wireless location accuracy using antenna arrays and interference cancellation," *Proc. of the IEEE Intl. Conf. on Acoustics, Speech, and Signal Processing* (ICASSP), 2003.

[52] W. C. Davidon, "Variance algorithm for minimization," *Computer Journal*, vol. 10, pp. 406–410, 1968.

[53] J. Riba and A. Urruela, "A non-line-of-sight mitigation technique based on ML-detection," *Proc. of the IEEE Intl. Conf. on Acoustics, Speech, and Signal Processing* (ICASSP), Montreal, Canada, May 17-21, 2004.

[54] J. Zangl and J. Hagenauer, "Large ad-hoc sensor networks with position estimation," *Proc. of the 10th Aachen Symposium on Signal Theory*, Aachen, Germany, pp. 115–118, 2001.

[55] D. Niculescu and B. Nath, "Ad hoc positioning system (APS)," *Proc. of IEEE Global Communications Conference* (GLOBECOM), 2001.

[56] S. Capkun, M. Hamdi and J. P. Hubaux, "GPS-free positioning in mobile ad-hoc networks," *Proc.* 34th *Hawaii Int. Conf. on System Sciences (HICSS)*, Jan. 2001.

[57] I. J. Mitola, "Cognitive radio for flexible mobile multimedia communications," *Proc. IEEE Intl. Workshop on Mobile Multimedia Communications 1999 (MoMuC 99)*, 1999; 3 10, doi:10. 1109/MOMUC.1999.819467.

[58] I. Akyildiz, W. Lee, M. Vuran, S. Mohanty, "Next generation/dynamic spectrum access/cognitive radio wireless networks: a survey," *Computer Networks*, 2006; 50(13):21272159.

[59] A. Bhattacharya, R. N. Ghosh, K. Sinha and B. P. Sinha, "Multimedia communication in cognitive radio networks based on sample division multiplexing," *Proc. 3rd IEEE Intl. Conf. on Communication Systems and Networks (COMSNETS)*, Bangalore, India, Jan. 4-8, 2011, doi: 10.1109/COMSNETS.2011.5716435.

Chapter 2

Mobility Management in Cellular Networks

In Chapter 1, we briefly introduced the two basic location management functions for call setup: location registration and call delivery, and also described the handoff management function when an MT moves from one cell to the other. We now discuss

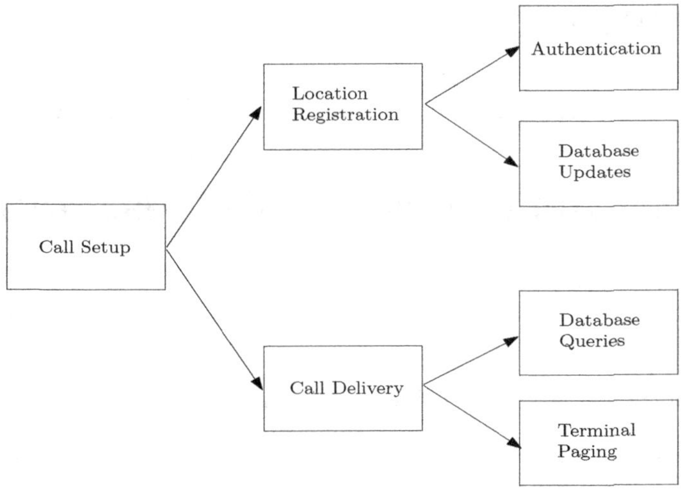

FIGURE 2.1: Call setup in PLMN

all these activities in more detail for a public land mobile network (PLMN) and a mobile IP (mobile Internet protocol) network.

2.1 Call Setup in Public Land Mobile Networks

A schematic diagram of the various operations to be performed for call setup in PLMN is shown in Figure 2.1. The *location registration* module in the Figure 2.1 updates the location databases of the mobile terminals (MTs) stored at various places (e.g., MSCs) of the network and is also responsible for authentication of a mobile terminal when a request for location update from an MT is received. The *call delivery* module shown in Figure 2.1 [1] performs the two functions: i) querying the databases to find the current MSC which keeps track of the called MT and ii) paging different LAs under that MSC to exactly locate the cell within which the called MT is currently residing. A detailed discussions on all these activities is given below.

2.1.1 Signaling Systems

In a public land mobile network, the two location management functions, i.e., location registration and call delivery required to set up a call are implemented by the exchange of signaling messages through a signaling network. Signaling System 7 (SS7) [2, 3, 4] is the protocol used for signaling exchange which is based on a two-level data hierarchy using two types of location registers, the home location register (HLR) and the visitor location register (VLR) associated with each MSC . A user or an MT is permanently associated with an HLR in its own subscribed network (where it is originally registered with all details about the user and the types of services to be provided). When an MT moves to another LA controlled by a different MSC, the information regarding the current location of that MT is kept in the VLR of the MSC controlling its current LA, and the current location information of the MT is also communicated to the HLR of the MT (by sending a signaling message to the MSC which manages its HLR). Figure 2.2 shows a hierarchical view of the corresponding SS7 signaling network where the intermediate nodes STPs stand for the signal transfer points used for routing the signaling messages between the HLR of the MT and the VLR of its current MSC.

2.1.2 Location Registration

Location registration updates the current location of the MT in the MSC of its current LA and also authenticates the MT with the help of signaling messages from the HLR of the MT. There are two commonly used standards for location registration and call delivery: i) IS-41 (Interim Standard-41) [5, 6] introduced by the Electronic and Telephone Industry Associations EIA/TIA, and ii) GSM MAP (Global System for Mobile Communications Mobile Application Part) [7, 6]. IS-41 is popular in North America, while GSM MAP is commonly used in Europe, and both these standards are based on the above two-level database hierarchy using HLR and VLR.

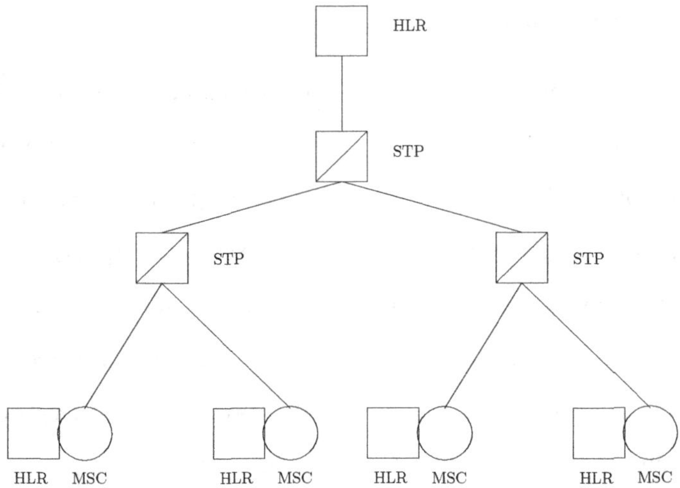

FIGURE 2.2: SS7 signaling network in PLMN

Both standards essentially follow the same steps in performing location registration and call delivery functions which are mentioned below.

The location registration process involves the following sequence of steps [1]:

1. The MT enters a new LA and transmits a location update message to the new (current) base station BS.
2. The current BS forwards this location update message to its MSC (controlling the current LA) which issues a registration query to its associated VLR.
3. The VLR updates its record on the location of the MT. If the new LA belongs to a VLR which is different from the VLR corresponding to the old LA of the MT, then the MSC corresponding to this new VLR determines the address of the HLR of the MT from its mobile identification number (MIN) through a global title translation process implemented by a table look-up mechanism. This MSC then sends a message to the home MSC of the MT (which holds the information of the MT in its HLR) for updating the current location information of the MT in its HLR.

4. The HLR then performs the required procedure for authentication of the MT and records the address of the new serving VLR of the MT and the HLR then sends a registration acknowledgment to the new VLR (through the respective MSCs).

5. The HLR also sends a registration cancellation message to the former serving VLR of the MT.

6. The former VLR cancels the registration of the MT from its record and then sends a registration cancellation message to the HLR.

Depending on the distance between the current and home locations of the MT, steps 3 to 6 above may need message communication through a number of STPs, which would require a lot of communication time and cause significant traffic in the signaling network. In particular, if a location update is initiated by an MT each time it crosses an LA boundary, excessive location updates may be performed by an MT which is located near an LA boundary as it makes frequent movements back and forth between two LAs across this boundary. Thus, with a large number of such MTs, the location update messages from various mobile terminals will generate very heavy traffic in the signaling network. The need for reducing the overhead of location updates leads to other possible schemes as mentioned below:

Time-based location update: An MT performs location updates periodically at a constant time interval, say ΔT. Figure 2.3 [1] shows this situation where the MT is at point A when a location update was initiated by the MT, say, at time $t = 0$. The next two successive location updates are then issued at time $t = \Delta T$ and $2\ \Delta T$ when the MT moves to points B and C, respectively. Note that while moving from A to B, the MT crossed one cell boundary while from B to C, it crossed three cell boundaries. One drawback of this scheme is that the MT may not move at the successive time instants of issuing the location update messages, i.e., unnecessary location

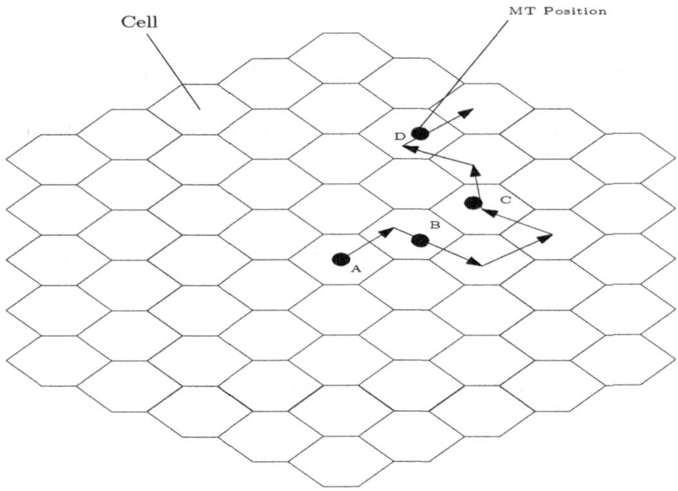

FIGURE 2.3: Time-based location update scheme

updates may have been issued.

Movement-based location update: The location of an MT is updated when the MT makes a predefined number of moves across the cell boundaries. The situation is illustrated by Figure 2.4 [1] assuming that the location update will be initiated after the MT crosses three cell boundaries. Thus, in this figure, the MT initially starts moving from point A and after reaching point B, when it has crossed three cell boundaries, a location update is initiated. The next location update is made at point C when it crosses another three cell boundaries from point B. The drawback of this approach is again that if the MT is situated around a cell boundary and moves back and forth across the same boundary without moving to any other distant cell, unnecessary location updates may be initiated.

Distance-based location update: The location update in this scheme is initiated when an MT moves through a predefined distance from the cell where it performed its last location update. The situation is illustrated in Figure 2.5 [1] with the assumed threshold of three-cell distance (two cells sharing a common boundary are assumed

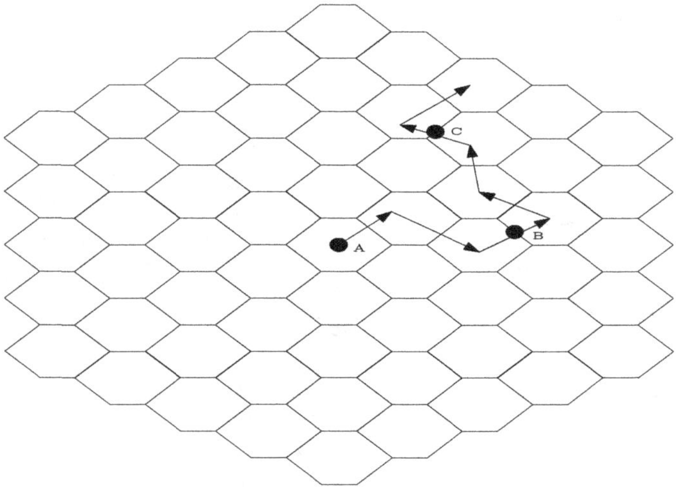

FIGURE 2.4: Movement-based location update scheme

to be one distance apart) for successive location updates. Thus, the MT first makes a location update at point A, then it moves along the path AB. At the position B where the distance of the MT is going to exceed the assumed threshold of three-cell distance, the next location update is initiated.

Dynamically changing size of LA: In this location update scheme, the LA size and hence the LA boundaries may be different for different MTs and may be dynamically changed even for the same MT depending on its recent mobility characteristics. For example, referring to Figure 2.6, [1] the LA size of an MT, say MT1 may consist of 2×2 square cells, whereas for another mobile terminal, say MT2, moving at a faster rate, the LA size may consist of 3×3 cells. Thus, essentially, the LA size is chosen on a per-user basis aiming at getting the optimum results in terms of the total overhead in time arising from location update and paging. Naturally, this scheme is more complicated than the others and is difficult to implement because of the dynamically changing LA size and LA boundaries for each individual MT. The scheme becomes even more complicated if the cells are hexagonal rather than square.

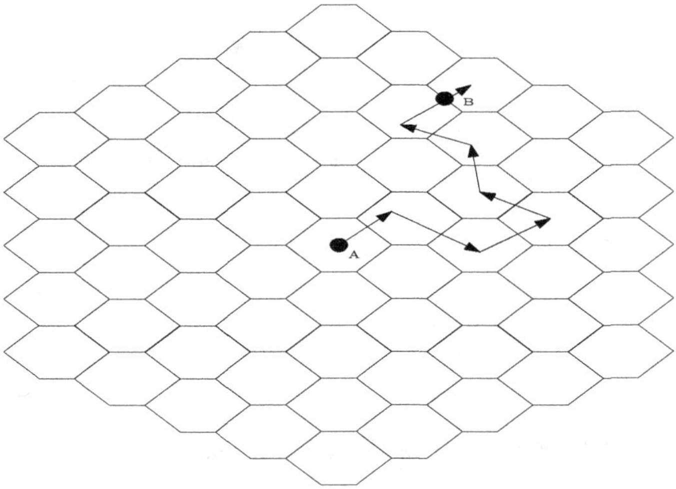

FIGURE 2.5: Distance-based location update scheme

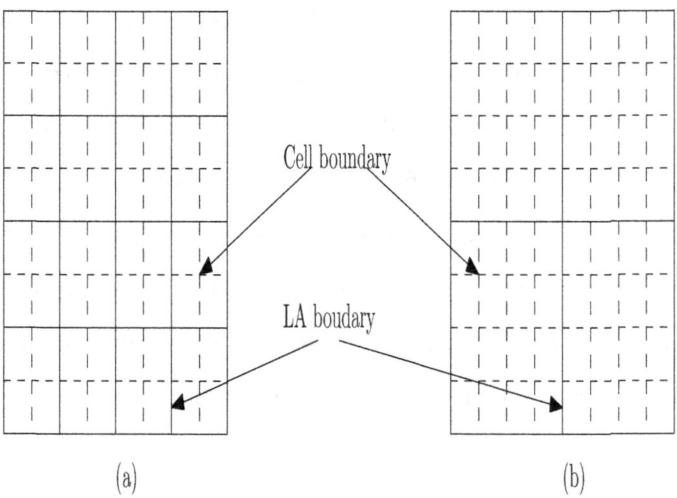

FIGURE 2.6: Dynamic location area-based update scheme

2.1.3 Call Delivery

The call delivery activity, in turn, involves two major functions [1]: a) determining the currently serving VLR of the called MT and b) finding the exact cell of the called MT within its current LA. The first function consists of the following sequence of steps:

1. The calling MT sends a call initiation message to its serving MSC through the base station.
2. The MSC finds the address of the HLR of the called MT through the global title translation process and sends a location request message to this HLR.
3. The HLR determines the currently serving VLR of the called MT and sends a route request message to this VLR through its corresponding MSC.
4. This MSC (corresponding to the VLR) then allocates a temporary identifier called the temporary local directory number (TLDN) to the called MT and sends a reply to the HLR along with this TLDN.
5. The HLR then forwards this information to the MSC of the calling MT.
6. The calling MSC requests a call setup to the called MSC through the SS7 network.

For the other function involved in call delivery, namely finding the exact cell of the called MT within its current LA, the network locates the MT by paging all cells within the LA. However, requiring the network to simultaneously poll all cells within the LA each time a call arrives may result in very heavy traffic in the network. It is, in general, very difficult to find the appropriate LA boundaries and the corresponding paging scheme which will lead to optimum results with minimum overhead in time for location update and paging for all MTs. In fact, an optimum call setup process with minimum overhead on location updates and minimum paging delay may be achieved on a per-user basis by optimally selecting the location areas for an MT depending on its mobility pattern and call arival rate.

If the paging delay is unconstrained, the message traffic arising due to polling can be reduced to some extent by sequentially searching the cells in decreasing order of probability that the MT is currently residing in that cell [8]. However, this requires some prior knowledge about the recent history of movements of the MT within that LA. Location updates and paging subject to delay constraints have been considered by various researchers [9, 10] and are essentially based on having an upper bound on the time required for paging. In [9], a distance-based location update mechanism has been assumed and the residing area of the called MT is partitioned into a number of smaller areas. When an incoming call for the MT arrives, all these subareas are polled sequentially. The number of polling areas is limited to some maximum value, say N, so that the total paging delay is limited to the time required for N polling operations. Depending on the mobility pattern of the MT (which may be decided by a mobility model), call arrival rate and maximum delay N, the optimal distance may be computed for distance-based location update. Thus, an optimal scheme for a constrained paging delay scheme with distance-based location update is complex to implement. However, as in [10], if a movement-based location update is used along with the maximum number N of polling operations to limit the paging delay, the implementation of the overall scheme for location update and paging becomes relatively simple.

2.1.4 Database Architectures

As discussed above, location registration involves updating the location databases with the most recent location information of an MT that is currently available, while call delivery requires querying these databases to find the current location of a called

MT first to the level of the MSC that stores the current information of the called MT and then to the cell level, i.e., the level of the base station (BS) of the cell where the called MT currently resides. If the called MT has moved far away from its home location, a lot of location update messages may be generated from its current VLR to the HLR and also to all the intermediate VLRs on the path from its current VLR to the HLR. Also, the acknowledgement and registration cancellation messages will be communicated among its HLR and the respective VLRs (see steps 4, 5 and 6 of the location registration process described above). The total overhead in terms of the time needed for the two basic operations of updating the databases and terminal paging as well as the message traffic in the network for the required signalling operations may be significant, as explained below, unless the location databases are properly managed.

For example, consider a situation that the called MT is originally registered in Kolkata, and has currently moved to Bangalore. Then a location update message is transmitted from the VLR in Bangalore to its HLR in Kolkata whenever the MT moves to a new LA that belongs to a different VLR. This may involve message communication through a number of intermediate MSCs. Further, when a call is generated from another MT in Bangalore to this roaming MT (which is also in Bangalore), the MSC of the calling MT must first query the HLR of the called MT in Kolkata, to discover that the called MT is currently residing in Bangalore. Due to an ever-increasing number of mobile users, this signalling traffic through the network can be very high, causing a significant reduction in the network's performance in handling the calls.

There are two basic approaches to design the architecture of the location databases to mitigate the above problem. The first is the *centralized database architecture* and the other is the *distributed database architecture*.

Centralized Database Architecture: One possible scheme under this category is a dynamic hierarchical database architecture which uses an additional level of databases called *directory registers (DRs)* along with the two-tier IS-41 database structure involving HLR and VLR. The basic idea is to use a DR that covers the service area of a number of MSCs so that it stores the location pointer configuration for the MTs roaming in its service area. There are three basic types of location pointers used by a DR [1]:

1. A *local pointer* is stored at an MT's serving DR pointing to the currently serving MSC of the MT..
2. A *direct remote pointer* is stored at a remote DR which points to the currently serving MSC of the MT.
3. An *indirect remote pointer* is stored at a remote DR which indirectly points to the currently serving MSC of the MT.

Along with this scheme of using the DRs, the HLR of the MT may now be configured to store a pointer to the serving DR and/or the currently serving MSC. In some cases, however, no such change in HLR may be made for a cost-effective operation and the original IS-41 scheme may be used in HLR. In view of these DRs, let us again consider the above example of an MT having its HLR in Kolkata, and currently roaming in Bangalore. Suppose it moves to Chennai and receives calls from the Bangalore area. If a significant number of incoming calls to this MT are detected as originating from Bangalore, a direct or indirect remote pointer can be set up for this MT in the DR at Chennai. When the next call is initiated for this MT from Chennai, the calling MSC first queries the DR and the call can be immediately forwarded to Bangalore, without the need for querying its HLR in Kolkata. This significantly reduces the signaling traffic in the network for call delivery. Likewise, signaling traffic in location registration can also be greatly reduced if its HLR records the ID of the serving DR (instead of the serving MSC) of the MT. If the MT moves to a new

LA under a different MSC (but under the same DR) in the Bangalore area, then only the local pointer at the serving DR of the MT has to be updated, without the need for accessing the HLR in Kolkata. Thus, such a scheme using DRs can be used for significant overhead reduction in both location registration and call delivery processes.

Another possible scheme for reducing the overhead in location registration and call delivery is through *per-user location caching* [11] which stores the current location information of an MT in a cache at a nearby signal transfer point (STP). Whenever an MT is accessed through the STP, an entry is added to the cache which refers to the mapping from the ID of the MT to that of its serving VLR. When a call is initiated to an MT, the STP first checks whether a cache entry exists for this MT. If no cache entry exists, the usual (e.g., IS-41) call delivery scheme as described above is used to locate the MT. If, however, a cache entry is found for this MT, the STP will query the VLR as specified by the cache entry. If the MT still resides under this VLR, a cache hit occurs and the MT is found easily without the need for querying its HLR. If, however, the MT has already moved to a another location not associated with this VLR, a miss occurs and the usual IS-41 call delivery scheme is followed in locating the MT. Figure 2.7 illustrates this scheme in which MT1 is trying to call MT2 and there is a cache entry at the signal transfer point STP1 corresponding to this MT2 that it is now mapped to VLR2. If the MT2 still resides under VLR2, the call is delivered to it immediately through VLR2 without querying the HLR of MT2. However, if MT2 moves to a new location under a VLR other than VLR2, there will be a cache miss and the usual IS-41 process will be applied to locate MT2. Thus, the overhead under this cache miss situation is more than that in the usual IS-41 scheme. For a better system performance, there should be a minimum cache hit ratio and the system should operate with a cache hit ratio always above this value. This minimum cache hit ratio depends on the local call-to-mobility ratio (LCMR) (LCMR) which is defined as ratio of the average number of calls to an MT from a given originating

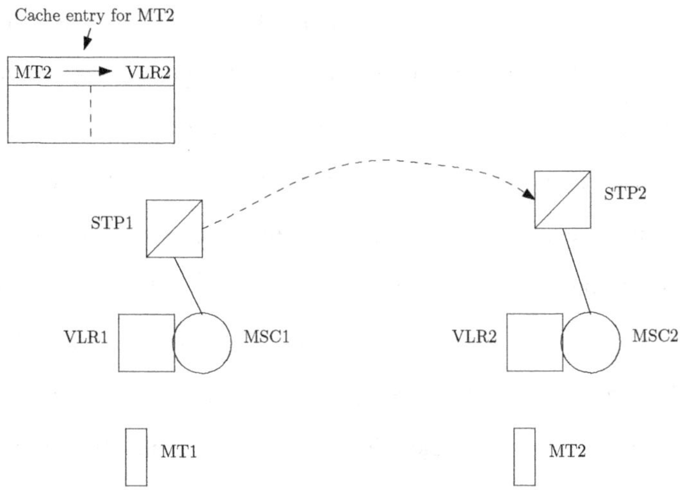

FIGURE 2.7: Location information caching scheme

STP to the average number of times the MT changes its VLR per unit time. It is recommended that some minimum value of LCMR is necessary to attain the minimum value of cache hit ratio. Also, to reduce the number of misses, it is suggested [35] that the cache entries be invalidated periodically after a certain interval of time.

There exist several other possible schemes using centralized databases for reducing the overhead in time and signalling traffic for location registration and call delivery, e.g., user profile replication [12], pointer forwarding [14], and local anchoring [13]. Brief discussions of all these schemes can also be found in [1].

Distributed Database Architecture: In contrast to the centralized database architecture described above, the distributed database architecture uses multiple databases distributed throughout a whole network. There are variants of this approach as described below:

Fully Distributed Tree-based Database Architecture: In this scheme [15], the HLR/VLR database architecture of the IS-41 standard is replaced by a large number of location databases organized as a tree as shown in Figure 2.8, in which the MTs are associated with the leaf location databases and each location database at a non-leaf node contains location information of all MTs residing in its subtree. Referring to Figure 2.8, suppose a mobile terminal MT1 is in a location area LA1. Entrpes for MT1 exist in all databases along the path from its current location to the root of the tree. If a call to MT1 is initiated by another mobile terminal MT2 under the database at node A, first the database at node A is queried if it contains location information of the called mobile terminal MT1. Since this database at A does not have an entry for MT1, the call request is forwarded up the tree to the nodes B and C, and then to the root node D. Since the database at node D contains location information of MT1, the called mobile terminal is located from this database information. If MT1 later moves to a different location corresponding to a different leaf node, all the databases from its current location up to the root are to be accordingly updated with new location information. Also, all the previous databases which contained the location information of MT1 and do not fall in the path from the current location of MT1 to the root need to be updated by deleting their corresponding entries for MT1. Thus, although this scheme reduces the signaling traffic for call delivery, the overhead in location updates increases due to a large number of databases to be updated in the tree, with the corresponding increases in time and signaling traffic for location registration.

Partitioning: This scheme [16] aims at reducing the overhead in location registration in the fully distributed tree-based architecture. Based on the mobility pattern of the MTs, the PLMN is partitioned into several groups, each group consisting of a set of location areas in the tree database structure as shown by the partitions $P1$, $P2$ and $P3$ in Figure 2.9. Suppose a mobile terminal MT1 enters a location area in partition $P2$ consisting of the five location areas shown in the figure. Its location is updated only in the location server $LS2$ which is the nearest common ancestor of all five nodes in the tree database architecture, indicating that MT1 is residing currently in the subtree rooted at $LS2$. If MT1 next moves to some other location area under the same subtree rooted at $LS2$, no location update is performed. This scheme min-

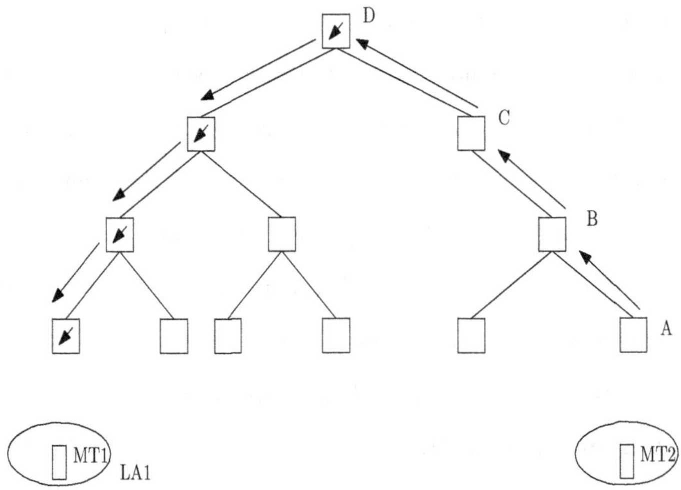

FIGURE 2.8: Distributed tree-based database architecture

imizes the overhead in location registration in areas (partitions) where the mobility rates of the MTs are very high within the same partition.

Variant of the distributed database hierarchy for improving upon the overhead in location registration and call delivery can be found in [17]. A summary of other possible schemes can be found in [1].

2.2 Call Setup in Mobile IP Networks

All the above discussions on call setup refer to the public land mobile network. The location registration and call delivery in a mobile IP network, however, works in a different manner *without the use of any location databases* as in a PLMN. The standards for communication in a mobile IP network have been developed by the IETF (Internet Engineering Task Force) [18]. In such a network, mobile terminals in the same subnetwork can send their packets directly, but the terminals belonging to

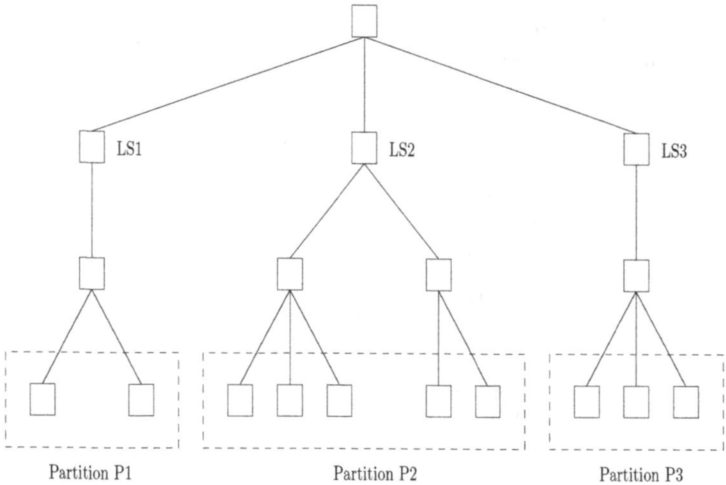

FIGURE 2.9: Partitioning scheme in database architecture

different subnetworks must send their packets through IP nodes or routers that perform the switching functions. Standards allow the packets to be sent by a mobile node (MN) without interruption to another destination MN in some other subnetwork even when the source MN is moving from one subnetwork to another. (Note that the term MN is used instead of MT in order to follow the mobile IP conventions.) When an MN resides in its own registered area, it accesses the Internet through a home agent (HA) which is a router in its own subnetwork (where it is originally registered). When the MN moves to a different subnetwork, it accesses the Internet through a foreign agent (FA) which is a router in its currently visiting subnetwork. The destination node which receives the packet from this MN is called the correspondent node (CN). The essence of the mobile IP network architecture is illustrated in Figure 2.10 [1].

As shown in the figure, the MN has a home agent in its own subnetwork (subnet A) through which it accesses the Internet. When the MN moves to a different subnetwork (subnet B in the figure), it will have a foreign agent in subnet B through which it will access the Internet. The communication between the MN and the correspondent node is now effected through the MN's HA and the current FA.

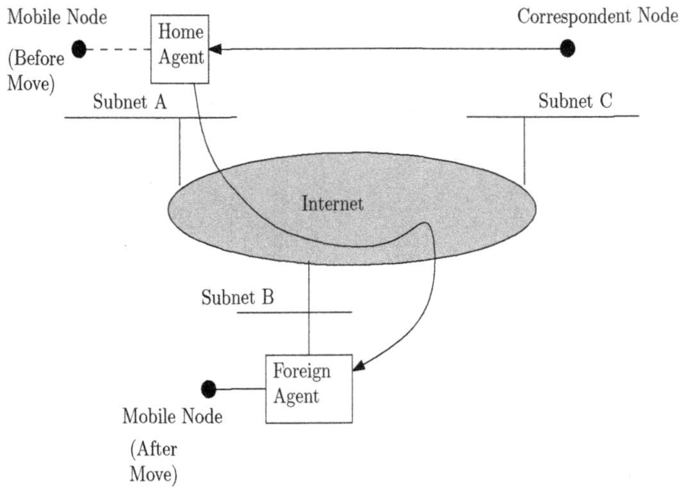

FIGURE 2.10: Mobile IP network architecture

When an MN moves from one subnet to another, it has to register its new IP address (in the newly visited subnet) with its home agent. Thus, in effect, there are two IP addresses associated with every MN: one is the home address for its identification and the other is for finding its current location. The latter address is known as its *care of address (CoA)*. The association between the current CoA and the MN's home address is maintained by some mobility binding rule with some lifetime period for this binding. Any packet destined for this MN is routed using the current CoA. The lifetime period of the binding with the CoA is determined at the time of MN's registration at its home agent and after this period, this registration is deleted. The MN must re-register with its HA before the expiry of this time period to ensure continuous service (e.g., without interrupting ongoing calls) using this CoA. Thus, the MN may need to send a registration request message even when it is in its home subnet, i.e., the registration message may be sent directly to its HA (when it is still in its own subnet) or through an FA it is currently visiting. According to the Mobile Internet Protocol Version 4 (IPv4), the following sequence of actions is performed in the location registration process of an MN, illustrated in Figure 2.11 [1].

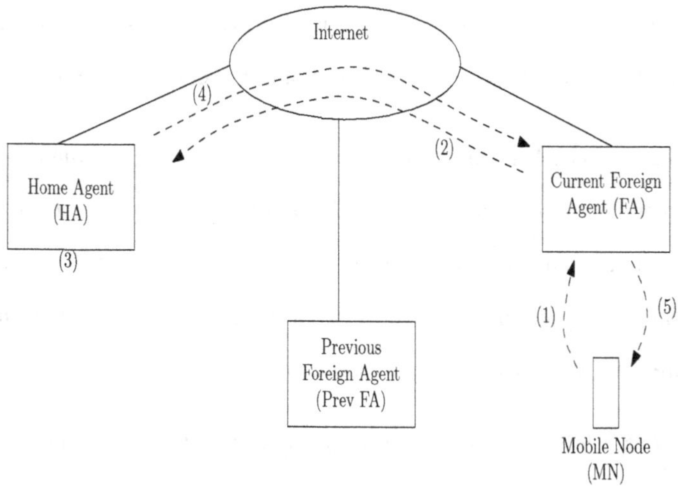

FIGURE 2.11: Mobile IP location registration

1. The MN registers its current IP address with the its HA using a registration request message which may be relayed to the HA by its current FA (arcs labeled (1) and (2) in Figure 2.11).

2. The HA creates or modifies a mobility binding with a new lifetime for the MN with this IP address (step (3) in Figure 2.11).

3. The HA sends a registration reply message to the MN (maybe also through the corresponding FA). This reply message contains the necessary code for the status of the registration request along with the binding lifetime information (arcs labeled (4) and (5) in Figure 2.11).

Revisions of the Mobile Internet Protocol include Version 4 (IPv4) and Version 6 (IPv6) [19]. IPv6 provides more addresses and mobility support than IPv4. In IPv6, the FAs in Figure 2.11) are thought of only as *access points (APs)*. IPv6 enables the use of some mobile agents to determine whether an MN has moved to a new subnet after changing its network APs. These mobile agents make themselves known by sending agent advertisement messages. There are two mechanisms which may be used by the MN to detect its movement from one subnet to another as given below:

1. **Advertisement Lifetime:** There is a *lifetime* field in the main body of the Internet Control Message Protocol (ICMP) router advertisement portion of the agent advertisement. The MN records this information for the lifetime of the binding with the current CoA, until the lifetime expires. When the lifetime expires, the MN must attempt to solicit a new agent.

2. **Network Prefix:** This scheme uses a network prefix consisting of some number of initial bits in the IP address to detect movement. If the prefix received through the agent advertisement differs from the prefix in the current CoA of the MN, the MN decides that it has moved to a different AP.

Once the MN discovers that it is on a new foreign network, it obtains a new CoA from the new AP and performs the location update procedure, which includes the creation or modification of a mobility binding at its home agent by associating the current CoA with the associated lifetime for the binding. This is in contrast to the location update process in PLMN which involves modifying location databases at various places of the network.

The steps involved in the location registration process in a mobile IP network are shown in Figure 2.12, consisting of the following sequence of operations [1]:

1. The MN registers its new CoA along with the associated lifetime at its home agent (HA) by sending a binding update (refer to the arc labeled (1) in the figure).

2. The MN also informs its current binding information to its correspondent node (CN) (arc (2) in the figure).

3. When the current binding is going to expire, the home agent and the correspon-

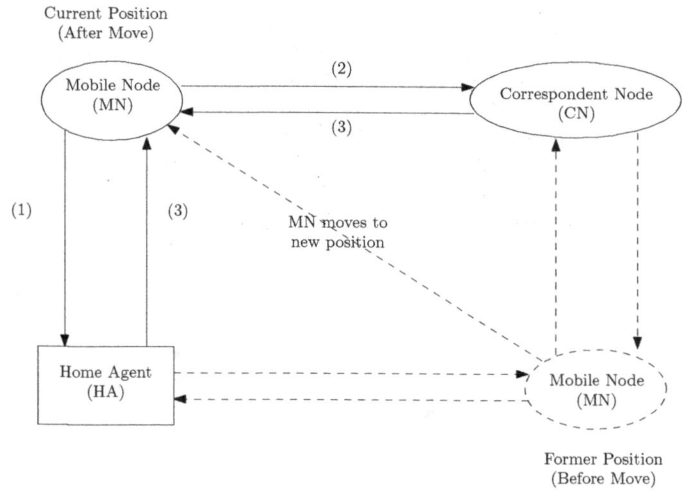

FIGURE 2.12: Sequence of operations in mobile IP location registration

dent node send a binding request to the MN for current binding information (arcs labeled (3) in the figure).

4. The MN responds to the binding requests from its HA and CN with its new binding information (binding update) (arcs labeled (4) in the figure).

5. Both HA and CN send a binding acknowledgement message (arcs labeled (5) in the figure) to the MN after receiving the binding update information from the MN.

The overall call setup process for Mobile IP networks is shown in Figure 2.13 [1].

After completing the location registration process, the requested call is delivered to the called MN through its current CoA obtained from its HA.

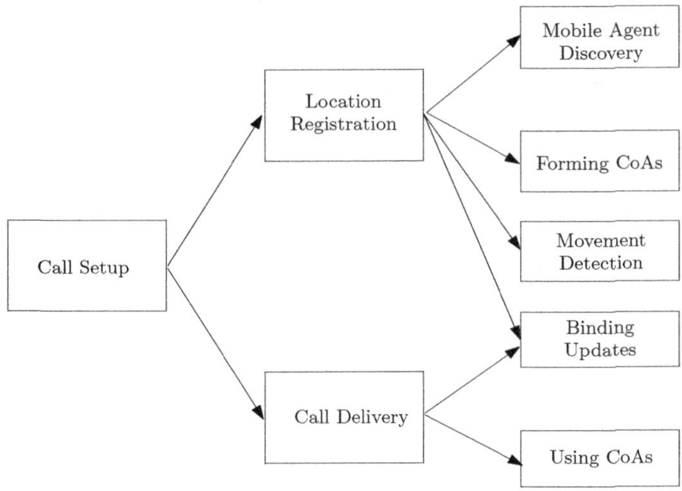

FIGURE 2.13: Call setup process in mobile IP network

2.3 Handoff Management

In Chapter 1, we mentioned three distinct stages for handoff: i) handoff initiation, ii) new connection generation and iii) data flow control. Actually, handoff intiation depends on the received signal strength by the MT from the BS and vice versa. This leads to two types of handoff situations: intracell handoff and intercell handoff. Intracell handoff may be initiated when the user moves even in the same cell, but the signal strength falls below the threshold value due to changed network surroundings (e.g., obstacles or reflecting walls, etc.) for reliable communication to be maintained between the BS and the MT. In this case, after the handoff, the ongoing call to the MT will be continued over a newly allocated channel through which the signal strength received by the MT from the BS and vice versa will be more to sustain a reliable communication. The intercell handoff is initiated when the MT moves from one cell to an adjacent cell so that the signal received from the old BS is of smaller strength than that from another BS in the direction of its movement.

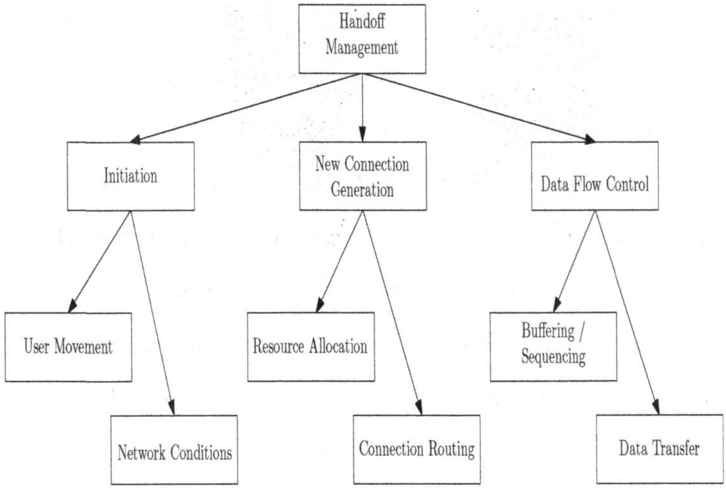

FIGURE 2.14: Handoff management process

When performing an intracell handoff, it is quite possible that an MT is simultaneously receiving strong signals from multiple new base stations because of physical proximity from all of them. In such a case, some kind of signaling diversity may be used by the MT to combine the multiple signals received from those base stations, to stay connected to all new base stations. This transient situation is termed *soft handoff*. Eventually, the MT will be handed over to only one of these new base stations as may be decided from the direction of movement of the MT and the received signal strengths. The connection with the former base station will be cleared under this situation after transferring the data packets of any ongoing call to the MT that might be buffered during the establishment of this handoff process. Also, the MT's required resources will be allocated from the new base station and the routing path for receiving the packets of the ongoing calls will be determined. At this point, a *hard handoff* of the MT from the old base station to the new one occurs. All these steps involved in handoff management are represented in Figure 2.14 [1].

For a smooth handoff process, it is essential that when a mobile user is changing its point of attachment (base station in a PLMN or an Internet access point (AP) in a mobile IP network), any ongoing call is allowed to continue for maintaining the required quality of service. To achieve this goal, any data packets to be communicated through the network on behalf of this ongoing call should be properly transferred to and from the mobile user so that this handoff is transparent. In case of a PLMN, the old base station may buffer the relevant data packets to and from the MT in the proper sequence, and eventually transfers these to the new base station (after the handoff) for delivering to the concerned MTs of the ongoing call. In a mobile IP network, the previous foreign agent maintains a binding for its former visitor MN along with the old CoA of the MN. As packets are sent to this old CoA, the former foreign agent forwards these packets to the home agent of the MN which, in turn, forwards them to the current CoA of the MN as newly registered in its home agent. However, this process of routing the data packets for the MN from its previous foreign agent to the current foreign agent via the home agent often results in a significantly longer path with associated increases in message traffic and routing delay. This can be avoided if the MN also sends information about its new CoA received from its current foreign agent to the immediate previous foreign agent. Thus, the former foreign agent can now forward these data packets directly to this new CoA of the MN, instead of sending them to the home agent of the MN. For more details, the reader is referred to [20, 21].

When a mobile terminal moves from one access point to another using a different network technology, the corresponding handoff is termed a *vertical handoff*, unlike the *horizontal handoff* in which the two access points before and after handoff use the same network technology, standards and protocols. For example, if a mobile terminal is currently under a wireless LAN and it moves to an access point of a cellular network, the corresponding handoff will be vertical. But if it moves from one access point of a cellular network to another access point of the same cellular network, the corresponding handoff will be a horizontal handoff. Transition from one access point to the other needs to be smooth in both horizontal and vertical handoff situations so that the user is transparent to any changes in the signal strength, network technologies, standards and protocols.

2.4 Mobility Models

In all the mobility management functions discussed above, lot of optimizations in location registration, call delivery and handoff management could be possible if the mobility patterns of each individual MT's or MN's were known or predictable. For example, if it is predicted with confidence that an MT will move to one particular adjacent cell in the next few seconds, the appropriate handoff management process might be started with the base station of that cell beforehand in a proactive manner, avoiding message traffic that would have been otherwise generated during the initiation of handoff. Likewise, paging cost in the call delivery phase could be minimized if the current location of the called MT could have been accurately predicted. Also, for the location registration activity, the choices of the location and size of the location databases (in distributed database architectures) can be optimized for reducing the overhead. The location registration request from an MT can also be made on a per-user basis as and when it is predicted that the MT will be in a new LA in the next few seconds. Of course, the success of all these optimization processes depends on the accuracy of the prediction. If the prediction is incorrect, the usual techniques (in absence of prediction) for performing the above functions should be adopted.

Several mobility models have been proposed in the literature to model the mobility patterns of individual users under different scenarios or environments. We give below a brief introduction to these models. Some of these mobility models can also be applicable to the nodes in a mobile ad hoc network (MANET) or a wireless sensor network (WSN).

2.4.1 Random Walk Mobility Model

According to the model proposed by Royer et al. [22], a mobile user chooses a random direction in the interval $[0, 2\pi]$ and a random velocity within some range, say $[v_{min}, v_{max}]$, to make a movement from a certain point in a closed region R. When the mobile user reaches the boundary of R, it again chooses some other random direction and velocity to move from that point, and so on. Thus, this model effectively assumes obstacle-free random movement for every individual MT (or MN) leading to a uniform spatial distribution of mobile nodes within region R during any simulation run.

There are some variants of this model proposed in the literature. The first proposes that a mobile node is bounced back from the boundary when it reaches a boundary point of the region R [23, 24]. A second variant allows a mobile node to move in a randomly chosen direction at a randomly chosen velocity for a time duration which is exponentially distributed and then the mobile node will again change its direction, velocity, and duration of movement with these chosen direction and velocity.

A third variant, known as the *Gauss-Markov mobility model* [25] involves the assignment of an initial speed and a direction of movement to each mobile node. After a fixed time interval, the speed and direction of each mobile node for the next interval, say n^{th} time interval, are determined from the speed and direction at the previous time interval (i.e., $(n-1)^{th}$ interval) using two statistically random variables S_x and D_x (for speed and direction, respectively) that follow Gaussian (normal) distribution and some tuning parameter α, $0 \leq \alpha \leq 1$ that determines the relative contribution by the current status and the variability in node movement. Specifically, the values of speed S_n and direction D_n at the nth time interval are calculated based on the values of speed S_{n-1} and D_{n-1} at the $(n-1)^{th}$ time interval with incremental random changes from the following equations:

$$S_n = \alpha * S_{n-1} + (1+\alpha) * \bar{S} + \sqrt{(1-\alpha^2)} * S_{x_{n-1}} \tag{2.1}$$

$$D_n = \alpha * D_{n-1} + (1+\alpha) * \bar{D} + \sqrt{(}1-\alpha^2) * D_{x_{n-1}}, \tag{2.2}$$

where $\bar{S}$ and $\bar{D}$ are the mean values of speed and direction, respectively, and $S_{x_{n-1}}$ and $D_{x_{n-1}}$ are values of two random variables S_x and D_x at the $(n-1)^{th}$ time interval.

2.4.2 Random Way Point Mobility Model

This is one of the most popular mobility models and is implemented in the $ns2$ simulator [18]. This model (also referred to as the *RWP model* in the literature) was proposed by Johnson and Maltz [23] to study the performance of the DSR protocol used for routing in ad hoc networks. Its main difference from the random walk model is that instead of choosing a random direction for movement, the mobile node chooses a destination point at random in a given region R. Then, with a randomly chosen velocity from a given interval, say $[v_{min}, v_{max}]$, the mobile node starts moving toward this randomly chosen destination point or *way point*. After reaching this way point, it remains stationary (pause time) for a predefined period (this is another point of difference from the random walk model), and then repeats the above steps to make another movement from that point. This RWP model also allows obstacle-free movements of individual mobile nodes in a given region R which may represent the movement pattern of a person moving in a shopping complex, market place, or convention site.

Figure 2.15 is a representative diagram showing the mobility pattern according

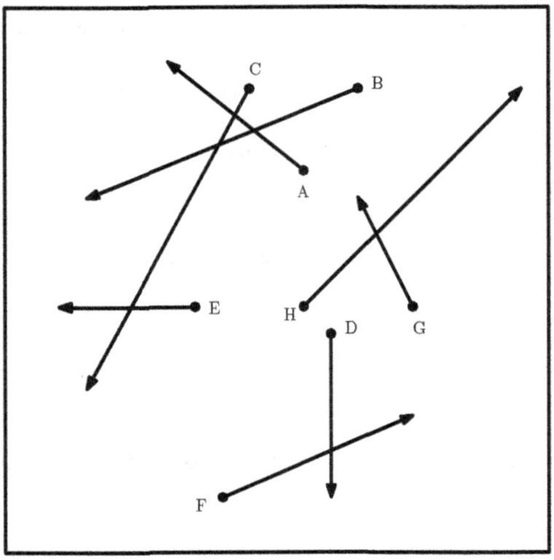

FIGURE 2.15: An example of random way point mobility

to this RWP model where A, B, C, D, E and F are different mobile terminals. The figure shows that the mobile nodes have a tendency to move across the central part of the deployment region. Several researchers [28, 29, 30] noted that the mobile nodes following the RWP model concentrate near the center of the deployment region in the long run. This is known as the *border effect* of the RWP model. The border effect is, however, not exhibited in the random walk model.

One small variation of the RWP model has been proposed by Bettsletter et al. [29] who suggests that the pause time between the time instants of reaching the destination point and starting the next move from there can also be randomly chosen following an arbitrary probability distribution. Also, some random fraction of the nodes in the network may be allowed to remain stationary for the entire duration of interest to reflect a more practical scenario.

2.4.3 Brownian Mobility Model

The above two models reflect the mobility pattern in a network where the nodes make intentional movements, with randomized direction and velocity of each movement at the next step. In contrast, the Brownian mobility model [31] allows non-intentional movements of the constituent nodes in which the position of a node at the next time step is dependent only on its position at the current time step. To be more specific, three parameters are used in this model: $prob_{stat}, prob_{move}$ and m. The $prob_{stat}$ parameter represents the probability that a node remains stationary for the entire time of interest. The $prob_{move}$ parameter represents the probability that a node moves at a given time step and finally m defines a square having each side of length $2m$ centered at the current position. If a node is at its current position at time step k, at the $(k+1)^{th}$ time step its position will be at some randomly chosen point within this square. In effect, the parameter m models the maximum velocity of the mobile node, for a given duration of each time step. Because of the short step movements of a node at each time step, the movement pattern appears like the movement of a drunken person and hence it appears that the nodes following this model make non-intentional movements.

2.4.4 Map-based Mobility Model

In the above three models, nodes are assumed to be free to move around within the deployment region R. In this model, we assume that the nodes move only along specified paths within the region R. Examples of such constrained movements include driving along roads or highways and people moving along sidewalks on the streets. Thus, a road map of the region R is needed in this model. The nodes will be randomly deployed on these roads and the movements of the nodes will be confined to these defined paths only. An example of a map-based model is the *freeway mobility model* proposed by Bai et al. [32] that assumes several freeways with varying numbers of lanes in both directions. Nodes will be deployed on the freeway in a random manner and they can move along these lanes in specified directions only, with randomly chosen velocities which are temporally dependent on their immediate previous velocities. However, if two nodes are moving in the same lane and within a certain minimum threshold distance (which may be directly proportional to the velocity of the following node), the velocity of the following node cannot exceed that of the preceding node.

A variant of freeway mobility model is the *Manhattan mobility model* also proposed in [32], in which the movements follow a Manhattan-like map in the region R consisting of horizontal and vertical streets as in a city grid. Nodes can move along these streets in both directions. After reaching a point of intersection, a node chooses the direction of its next movement: along the same direction or along the left road or the right road. The velocity of a node will follow the same rules discussed above for the freeway mobility model.

2.4.5 Group-based Mobility Model

This model [33, 34] is used to mimic certain real-life situations like movements of a group of tourists in a city or a group of trekkers on a mountain. The idea is that the set of nodes, after being randomly deployed, are divided into clusters or groups with a leader for each such group. The group leaders follow one of the mobility models as discussed above (e.g., random way point mobility model) and the other group members follow the leader with velocities which are small perturbations of that of the leader. However, the model adheres to the provision that when two groups cross each other, any group member may leave the current group and join the other with a certain probability.

2.5 Exercises

1. What are the two important functions to be executed for call setup in a PLMN? Describe the SS7 signalling network architecture used in a PLMN.

2. List the sequence of steps to be followed in the location registration process as used in a PLMN. How are the databases maintained and updated for this purpose?

3. Describe the relative advantages and disadvantages of time-based, movement-based and distance-based location update mechanisms. Which one do you think is the most efficient and why?

4. What are the steps followed during the call delivery process in a PLMN? Explain all these steps with the help of a suitable illustrative example.

5. What are the basic functions of the directory registers used in a centralized database architecture for a PLMN? How are the pointers managed by such directory registers?

6. Describe the partitioning scheme used in a distributed database architecture in a PLMN.

7. Define the terms *home agent, foreign agent, correspondent node* and *care of address* in a mobile IP network.

8. Describe the steps involved in the location registration process used in a mobile IP network architecture.

9. What are the possible ways of tracking the movement of a mobile node in a mobile IP network architecture?

10. State the differences between horizontal handoff and vertical handoff.

11. Distinguish between a soft handoff and a hard handoff.

12. Define intracell handoff and intercell handoff. Can an intracell handoff be also a vertical handoff?

13. What are the different ways a handoff can be initiated and handled by a network?

14. Explain the RWP mobility model used for characterizing the mobility patterns of nodes in a mobile network. Under what situation will the group-based mobility model be better suited than this model?

Bibliography

[1] I. F. Akyildiz, J. McNair, J. S. M. Ho, H. Uzunalioglu and W. Wang, "Mobility management in next generation wireless systems," *Proc. IEEE*, Vol. 87, pp. 1345-1382, August 1999.

[2] A. R. Modarressi and R. A. Skoog, "Signaling System 7: A tutorial," *IEEE Commun. Mag.*, Vol. 28, pp. 19-35, July 1990.

[3] D. R. Wilson, "Signaling system No. 7, IS-41 and cellular telephony networking," *Proc. IEEE*, Vol. 80, pp. 664-652, April 1992.

[4] Y. B. Lin and S. K. DeVries, "PCS network signaling using SS7," *IEEE Commun. Mag.*, Vol. 33, pp. 44-55, June 1995.

[5] EIA/TIA, "Cellular radio-telecommunications intersystem operations," *EIA/TIA Tech. Rep.*, IS-41 Revision C, 1995.

[6] S. Mohan and R. Jain, "Two user location strategies for personal communication services," *IEEE Personal Commun.*, Vol. 1, pp. 42-50, 1994.

[7] ETSI/TC, "Mobile Application Part (MAP) Specification, Version 4.8.0," *Tech. Rep., Recommendation GSM 09.02*, 1994.

[8] C. Rose and R. Yates, "Minimizing the average cost of paging under delay constraints," *ACM-Baltzer J. Wireless Networks*, Vol. 1, No. 2, pp. 211-219, July 1995.

[9] J. S. M. Ho and I. F. Akyildiz, "A mobile user location update and paging mechansim," *ACM-Baltzer J. Wireless Networks*, Vol. 1, No. 4, pp. 413-425, Dec. 1995.

[10] I. F. Akyildiz, J. S. M. Ho and Y. B. Lin, "Movement-based location update and selective paging for PCS networks," *IEEE-ACM Trans. Networking*, Vol. 4, pp. 629-636, Aug. 1996.

[11] R Jain, Y. B. Lin and S. Mohan, "A caching strategy to reduce network impacts of PCS," *IEEE J. Select. Areas Commun.*, Vol. 12, pp. 1434-1444, Oct. 1994.

[12] N. Shivakumar and J. Widom, "User profile replication for faster location lookup in mobile environments," *Proc. ACM/IEEE MOBICOM '95* pp. 161-169, 1995.

[13] J. S. M. Ho and I. F. Akyildiz, "Local anchor scheme for reducing signaling cost in personal communication networks," *IEEE-ACM Trans. Networking*, Vol. 4, No. 5, pp. 709-726, Oct. 1996.

[14] R. Jain and Y. B. Lin, "An auxiliary user location strategy employing forwarding pointers to reduce network impact of PCS," *ACM-Baltzer J. Wireless Networks*, Vol. 1, No. 2, pp. 197-210, July 1995.

[15] J. Z. Wang, "A fully distributed location registration strategy for universal personal communication systems," *IEEE J. Select. Areas Commun.*, Vol. 11, pp. 850-860, Aug. 1993.

[16] B. R. Badrinath, T. Imielinski and A. Virmani, "Locating strategies for personal communication network," *Proceedings of Workshop on Networking of Personal Communications Applications*, Dec. 1992.

[17] V. Anantharam, M. L. Honig, U. Madhow and V. K. Wei, "Optimization of a database hirerarchy for mobility tracking in a personal communication network," *Performance Evaluation*, Vol. 20, No. 1-3, pp. 287-300, May 1994.

[18] C. Perkins, "Mobile IP," *IEEE Commun. Mag.*, Vol. 35, pp. 84-99, May 1997.

[19] D. B. Johnson and C. Perkins, "Mobility support in IPv6," *Internet Engineering Task Force, Internet Draft, draft-ietf-mobileip-ipv6-04.txt*, Nov. 1997.

[20] C. Perkins and D. Johnson, "Route optimization in mobile IP," *Internet Engineering Task Force, Internet Draft, draft-ietf-mobileip-optom-07.txt*, Nov. 20, 1997.

[21] P. Calhoun and C. Perkins, "Tunnel establishment protocol," *Internet Engineering Task Force, Internet Draft, draft-ietf-mobileip-calhoun-tep-00.txt*, Nov. 21, 1997.

[22] E. Royer, P. Melliar-Smith and L. Moser, "An analysis of the optimum node density for ad hoc mobile networks," *Proc. IEEE Intl. Conf. Communications*, Helsinki, pp. 857-861, 2001.

[23] Z. Haas and M. Pearlman, "The performance of query control schemes for the zone routing protocol," *Proc. ACM Sigcomm*, Vancouver, pp. 167-177, 1998.

[24] M. Pearlman, Z. Haas, P. Sholander and S. Tabrizi, "On the impact of alternate path routing for load balancing in mobile ad hoc networks," *Proc. ACM MobiHoc*, Boston, MA, pp. 3-10, 2000.

[25] T. Camp, J. Boleng, and V. Davies, "A survey of models for ad hoc network research," *Wireless Communication and Mobile Computing (WCMC): Special issue on Mobile Ad Hoc Networking: Research, Trends, and Applications*, Vol. 2, pp. 483-502, 2002.

[26] NS2 2002, "The network simulator – ns-2," *http://www.isi.edu/nsnam/ns/.*

[27] D. Johnson and D. Maltz, "Dynamic source routing in ad hoc wireless networks," *Mobile Computing*, Kluwer Academic Publishers, pp. 153-181.

[28] C. Bettsletter and O. Krause, "On border effects in modelling and simulation of wireless ad hoc networks," *Proc. IEEE Intl. Conf. Mobile and Wireless Commun. Network (MWCN)*, Recife, 2001.

[29] C. Bettsletter, G. Resta and P. Santi, "The node distribution of the random waypoint mobility model for wireless ad hoc networks," *IEEE Trans. Mobile Computing*, Vol. 2, pp. 257-269, 2003.

[30] D. Blough, G. Resta and P. Santi, "A statistical analysis of the long-run node spatial distribution in mobile ad hoc networks," *Wireless networks*, Vol. 10, pp. 543-554, 2004.

[31] P. Santi and D. Blough, "The critical transmitting range for connectivity in sparse wireless ad hoc networks," *IEEE Trans. Mobile Computing*, Vol. 2, pp. 25-39, 2003.

[32] F. Bai, N. Sadagopan and A. Helmy, "Important: A framework to systematically analyze the impact of mobility on performance of routing protocols for ad hoc networks ," *Proc. IEEE Infocom*, San Francisco, pp. 825-835, 2003.

[33] X. Hong, M. Gerla, G. Pei and C. Chiang, "A group mobility model for ad hoc wireless networks," *Proc. ACM MSWiM*, Seattle, pp. 53-60, 1999.

[34] K. Wang and B. Li, "Group mobility and pattern prediction in wireless ad hoc networks," *Proc. IEEE Intl. Conf. Commun.*, New York, pp. 1017-1021, 2002.

[35] Y. B. Lin, "Determining the user locations for personal communications services networks," *IEEE Trans. Veh. Technol.*, vol. 43, pp. 466473, Aug. 1994.

Chapter 3

Bandwidth Management in Cellular Networks

3.1 Introduction

Mobile computing, particularly over cellular networks, has emerged as an important topic of research because of the need for computing capabilities even when people are on the move. Over the past decade, the number of such cellular network dependent mobile users has increased almost exponentially. However, the available bandwidth for providing services to those users is very limited. Such a limited availability of the radio spectrum, signal distortion, and the interference caused by the environment and other mobile users impose inherent limitations on the capacity of wireless networks. Therefore, developing methods to utilize the scarce radio spectrum efficiently is more critical than ever.

3.1.1 Channel Assignment Problem and Variants

For cellular networks, the available radio spectrum is divided into a number of non-overlapping frequency bands. We assume that the frequency bands are of equal length and numbered $0, 1, 2, ...$ from the lower end. Each such frequency band is called a channel. The highest numbered channel required in an assignment problem is known as the required bandwidth or *span* of the assignment. When a mobile cellular network is designed, each cell of the network is assigned a set of channels to provide services to the individual calls. The same frequency channel may be assigned to different cells (reused) if they are at a sufficient distance without causing any interference. However, the assignment of channels to nearby cells should satisfy certain constraints to avoid channel interference. Interference is a function primarily of transmitter power, receiver sensitivity, antenna gains, and channel loss. Channel loss is a function of distance, frequency, and weather, and is quantified by the minimum acceptable signal-to-noise ratio (SNR). Assuming that most of these factors are already determined or beyond the ability to influence, the interference can be defined directly as a function of frequency and distance. Based on that, three types of interference are generally considered: 1) co-channel constraint, due to which the same channel cannot be assigned to certain pairs of cells simultaneously, 2) adjacent channel constraint, for which adjacent channels cannot be assigned to certain pairs of cells simultaneously, and 3) co-site constraint, which implies that any pair of channels assigned to the same cell must be separated by a certain number of channels [20].

The channel assignment problem (CAP) deals with assigning frequency channels to cells to satisfy the frequency separation constraints with a view to avoiding channel interference and using as little bandwidth as possible. This variant of the channel assignment problem is known as the *minimum span* channel assignment problem.

For another variant of the CAP, consider the channel assignment when the bandwidth B of the system is given, which may even be smaller than the required lower bound on bandwidth for the given problem. Depending on B, it may or may not be possible to satisfy all the channel demands of each cell unless B is sufficiently large.

Thus, a solution to this variant of CAP may cause blocked calls. However, the objective is to minimize call blocking. This class of assignment problem is known as the *fixed bandwidth* channel assignment problem.

There is, however, another class of problem of real-life importance known as the *perturbation-minimizing* frequency assignment problem (PMFAP) [17]. Assume that an assignment has been obtained for a given network to satisfy the required channel demands. After some time, demands of some of the cells may be changed due to either 1) newly generated calls or 2) a handoff situation or 3) completion of some ongoing calls. The objective of this variant of CAP is to accommodate these small changes in demands; but while doing so, the number of changes in the existing assignment should be minimized while meeting the desired QoS. This variant is useful for addressing the short-term demand fluctuations that often arise in real-life scenarios.

Channel assignment schemes can be classified into several categories. In the fixed (or static) channel assignment problem (FCA), the number of calls to each cell is known a priori and a number of nominal channels is permanently allocated to each cell for its exclusive use. In the simple uniform FCA, the same number of nominal channels is allocated to each cell. Since the traffic in cellular systems is mostly nonuniform, uniform allocation may result in poor channel utilization [34]. To achieve better channel utilization, the number of channels allocated to each cell should depend on the expected traffic load of the cell. This technique is called nonuniform channel allocation [35, 36]. Simple FCA schemes are not adaptive to temporal and spatial fluctuations of user demands and to overcome this drawback, dynamic channel assignment (DCA) schemes have been developed [37]. In DCA, all channels are available to every cell and channels are assigned to newly generated calls as and when required, provided they satisfy the interference constraints. Though DCA schemes provide flexibility and traffic adaptability at the cost of higher complexity, they are less efficient than FCA under high load situations [34, 38, 40]. To address this drawback, many hybrid schemes combining the benefits of both FCA and DCA have been developed. Both the FCA and DCA schemes can be implemented in centralized or distributed fashion. In centralized schemes, a channel is assigned by the central controller whereas in the distributed schemes, a channel is selected by the base station at which the call is initiated or by the mobile [34]. If the base stations select the channel, each base station keeps the information about the currently available channels in its vicinity and this information is updated by exchanging status information among these stations. In the other situation where the mobile terminals are allowed to select the channel, the mobile terminals choose the channels based on the local signal-to-interference measurements [34].

3.1.2 Mathematical Formulation of CAP

The channel assignment problem can be represented as a mathematical optimization problem. The model used in [1], [8], [14] to represent a CAP can be described by the following components:

1. A set of n distinct cells $X = \{0, 1, \cdots, n-1\}$.
2. A demand vector $W = (w_i)(0 \leq i \leq n-1)$ where w_i represents the number of channels required for cell i.
3. A frequency separation matrix $C = (c_{ij})$ where c_{ij} represents the frequency separation requirement between a call in cell i and a call in cell j $(0 \leq i, j \leq n-1)$.
4. A frequency assignment matrix $\Phi = (\phi_{ij})$, where ϕ_{ij} represents the frequency assigned to call j in cell i $(0 \leq i \leq n-1, 0 \leq j \leq w_i - 1)$.
5. A set of frequency separation constraints specified by the frequency separation

matrix :

$|\phi_{ik} - \phi_{jl}| \geq c_{ij}$ for all i, j, k, l (except when both $i = j$ and $k = l$).

Based on this model a channel assignment problem P can be characterized by the triplet (X, W, C). A frequency assignment Φ for P is said to be *admissible* if ϕ_{ij}'s satisfy the component 5 above for all i, j, where $0 \leq i \leq n-1$ and $0 \leq j \leq w_i - 1$. The *span* $S(\Phi)$ of a frequency assignment Φ is the maximum frequency assigned to the system. That is,

$$S(\Phi) = \max_{i,j} \phi_{ij}.$$

Thus, the objective of the *minimum span* frequency assignment is to find an admissible frequency assignment with the minimum span $S_{min}(P)$, where $S_{min}(P) = \min\{S(\Phi)|\ \Phi \text{ is admissible for } P\}$. The advantage of this category of algorithms is that the derived channel assignment always fulfills all the interference constraints for a given demand. However, it may be hard to find optimal solutions for large and difficult problems, even with powerful optimization tools [29].

For the *fixed bandwidth* channel assignment, let B be the given bandwidth of the system. Now suppose, due to the bandwidth constraint B, only w'_i channels are assigned to cell i instead of w_i in an admissible frequency assignment $\Phi = (\phi_{ij})$ $(0 \leq i \leq n-1, 0 \leq j \leq w'_i - 1)$ for P where $w'_i < w_i$ for some or all i. Let $b_i = (w_i - w'_i)$ when $w'_i < w_i$ and 0 when $w'_i = w_i$. Then we say that b_i calls are blocked in cell i by the assignment Φ. Then the total number of blocked calls $BL(\Phi)$ produced by Φ is defined as

$$BL(\Phi) = \sum_{i=0}^{n-1} b_i.$$

Given bandwidth B, the objective of the fixed bandwidth channel assignment is to find an admissible frequency assignment Φ for P with minimum blocking $BL_{min}(P)$, where $BL_{min}(P) = \min\{BL(\Phi)|\ \Phi \text{ is admissible for } P\}$. The principal disadvantage of this formulation is that it is almost impossible to minimize the blocking to the desired value of zero taking the minimum number of required channels as the given bandwidth for very difficult problems [29].

Consider a CAP $P = (X, W, C)$ and an admissible frequency assignment Φ for P. After some time, demands of some of the cells may be increased due to fluctuations that often arise in real-life scenarios and let $W' = (w_i')$ be this changed demand vector where $w_i' > w_i$ for some or all i. Given Φ, an initial admissible frequency assignment, the objective of the *perturbation-minimizing* frequency assignment [17] is to accommodate the additional requirements as specified in W' with minimum changes in the existing assignment.

3.1.3 CAP and Generalized Graph Coloring

In its most general form, the channel assignment problem is equivalent to the generalized graph coloring which is a well-known NP-complete problem [3]. The minimum span channel assignment problem can be represented as a graph coloring problem defined on a *CAP graph.* Call k of cell i is represented as node (ik) of a CAP graph and the nodes (ik) and (jl) are connected by an edge with weight c_{ij}, if $c_{ij} > 0$. Let us assume that the channels are assigned to the nodes of the CAP graph in a specific order and a node will be assigned the channel corresponding to the smallest integer that will satisfy the frequency separation constraints with all the previously assigned nodes. It can be shown that the ordering of the nodes has strong impact on the required channel bandwidth. Suppose there are m nodes in the CAP graph. Therefore, the nodes can be ordered in $m!$ ways and hence for sufficiently large m, it is impractical to find the best ordering by an exhaustive search.

Example 1. *Figure 3.1(a) shows a CAP graph with three cells having demands* w_0 *= 1,* w_1 *= 2, and* w_2 *= 2 respectively. Each node is labeled* (rs) *where* r *is the cell number at which a call is generated and* s *is the call number to this cell* r*. That is,* *the node (10) represents call 0 in cell 1. The frequency separation requirements for this example is given by the following matrix:*

$$
\begin{array}{cccccc}
 & cell & no. \rightarrow & 0 & 1 & 2 \\
 & \downarrow & & & & \\
C = & 0 & & 7 & 3 & 2 \\
 & 1 & & 3 & 7 & 4 \\
 & 2 & & 2 & 4 & 7.
\end{array}
$$

The edges of the CAP graph are labeled with weights according to this matrix $C = (c_{ij})$. *The label* $[\alpha]$ *associated with each node of the CAP graph of Figures 3.1(b) and (c) indicates that the frequency channel* α *is assigned to that node. If the channels are assigned to nodes in the order ((21), (00), (10), (11), (20)), as shown in Figure 3.1(b), the minimum bandwidth required will be 16. But, if the channels are assigned to nodes in the order ((20), (00), (10), (21), (11)), as shown in Figure 3.1(c), the minimum bandwidth required will be just 13.*

3.2 Benchmark Instances

To compare the performances of these algorithms for channel assignment, some well-known benchmark instances, commonly known as Philadelphia benchmarks, are widely cited in the literature [24, 26, 15, 23, 19, 22, 16, 8, 20, 17, 21, 9]. These benchmarks are defined on a 21-node cellular network shown in Figure 3.2. Here, each node represents a cell, and two nodes are connected by an edge if the corresponding cells share a common boundary. The demands of the cells are represented by any one of the two non-homogeneous demand vectors D_1 and D_2 shown in Table

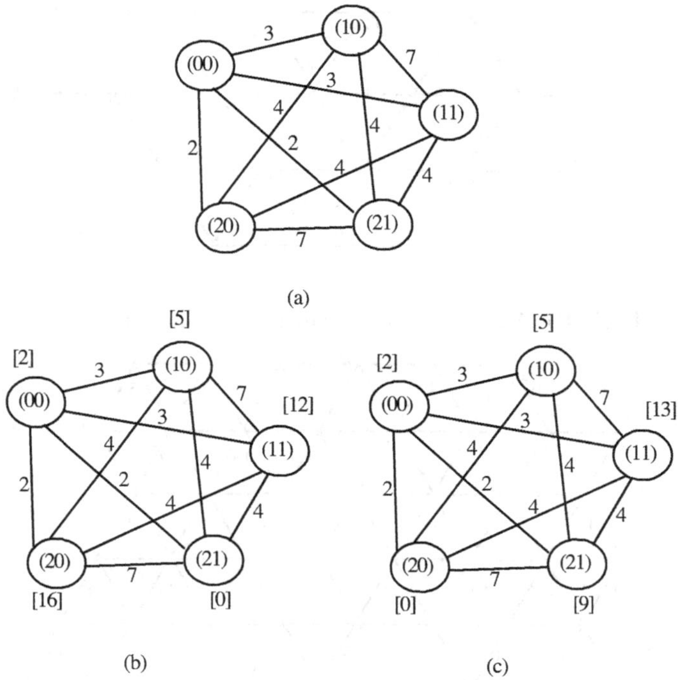

FIGURE 3.1: (a) Typical CAP graph; (b) and (c) two frequency assignments

3.1. The table indicates the channel demand from cell i corresponding to D_1 or D_2. These benchmark instances have been defined on the hexagonal cellular network assuming a two-band buffering restriction i.e., the interference does not extend beyond two cells from the call originating cell. It is assumed that to avoid channel interference, the calls in the same cell should be separated by at least s_0 channels and the calls in the cells that are distances of one and two cells apart should be separated by at least s_1 and s_2 channels, respectively. Table 3.2 shows the specifications of the benchmark problems 1 through 8 in terms of the specific values of s_0, s_1 and s_2 for the two-band buffering system, and the corresponding demand vector used for each one.

Apart from the Philadelphia benchmarks described above, we also consider a

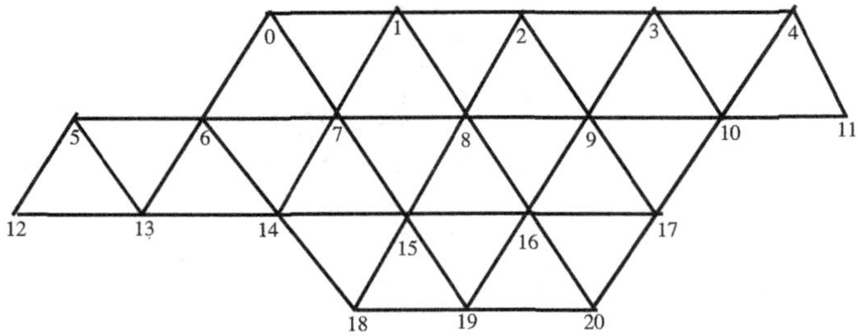

FIGURE 3.2: The 21-node benchmark cellular network

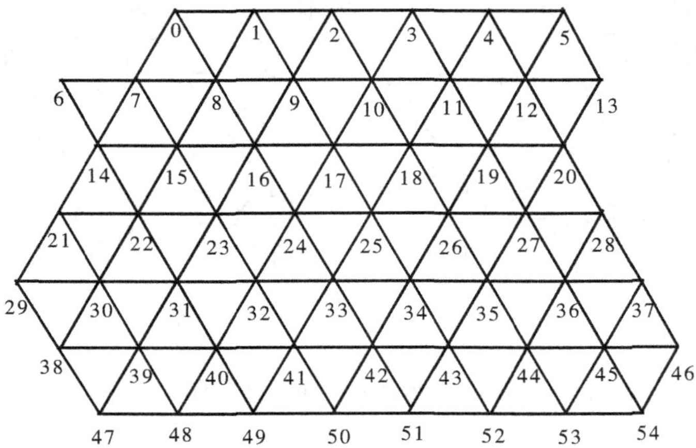

FIGURE 3.3: The 55-node benchmark cellular network

practical assignment problem from Helsinki, Finland defined on a 25-node benchmark network [26, 24].

Problem 9 has been formulated on this 25-cell network, whose frequency separation matrix C and demand vector D_3 are shown in Tables 3.3 and 3.4, respectively. The entry corresponding to the i-th row and j-th column of Table 3.3, i.e., c_{ij}, represents the minimum frequency separation requirement between a call in cell i and a call in cell j ($0 \leq i, j \leq 24$). Row D_3 in Table 3.4 indicates the channel demand from cell i.

In addition to the Philadelphia benchmarks and the 25-cell benchmark discussed above, we also considered two other benchmarks defined on the 55-node cellular network [26] shown in Figure 3.3. These two benchmarks also have been defined on a two-band buffering system where s_0, s_1 and s_2 are given as 7, 1 and 1, respectively. The demand vectors of these two problems (Problems 10 and 11) are given by D_4 and D_5 respectively, as shown in Table 3.5.

TABLE 3.1: Two different demand vectors for Philadelphia benchmark problems

Cell nos.	0	1	2	3	4	5	6	7	8	9	10	11	12	13	14	15	16	17	18	19	20
D_1	8	25	8	8	8	15	18	52	77	28	13	15	31	15	36	57	28	8	10	13	8
D_2	5	5	5	8	12	25	30	25	30	40	40	45	20	30	25	15	15	30	20	20	25

TABLE 3.2: Specifications of benchmark problems

Problems		1	2	3	4	5	6	7	8
Frequency	s_0	5	5	7	7	5	5	7	7
separation	s_1	1	2	1	2	1	2	1	2
constraints	s_2	1	1	1	1	1	1	1	1
Demand vector		D_1	D_1	D_1	D_1	D_2	D_2	D_2	D_2

TABLE 3.3: Frequency separation matrix for Problem 9

$$C = \begin{bmatrix}
2&1&1&0&1&0&1&1&1&1&0&1&1&1&1&0&0&0&0&0&0&0&0&0&0\\
1&2&1&0&1&0&1&1&0&1&0&1&1&1&1&0&0&0&0&0&0&0&0&0&0\\
1&1&2&1&1&1&1&1&1&1&1&1&1&1&0&0&0&0&0&0&0&0&0&0&0\\
0&0&1&2&0&0&1&1&1&1&1&1&1&0&0&0&0&0&0&0&0&0&1&1&1\\
1&1&1&0&2&0&0&0&0&1&1&1&1&1&1&1&0&0&0&0&0&0&0&0&0\\
0&0&1&0&0&2&1&1&1&1&0&0&0&0&0&0&0&0&0&0&0&0&0&0&0\\
1&1&1&1&0&1&2&1&1&1&1&1&1&0&0&0&0&0&0&0&0&0&0&0&0\\
1&1&1&1&0&1&1&2&1&1&1&1&1&0&0&0&0&0&0&0&0&0&0&1&0\\
1&0&1&1&0&1&1&1&2&1&1&1&0&0&0&0&0&0&0&0&0&0&0&1&1\\
1&1&1&1&1&1&1&1&1&2&1&1&1&1&1&1&0&0&0&0&0&1&0&1&0\\
0&0&1&1&1&0&1&1&1&1&2&0&1&1&1&1&0&1&1&1&1&1&1&1&1\\
1&1&1&1&1&0&1&1&1&1&0&2&1&1&0&0&0&0&0&0&0&0&0&0&0\\
1&1&1&1&1&0&1&1&0&1&1&1&2&1&1&1&1&1&1&1&0&0&0&0&0\\
1&1&1&0&1&0&0&0&0&1&1&1&1&2&1&1&1&1&1&1&0&0&0&0&0\\
1&1&0&0&1&0&0&0&0&1&1&0&1&1&2&1&1&1&1&1&1&1&0&0&0\\
0&0&0&0&1&0&0&0&0&1&1&0&1&1&1&2&1&1&1&1&0&0&0&0&0\\
0&0&0&0&0&0&0&0&0&0&0&0&1&1&1&1&2&1&1&0&0&0&0&0&0\\
0&0&0&0&0&0&0&0&0&0&1&0&1&1&1&1&1&2&1&1&0&0&0&0&0\\
0&0&0&0&0&0&0&0&0&0&1&0&1&1&1&1&1&1&2&1&1&1&1&0&0\\
0&0&0&0&0&0&0&0&0&0&1&0&1&1&1&1&0&1&1&2&1&1&1&0&0\\
0&0&0&0&0&0&0&0&0&0&1&0&0&0&1&0&0&0&1&1&2&1&1&0&0\\
0&0&0&0&0&0&0&0&0&1&1&0&0&0&1&0&0&0&1&1&1&2&1&1&1\\
0&0&0&1&0&0&0&0&0&0&1&0&0&0&0&0&0&0&1&1&1&1&2&1&1\\
0&0&0&1&0&0&0&1&1&1&1&0&0&0&0&0&0&0&0&0&0&1&1&2&1\\
0&0&0&1&0&0&0&0&1&0&1&0&0&0&0&0&0&0&0&0&0&1&1&1&2
\end{bmatrix}$$

Among the eight Philadelphia benchmark instances, it is relatively easy to derive the

TABLE 3.4: Demand vector for Problem 9

Cell nos.	0	1	2	3	4	5	6	7	8	9	10	11	12	13	14	15	16	17	18	19	20	21	22	23	24
D_3	10	11	9	5	9	4	5	7	4	8	8	9	10	7	7	6	4	5	5	7	6	4	5	7	5

TABLE 3.5: Two different demand vectors for 55-node benchmark problems

Cell nos.	0	1	2	3	4	5	6	7	8	9	10	11	12	13	14	15	16	17	18	19	20	21
D_4	5	5	5	8	12	25	30	25	30	40	40	45	20	30	25	15	15	30	20	20	25	8
D_5	10	11	9	5	9	4	5	7	4	8	8	9	10	7	7	6	4	5	5	7	6	4

Cell nos.	22	23	24	25	26	27	28	29	30	31	32	33	34	35	36	37	38	39	40	41	42	43
D_4	5	5	5	5	5	5	8	12	25	30	25	30	40	40	45	20	30	25	15	15	30	20
D_5	5	7	5	10	11	9	5	9	4	5	7	4	8	8	9	10	7	7	6	4	5	5

Cell nos.	44	45	46	47	48	49	50	51	52	53	54
D_4	20	25	8	5	5	5	25	8	5	5	5
D_5	7	6	4	5	7	5	6	4	5	7	5

optimal solutions for all the problems except 2 and 6, because in all six cases the required number of channels is primarily limited by the co-site interference constraint only. More difficult is finding optimal solutions for the other two Philadelphia instances: Problems 2 and 6 [15], [24]. For example, the assignment algorithm given in [16] required 165 hours of computing time for Problem 6 on an unloaded HP Apollo 9000/700 workstation, but produced only a non-optimal solution with 268 channels (optimal solution requires only 253 channels). Later, however, the authors in [15] proposed an algorithm which provided optimal solutions for Problems 2 and 6 with running times of 8 and 10 minutes, respectively on the same workstation. Among the later works, the FESR (frequency exhaustive strategy with rearrangement) algorithm in [17] and the randomized saturation degree (RSD) heuristic presented in [24] also produced only non-optimal solutions to the benchmark Problems 2 and 6. However, combining their RSD heuristic with a local search (LS) algorithm, the authors in [24] were able to find an optimal solution for Problem 2 but not for Problem 6. Recently, an efficient heuristic algorithm has been proposed in [26] which also produced non-optimal results for Problems 2 and 6 with 463 and 273 channels, respectively. Given the concept of a *critical block* of the hexagonal cellular network, the authors in [28] proposed a novel algorithm which provides an optimal assignment for Problems 2 and 6 with relatively less computation time than that cited in [15]. The critical block approach [28] requires only a few seconds for optimal channel assignment of the other six benchmark instances, on an unloaded Sun Ultra 60 workstation. For the benchmark Problems 2 and 6, however, the approach required only around 60 and 72 seconds, respectively, on the same workstation.

3.3 Lower Bounds on Bandwidth

As mentioned in the earlier subsection, because of the NP-completeness nature of the CAP, researchers attempted to solve the problem based on heuristic approaches or approximation algorithms. Many of these algorithms do not assess how far their results are away from optimality and some of them over-estimate the frequencies by more than 100% [41, 42]. As a result, even if by some lucky coincidence, an optimal assignment is found, it may not be recognized as optimal unless one has a clear idea about the lower bound on the required bandwidth. In heuristic approaches like the genetic algorithm (GA), the process usually terminates after a certain number of iterations. Hence, a prior idea about the lower bound on the required bandwidth will be very useful in deciding the termination conditions of these algorithms. Furthermore, in the approaches based on neural networks, simulated annealing, and tabu search, the techniques typically start from some known lower bounds and then gradually improve the results through each iteration.

The problem of determining the lower bounds on bandwidth has been extensively studied by several authors [20, 1, 29, 28, 42, 43, 7, 45, 46]. In [43], a lower bound on bandwidth developed for CAP considered only the co-channel interference constraints. In [42], Gamst et al. presented some lower bounds on bandwidth for CAP that took the additional constraints for adjacent channel interferences and co-site interferences into account. Improving these results, Tcha et al. [7] presented some results on the lower bound which, however, did not explicitly include the co-site constraints. In [20], the authors derived a lower bound which, in some cases, was tighter than those presented in [42] considering all the co-channel, adjacent channel, and co-site constraints. In [44], new lower bounds were generalizations of the bounds presented in [20, 42]. All these lower bounds are defined on a general network of arbitrary cell structure.

For the special case of hexagonal cellular networks, the problem of determining the lower bounds has been studied by several authors [29, 28]. In this special type of network, a k-band buffering exists in the network if the channels assigned to any two calls in two cells which are more than distance k apart do not interfere with each other. In other words, for a k-band buffering network we need to define $k+1$ constants $s_0, s_1, s_2, \cdots, s_k$ to specify the constraints for avoiding channel interferences where $s_i, 0 \leq i \leq k$ denotes the minimum gap between two channels assigned to two calls in two cells which are distance i apart. Clearly, $s_0 \geq s_1 \geq s_2 \geq \cdots \geq s_k$. Authors in [29, 28] studied the problem of determining the lower bounds for a two-band buffering situation in hexagonal cellular networks using only the three parameters s_0, s_1, and s_2, where s_0, s_1, and s_2 are the minimum frequency separations required to avoid interference for calls in the same cell, cells at distances 1 and 2, respectively. They derived expressions for the lower bounds on bandwidths for different relative values of s_0, s_1, and s_2 for both uniform and non-uniform demand situations. These lower bounds have been computed based on the minimum bandwidth requirement for assigning frequencies to a seven-node subgraph as shown in Figure 3.4. Note that all nodes in this subgraph are within distance 2 of each other. Therefore, under the two-band buffering scenario, the channels assigned to these nodes may interfere with each other if the channels are not sufficiently far apart. Most importantly, no frequency re-use is possible within this subgraph. Hence, the bandwidth requirement of this subgraph will give a lower bound on the bandwidth requirement for the whole cellular network which is not necessarily a tight one.

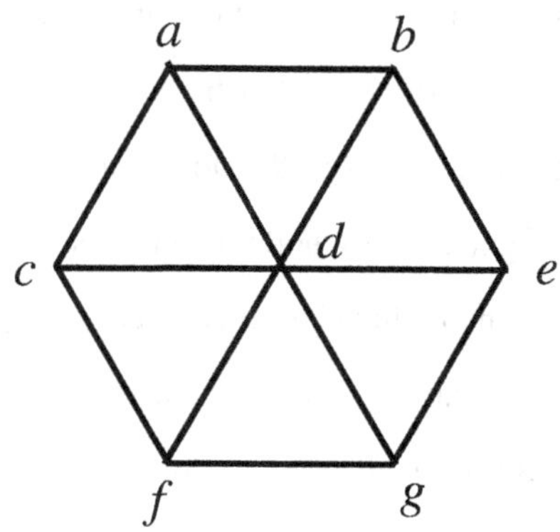

FIGURE 3.4: Seven-node subgraph of hexagonal cellular network

3.3.1 Lower Bounds on Bandwidth for Homogeneous Demand

For the uniform demand of only one channel per cell, the lower bound derived in [29] is stated below.

Lemma 1. *The minimum bandwidth required for assigning channels to the seven-node subgraph of a hexagonal cellular network having homogeneous demand of single channel and two-band buffering with frequency separation* $s_1 \geq s_2$, *is* $(s_1 + 5s_2)$ *when* $s_2 \leq s_1 \leq 2s_2$, *and* $(2s_1 + 3s_2)$ *when* $s_1 \geq 2s_2$.

Proof: Since any two nodes of the subgraph shown in Figure 3.4 are Distance 1 (adjacent) or Distance 2 (two-hop away) apart from each other, the frequencies assigned on any two nodes of this subgraph must be separated by at least s_1 or s_2 depending on whether they are Distance 1 or Distance 2 apart, respectively. In Figure

3.4, d is the central node and all the remaining nodes a, b, c, e, f and g of this subgraph are Distance 1 apart from d. We divide nodes a, b, c, e, f and g into two sets A and B in such a way that any two nodes of a set are Distance 2 apart from each other. Let $A = \{a, e, f\}$ and $B = \{b, c, g\}$. The minimum frequency that can be assigned to any node of this subgraph is $(0, 0)$. We now have the following cases.

Case 1: Let frequency $(0, 0)$ be assigned to the central node d, then assign frequencies to the nodes of set A (or B). After the assignment of all the nodes of set A (or B), we will assign frequencies to the nodes of set B (or A). The minimum frequency that can be assigned to any node of set A is $(1, 0)$ because any node of set A is Distance 1 apart from d which has already been assigned a frequency $(0, 0)$. Without loss of generality (w.l.o.g), let $(1, 0)$ be assigned to node a. The minimum frequency that can be assigned to either node f or node e is $(1, 1)$, because both the nodes e and f are Distance 2 apart from a (which has already been assigned a frequency $(1, 0)$) and Distance 1 apart from d with assigned frequency $(0, 0)$. W.l.o.g, let $(1, 1)$ be assigned to f. It follows from a similar argument that the minimum frequency which can be assigned to e is $(1, 2)$. We can now start assigning frequencies to the nodes of set B. Note that e has been assigned the highest frequency among all the previously assigned nodes d, a, f and e. Node c of set B is at Distance 2 from the node e, i.e., it requires less frequency separation from e. To satisfy the frequency separation requirements with all the previously assigned nodes d, a, f and e, the minimum frequency that can be assigned to the node c of set B is i) $(1, 3)$ when $s_2 \leq s_1 \leq 2s_2$, and ii) $(2, 1)$ when $s_1 \geq 2s_2$. By similar arguments, the minimum frequencies which can be assigned to the nodes b and g are : i) $(1, 4)$ and $(1, 5)$ respectively, when $s_2 \leq s_1 \leq 2s_2$, and ii) $(2, 2)$ and $(2, 3)$ respectively, when $s_1 \geq 2s_2$. The corresponding assignments for $s_2 \leq s_1 \leq 2s_2$ and $s_1 \geq 2s_2$ are shown in Figures 3.5(a) and (b), respectively.

Note in Figure 3.5(a), the inequality $s_1 \leq 2s_2$ has been used because the frequencies (1, 1) and (1, 3) are assigned to nodes f and c respectively, which are Distance 1 apart from each other.

Similarly, in Figure 3.5(b), the inequality $s_1 \geq 2s_2$ has also been used because the frequencies (1, 2) and (2, 1) are assigned to nodes e and c respectively, which are Distance 2 apart from each other.

It shows clearly that $s_1 = 2s_2$ is the critical point for determining the minimum bandwidth requirement for the seven-node subgraph.

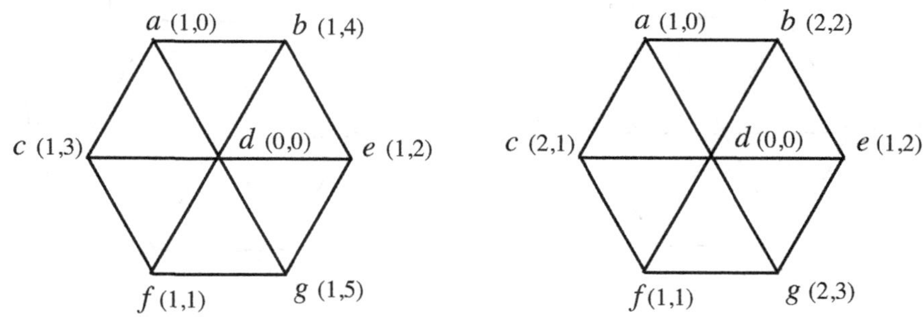

(a) (b)

FIGURE 3.5: Different frequency assignments to seven-node subgraph with minimum channel at central node for (a) $s_2 \leq s_1 \leq 2s_2$, and (b) $s_1 \geq 2s_2$

Case 2: Let channel $(0, 0)$ be assigned to any node other than the central node. W.l.o.g, let $(0, 0)$ be assigned to the node e. As $e \in$A, our next step is to assign frequencies to a or f. The minimum frequency that can be assigned to a or f is $(0, 1)$ because both of them are Distance 2 apart from e which has already been assigned a frequency $(0, 0)$. W.l.o.g, let $(0, 1)$ be assigned to a. Then the minimum frequency that can be assigned to f is $(0, 2)$ because f is Distance 2 apart from both a and e. We now need to assign frequencies to the central node d and the nodes of the set B. Note that f has been assigned the highest frequency among all the previously assigned nodes a, e and f. Node b of set B is at Distance 2 from node f, i.e., it requires minimum frequency separation from f. Now, satisfying the frequency separation requirements with all the previously assigned nodes a, e and f, the minimum frequency that can be assigned to the node b is i) $(0,3)$ when $s_2 \leq s_1 \leq 2s_2$, and ii) $(1,1)$ when $s_1 \geq 2s_2$. By similar arguments, the minimum frequencies that can be assigned to the nodes g and c are : i) $(0,4)$ and $(0,5)$ respectively, when $s_2 \leq s_1 \leq 2s_2$, and ii) $(1,2)$ and $(1,3)$ respectively, when $s_1 \geq 2s_2$.

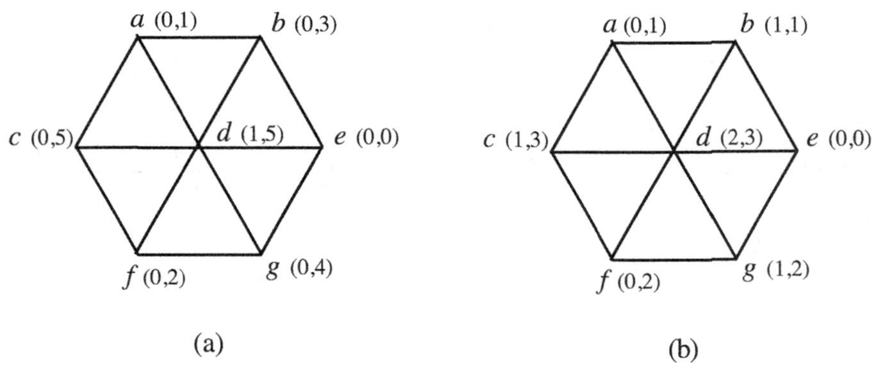

FIGURE 3.6: Different frequency assignments to seven-node subgraph with maximum channel at central node for (a) $s_2 \leq s_1 \leq 2s_2$, and (b) $s_1 \geq 2s_2$

Since the central node d is Distance 1 apart from all the remaining nodes, the minimum frequency that can be assigned now to d is $(1,5)$ if $s_2 \leq s_1 \leq 2s_2$, and $(2,3)$ if $s_1 \geq 2s_2$.

If, after assigning frequencies to the nodes of set A, we assign frequencies to the central node d followed by the nodes of set B, by proceeding in a similar way we would see that the minimum bandwidth required will be $(2,4)$ for both the cases of i) $s_2 \leq s_1 \leq 2s_2$ and ii) $s_1 \geq 2s_2$.

Figure 3.6(a) shows the assignment with a bandwidth of $(1,5)$ for $s_2 \leq s_1 \leq 2s_2$, and Figure 3.6(b) shows the assignment with a bandwidth of $(2,3)$ for $s_1 \geq 2s_2$.

Hence all the nodes of the subgraph can be assigned frequency channels without interference by using a bandwidth of $(s_1 + 5s_2)$ when $s_2 \leq s_1 \leq 2s_2$, or $(2s_1 + 3s_2)$ when $s_1 \geq 2s_2$. □

Remark 1. *The results of Lemma 1 show that the minimum bandwidth requirement for assigning channels to the seven-node subgraph of Figure 3.4 is actually lower (except when* $s_1 = s_2$*) than* $(2s_1 + 4s_2)$ *reported in [1].*

Generalizing Lemma 1, the following two theorems on the minimum bandwidth requirement for assigning w (≥ 2) frequency channels to each of the nodes of the subgraph were reported in [29].

Theorem 1. *The minimum bandwidth required for assigning channels to the seven-node subgraph of a hexagonal cellular network with homogeneous demand* w (≥ 2) *and two-band buffering with frequency separation* $s_2 \leq s_1 \leq 2s_2$*, is*

1. $(2s_1 + 5s_2) + (w - 2)(s_0 + 6s_2) + 6s_2$, *when* $s_1 \leq s_0 \leq (2s_1 - s_2)$,
2. $(w - 1)(2s_1 + 5s_2) + 6s_2$, *when* $(2s_1 - s_2) \leq s_0 \leq 6s_2$,
3. $(w - 1)(2s_1 + 5s_2) + s_0$, *when* $6s_2 \leq s_0 \leq (s_1 + 5s_2)$,
4. $(w - 1)(2s_1 + 5s_2) + (s_1 + 5s_2)$, *when* $(s_1 + 5s_2) \leq s_0 \leq (2s_1 + 5s_2)$, *and*
5. $(w - 1)s_0 + (s_1 + 5s_2)$, *when* $s_0 \geq (2s_1 + 5s_2)$.

Proof: Let us consider the seven-node subgraph of a hexagonal cellular network as shown in Figure 3.4. Suppose, using the frequency channels within the closed interval $[0, p]$, it is possible to assign w frequency channels to each of the nodes of the subgraph while satisfying all other constraints. Therefore, our objective is to find *minimum* p.

Note that any two nodes of the subgraph are within Distance 2 from each other.

Therefore, any two frequencies assigned to two nodes of the subgraph must be separated by at least s_2. Suppose the frequency channel y ($\in [0, p]$) is assigned to any node of the subgraph. Then to satisfy the interference criteria, no frequency channel within the open interval $(y - s_2, y)$ and $(y, y + s_2)$ can be assigned to any node of the subgraph (including the node itself, since $s_0 \geq s_2$). That is, there would be an unusable gap of $(y - s_2, y)$ before y, and another such gap of $(y, y + s_2)$ after y on the frequency spectrum line where gap (a, b) implies that the integers within the open interval (a, b) cannot be assigned to any node of the subgraph. We refer to $(b - a)$ as the length of gap (a, b). Obviously, for $y = 0$ (or, p), one of these gaps of unusable frequencies, e.g., $(y - s_2, y)$ or $(y, y + s_2)$ will be beyond the interval $[0, p]$.

In addition, the central node d is Distance 1 apart from each of the other nodes of the subgraph. Suppose, during the assignment, a frequency channel x ($\in [0, p]$) is assigned to the central node d. To satisfy the interference criteria, no frequency channel within the open intervals $(x - s_1, x)$ and $(x, x + s_1)$ can be assigned to any node of the subgraph including the central node itself, since $s_0 \geq s_1$. That is, there must be unusable gaps $(x - s_1, x)$ and $(x, x + s_1)$ before and after x, respectively. If $x = 0$ (or p), one of these gaps of unusable frequencies, e.g., $(x - s_1, x)$ or $(x, x + s_1)$ falls beyond the interval $[0, p]$.

Since $s_1 \geq s_2$, we can have more usable frequency channels if the minimum and maximum frequencies are assigned to the central node d, rather than to any other node.

It is always possible to assign either the minimum frequency or the maximum frequency to the central node d, but depending on the relative values of s_0, s_1 and s_2, it may or may not be possible to assign both of these to the central node for keeping p minimum.

Let us now consider the cases where we can do that and where we cannot.

Lemma 1 states that $(s_1 + 5s_2)$ is the minimum bandwidth required to assign only one frequency channel to each of the nodes of the subgraph when $s_2 \leq s_1 \leq 2s_2$. For this assignment, the minimum frequency 0 (or the maximum frequency $(s_1 + 5s_2)$) is assigned to the central node d as shown in Figure 3.5(a) (or Figure 3.6(a)). W.l.o.g let us start with the single channel assignment of Figure 3.5(a). Next, we are left to assign $(w - 1)$ more frequency channels to each, and see whether the maximum frequency channel can also be assigned to the central node d, keeping p minimum.

We see in Figure 3.5(a) that the node g has been assigned the frequency $(s_1 + 5s_2)$. The frequency assigned to node a is the minimum of all assigned frequencies to the nodes Distance 2 apart from node g, and the frequency assigned to node d is the minimum of all assigned frequencies to the nodes Distance 1 apart from g. To assign the second channel to all the nodes, we can start by assigning the next available minimum frequency either to a (a peripheral node) or to d (the central node). For the first case, the minimum frequency that can be assigned to a is $(s_1 + 5s_2) + s_2$, if $s_0 \leq 6s_2$ (to avoid interference between two channels assigned to a), and $(s_0 + s_1)$ otherwise. For the second case, the minimum frequency that can be assigned to d is $(s_1 + 5s_2) + s_1$, if $s_0 \leq (2s_1 + 5s_2)$, and s_0 otherwise. It follows from these observations that for $6s_2 \leq s_0 \leq (s_1 + 5s_2)$, it is more economical to start with assigning the minimum frequency for the second channel to the node a, rather than to the node d. We now have the following cases.

Case 1 : When $s_0 \leq 6s_2$ (or $6s_2 \leq s_0 \leq (s_1 + 5s_2)$):

In this case, to assign the second and successive channels to the nodes in Figure 3.5(a), we start by assigning the frequency $(s_1 + 6s_2)$ (for $s_0 \leq 6s_2$) or $(s_0 + s_1)$ (for $6s_2 \leq s_0 \leq (s_1 + 5s_2)$) to the node a. Our objective is to end up with the maximum frequency at the central node d along with minimum gaps between successively assigned frequencies to keep the required p minimum. Let us now consider the following two sub cases:

a) When $s_1 \leq s_0 \leq (2s_1 - s_2)$:

After the completion of second round channel assignment to all peripheral nodes (a, b, c, e, f, g), if we assign a channel to the central node, the minimum gap would be s_1. For the next round, we are to keep another gap of length s_1 before assigning to node a. It would be more economical to complete the assignments of all w channels to the peripheral nodes only, and assign the remaining $(w-1)$ channels to the central node.

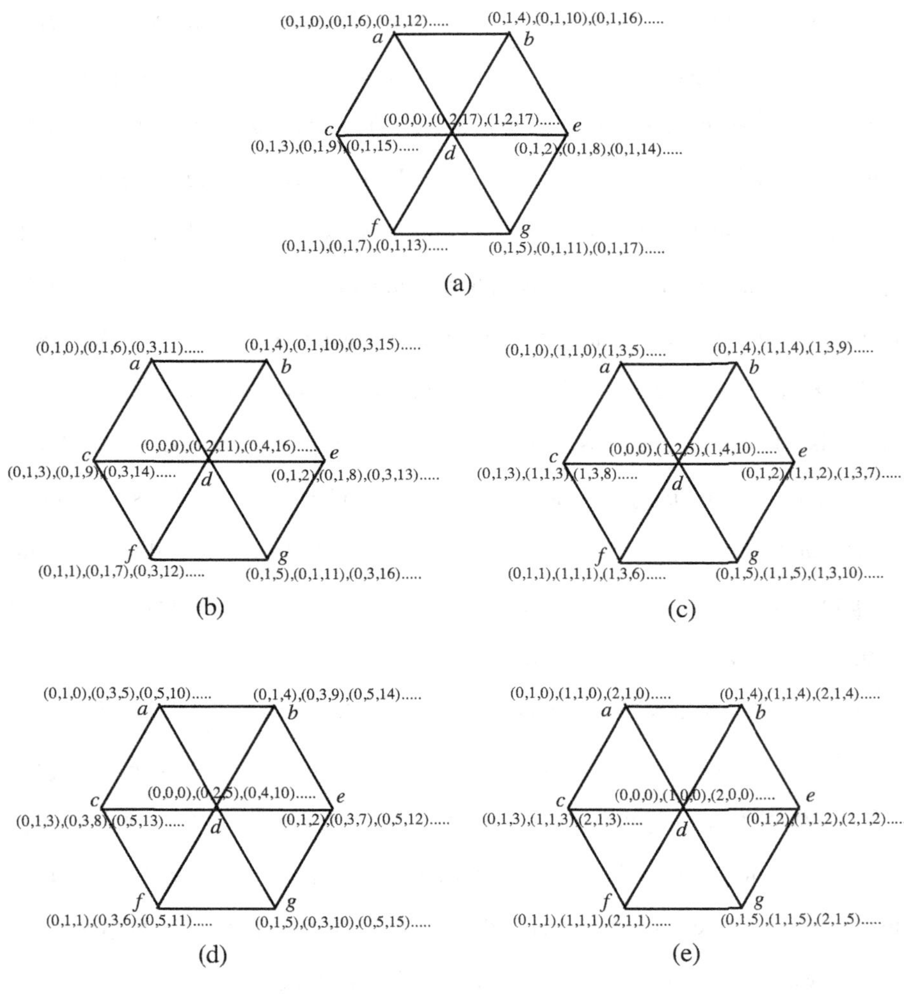

FIGURE 3.7: Different frequency assignments to seven-node subgraph for $s_2 \leq s_1 \leq 2s_2$ when (a) $s_1 \leq s_0 \leq (2s_1 - s_2)$, (b) $(2s_1 - s_2) \leq s_0 \leq 6s_2$, (c) $6s_2 \leq s_0 \leq (s_1 + 5s_2)$, (d) $(s_1 + 5s_2) \leq s_0 \leq (2s_1 + 5s_2)$, and (e) $s_0 \geq (2s_1 + 5s_2)$

As shown in Figure 3.7(a), we assign the second round channels to all the nodes except the central node as (0, 1, 7), (0, 1, 8), (0, 1, 9), (0, 1, 10) and (0, 1, 11) to the nodes f, e, c, b and g, respectively. To assign the remaining $(w-2)$ channels to each of the nodes (a, b, c, e, f, g) of the subgraph, we follow the same ordering of nodes, i.e., a, f, e, c, b and g so that each node is assigned with the next minimum frequency available at that time. Proceeding similarly for the third, fourth, ..., wth channel, we end at node g with a frequency of $(s_1+5s_2)+(w-1)(6s_2)$ for $s_1 \leq s_0 \leq (2s_1-s_2)$. Now we are left to assign $(w-1)$ frequencies to the central node d. For that, the minimum frequency by which we can start at node d is $(s_1+5s_2)+(w-1)(6s_2)+s_1$. Therefore, after the assignment of remaining $(w-2)$ frequencies, we can end up with a frequency of $(s_1+5s_2)+(w-1)(6s_2)+s_1+(w-2)s_0$ at node d. Since, we assign both the minimum and maximum frequencies to the central node d along with minimum gaps between successively assigned frequencies, the required p will be a minimum whose value is given by $(s_1+5s_2)+(w-1)(6s_2)+s_1+(w-2)s_0 = (2s_1+5s_2)+(w-2)(s_0+6s_2)+6s_2$, *when* $s_1 \leq s_0 \leq (2s_1-s_2)$. Figure 3.7(a) illustrates the assignments up to the third channel to each node following this method.

b) When $(2s_1-s_2) \leq s_0 \leq 6s_2$(or $6s_2 \leq s_0 \leq (s_1+5s_2)$):
In this case, it is more economical to complete the assignments of channels in each round to all the nodes, including the central node.

As shown in Figure 3.7(b), for $(2s_1-s_2) \leq s_0 \leq 6s_2$, we assign the second round frequencies (0, 1, 7), (0, 1, 8), (0, 1, 9), (0, 1, 10), (0, 1, 11) and (0, 2, 11) to the nodes f, e, c, b, g and d, respectively. To assign the remaining $(w-2)$ channels to each of the nodes of the subgraph, we follow the same ordering of nodes, i.e., a, f, e, c, b, g and d so that each node is assigned the next available minimum frequency. Figure 3.7(b) illustrates the assignments up to the third channel to each node following this method. Since we finally achieve the assignment of both the minimum and maximum frequencies to the central node d along with minimum gaps between successively assigned frequencies, the required p will be minimum and its value is given by $(s_1 + 5s_2) + (s_1 + 6s_2) + (w - 2)(s_1 + (s_1 + 5s_2)) = (w - 1)(2s_1 + 5s_2) + 6s_2$, *for* $(2s_1 - s_2) \leq s_0 \leq 6s_2$, $(s_1 + 5s_2) + (s_0 + s_1) + (w - 2)(s_1 + (s_1 + 5s_2)) = (w - 1)(2s_1 + 5s_2) + s_0$, *for* $6s_2 \leq s_0 \leq (s_1 + 5s_2)$. Figure 3.7(c) illustrates the assignment when $6s_2 \leq s_0 \leq (s_1 + 5s_2)$ for $w = 3$.

Case 2 : When $(s_1 + 5s_2) \leq s_0 \leq (2s_1 + 5s_2)$ (or $s_0 \geq (2s_1 + 5s_2)$):
To assign the second and successive channels to the nodes in Figure 3.5(a), we start by assigning the frequency $(2s_1 + 5s_2)$ (for $(s_1 + 5s_2) \leq s_0 \leq (2s_1 + 5s_2)$), or s_0 (for $s_0 \geq (2s_1 + 5s_2)$) to the central node d. However, after the assignment of second channel to every node, we assign the maximum frequency to a non-central node (say, node g). This is illustrated in Figure 3.7(d) (for $(s_1 + 5s_2) \leq s_0 \leq (2s_1 + 5s_2)$) and Figure 3.7(e) for $s_0 \geq (2s_1 + 5s_2)$, respectively, for $w = 3$. Proceeding similarly for the third, fourth, ..., wth channel, we can end up at node g with a frequency of $(s_1 + 5s_2) + (2s_1 + 5s_2) + (w - 2)(s_1 + (s_1 + 5s_2)) = (w - 1)(2s_1 + 5s_2) + (s_1 + 5s_2)$, *for* $(s_1 + 5s_2) \leq s_0 \leq (2s_1 + 5s_2)$, $(s_1 + 5s_2) + (s_0) + (w - 2)(s_0) = (w - 1)(s_0) + (s_1 + 5s_2))$, *for* $s_0 \geq (2s_1 + 5s_2)$. Hence the proof. □

Theorem 2. *The minimum bandwidth required for assigning channels to the seven-node subgraph of a hexagonal cellular network with homogeneous demand* w (≥ 2) *and two-band buffering with frequency separation* $s_1 \geq 2s_2$, *is*

1. *(a)* $(3s_1 + 3s_2) + (w - 2)(s_0 + 6s_2) + 6s_2$, *when* $s_1 \leq s_0 \leq 3s_2$,

   (b) $(3s_1 + 3s_2) + (w - 2)(3s_0) + 2s_0$, *when* $3s_2 \leq s_0 \leq (s_1 + s_2)$,

2. $(w - 1)(3s_1 + 3s_2) + (2s_1 + 2s_2)$, *when* $(s_1 + s_2) \leq s_0 \leq (2s_1 + 2s_2)$,

3. $(w - 1)(3s_1 + 3s_2) + s_0$, *when* $(2s_1 + 2s_2) \leq s_0 \leq (2s_1 + 3s_2)$,

4. $(w - 1)(3s_1 + 3s_2) + (2s_1 + 3s_2)$, *when* $(2s_1 + 3s_2) \leq s_0 \leq (3s_1 + 3s_2)$, *and*

5. $(w-1)s_0 + (2s_1 + 3s_2)$, *when* $s_0 \geq (3s_1 + 3s_2)$.

Proof: We start from Figure 3.5(b) and proceed in a similar way as in the proof of Theorem 1 to get the desired result. □

Note that the results presented in theorems above assumed homogeneous demand.

3.3.2 Lower Bounds on Bandwidth for Non-Homogeneous Demand

The lower bounds for the non-uniform demands have been derived in [28]. These lower bounds are also based on the minimum bandwidth requirement for assigning frequencies to the seven-node subgraph shown in Figure 3.4. Let us now consider the seven-node subgraph G with non-homogeneous demand vector $W = (w_i)$ (w_i being the channel requirement for cell i) where $w = max(w_i)$, $i = 1, 2, \cdots, 7$. It is evident that the minimum bandwidth ($B^{(h)}_{min}$) required to satisfy the homogeneous demand w is an *upper bound* on the minimum bandwidth requirement B_{min} of G with the demand vector $W = (w_i)$. It is also evident that a *lower bound* on B_{min} for G is $(w-1)s_0$. However, this lower bound is not always tight for all values of s_0, s_1, s_2, and W. We find a tighter *lower bound* on bandwidth for the general case in the following way.

Lemma 2. *A lower bound on B_{min} for G with demand vector $W = (w_i)$, where $w = max(w_i)$, $i = 1, 2, \cdots, 7$, is:*

1. $max((w-1)s_0, (\sum_{i=1}^{7} w_i - 1)s_2 + (s_0 - s_2)(w_4 - 2) + 2(s_1 - s_2))$ *for* $s_1 \leq s_0 \leq (2s_1 - s_2)$, *and*

2. $max((w-1)s_0, (\sum_{i=1}^{7} w_i - 1)s_2 + 2(s_1 - s_2)(w_4 - 2) + 2(s_1 - s_2))$ *for* $s_0 \geq (2s_1 - s_2)$.

Proof: Let us consider the Distance 2 clique G with the demand vector $W = (w_i)$, $i = 1, 2, \cdots, 7$, as shown in Figure 3.4. Suppose, using the frequency channels within the closed interval $[0, p]$, it is possible to assign frequencies to the nodes of G, satisfying all other constraints. Therefore, our objective is to find a lower bound on *minimum p*.

The total demand for G is $\sum_{i=1}^{7} w_i$. Any two nodes of G are within Distance 2 from each other. Therefore, any two frequencies assigned to this clique G must be separated by at least s_2. Suppose the frequency channel y ($\in [0, p]$) is assigned to any node of G. To satisfy the interference criteria, no frequency channel within the open interval $(y - s_2, y)$ and $(y, y + s_2)$ can be assigned to any node of G (including the node itself since $s_0 \geq s_2$). That is, there would be an unusable gap of $(y - s_2, y)$ before y and another such gap of $(y, y + s_2)$ after y on the frequency spectrum line where the gap (a, b) implies that the integers within the open interval (a, b) cannot be assigned to any node of the subgraph. We call $(b - a)$ the length of gap (a, b). Obviously, for $y = 0$ (or p), one of these gaps of unusable frequencies, e.g., $(y - s_2, y)$ or $(y, y + s_2)$ will be outside the interval $[0, p]$.

In addition, the central node 4 is Distance 1 apart from every other node of G. Suppose, during the assignment, a frequency channel x ($\in [0, p]$) is assigned to the central node 4. Then, to satisfy the interference criteria, no frequency channel within the open interval $(x - s_1, x)$ and $(x, x + s_1)$ can be assigned to any node of G (including the central node itself since $s_0 \geq s_1$). That is, there must be unusable gaps $(x - s_1, x)$ and $(x, x + s_1)$ before and after x, respectively. If $x = 0$ (or p), one of these gaps of unusable frequencies, e.g., $(x - s_1, x)$ or $(x, x + s_1)$ falls beyond the interval $[0, p]$.

Since $s_1 \geq s_2$, it follows that we can have more usable frequency channels if the minimum and maximum frequencies are assigned to the central node 4 rather than to any other node. Depending on the relative values of s_0, s_1, and s_2, we now have the following two cases.

Case 1) When $s_1 \leq s_0 \leq (2s_1 - s_2)$:

After the assignment to a peripheral node, to assign a channel to the central node, the minimum gap length would be s_1. And also we are to keep another gap of length s_1 before assigning a frequency to a peripheral node after the central node. Instead, it would be more economical to first complete the assignments of all the required channels to the peripheral nodes and then assign channels to the central node.

If, however, the minimum and maximum frequency channels are assigned to the central node, the lower bound on *minimum p* as follows from the above discussion is $(\sum_{i=1}^{7} w_i - 1)s_2 + (s_0 - s_2)(w_4 - 2) + 2(s_1 - s_2)$, w_4 being the demand of the central node 4.

Case 2) When $s_0 \geq (2s_1 - s_2)$:

It would be more economical to assign to the central node after the assignment of a peripheral node, and after that to assign to a peripheral node from the central node.

If the minimum and maximum frequency channels are assigned to the central node, the lower bound on *minimum p* would be $(\sum_{i=1}^{7} w_i - 1)s_2 + 2(s_1 - s_2)(w_4 - 2) + 2(s_1 - s_2)$. □

It does not, however, necessarily mean that there always exists a conflict-free assignment of G with the lower bound given above.

Several lower bounds on the minimum bandwidth requirement $S_0(P)$ for a given CAP P have been described in [42, 7, 20] for non-homogeneous demands. These bounds are based on solving $CAPs$ which are simpler than P defined in the following subsets of X [7].

Definition 1. *Let v and v' be two positive integers such that $v' > v$. Then,*

- *A subset Q of X is said to be v-complete if $c_{ij} \geq v$ for all $i, j \in Q$.*
- *A subset pair (Q, Q') of X is said to be (v, v')-complete if Q is a v-complete subset of X and Q' is a v'-complete subset of Q.*

The following is a lower bound on the minimum number of channels needed for the solution of a given channel assignment problem P from Gamst [42].

Result 1. *(Lemma 9 in [42]): Let* $P = (X, W, C)$ *be a channel assignment problem (CAP), and assume there exists a* (v, v')*-complete subset pair* (Q, Q') *of* X *such that*

$$c_{ij} \geq v' > v, \quad for\ all\ i \in Q'\ and\ j \in Q. \tag{3.1}$$

$$If \sum_{i \in Q \backslash Q'} w_i > 0,$$

let

$$L_1 = 1 + v' \sum_{i \in Q'} w_i + v\Big(\sum_{i \in Q \backslash Q'} w_i - 1 \Big).$$

Then, $S_0(P) \geq L_1$.

Note that, Result 1 is particularly useful when applied to combinations of co-channel ($v = 1$) and adjacent channel ($v' = 2$) constraints (here represented by $s_2 = 1$ and $s_1 = 2$, respectively)[42]. Besides that, Result 1 is of little practical use due to the difficulty of finding a (v, v')-complete subset pair satisfying condition (3.1) that each cell in Q' is at least v'-compatible with every cell in Q [7]. Motivated by this observation, in [7], Tcha et al. derived a new lower bound L_2 for P, when the strict condition (3.1) is relaxed by devising a procedure of *frequency insertion* which allows lower-than-v'-compatibility for some cell pairs between Q' and Q.

Result 2. *(Theorem 2 in [7]): Let* $P = (X, W, C)$ *be a CAP having a* (v, v')*-complete subset pair* (Q, R) *of* X *with a maximal* v'*-complete subset* R *of* Q*, from which a CAP* $P' = (X, W', C')$ *is defined, wherein* $W' = (w_i')$ *and* $C' = (c_{ij}')$ *are given as*

$$w'_i = \begin{cases} w_i, & if\ i \in Q, \\ 0, & otherwise \end{cases}$$

$$c'_{ij} = \begin{cases} v' I_{[v',\infty]}(c_{ij}) + v I_{[v,v']}(c_{ij}), & if\ i \in R\ or\ j \in R, \\ min\{v, c_{ij}\}, & otherwise \end{cases}$$

where $I_A(e)$ is an indicator function which takes value as 1 when e is contained in A, 0 otherwise. Let a_0 and a_{v-v_0} be the number of times the procedure of frequency insertion (PFI) (see [7] for details) takes the 0-insertion and $(v - v_0)$-insertion, respectively, where v_0 is a nonnegative integer and q is a positive integer such that $v' = qv + v_0$, $0 \leq v_0 < v$. Let α, β be some upper bounds on a_0 and $a_0 + a_{v-v_0}$ respectively for P'. Define

$$L_2 = 1 + v'\left(\sum_{i \in R} w_i - 1\right) + (v - v_0)(\beta - \alpha) + v\left(\sum_{i \in Q \setminus R} w_i - \beta\right).$$

Then, $S_0(P) \geq S_0(P') \geq L_2$.

The difficulty with Result 2, however, is that both a_0 and a_{v-v_0} depend on the initialization of the PFI, i.e., the initial frequency assignment of the requirements in R. Since there exist as many as $\frac{(\sum_{i \in R} w_i)!}{\prod_{i \in R} w_i!}$ possible initializations, it is computationally prohibitive to obtain the exact values of $a_0 = a_0^*$ and $a_{v-v_0} = a_{v-v_0}^*$ which maximize $va_0 + v_0 a_{v-v_0}$ by enumerating all such cases [7]. Instead the authors derived upper bounds on a_0 and $a_0 + a_{v-v_0}$, which finally produced an upper bound on $va_0 + v_0 a_{v-v_0}$. Two maximal flow problems (MFP's), $\mathcal{P}_1$ and $\mathcal{P}_2$, are defined for the purpose of generating upper bounds on a_0 and $a_0 + a_{v-v_0}$, respectively. In [7], it has also been proved that $L_2 \geq L_1$.

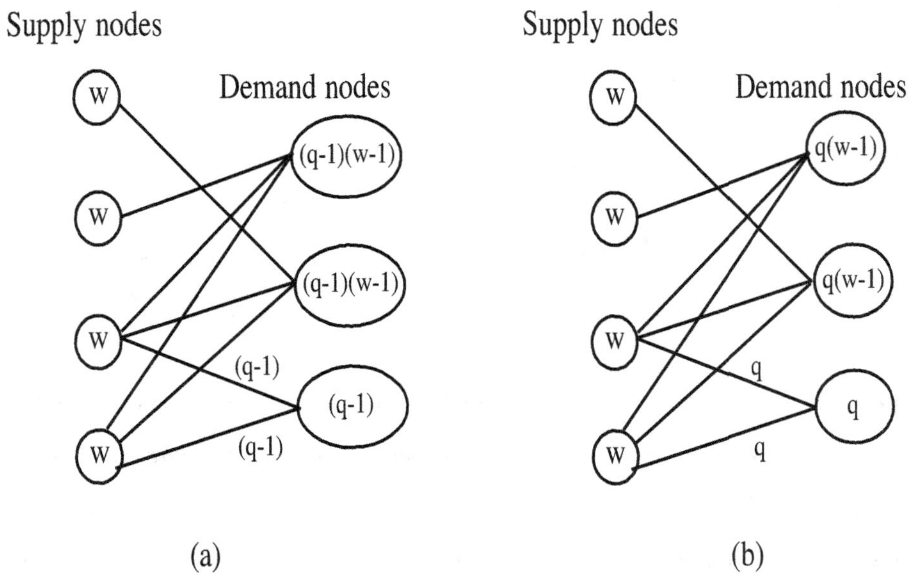

FIGURE 3.8: (a) MFP $\mathcal{P}_1$ to find α (b) MFP $\mathcal{P}_2$ to find β

Here we consider another lower bound presented in [42] in the presence of co-site constraints (s_0).

Result 3. *(Lemma 10 in [42]): Let $P = (X, W, C)$ be a CAP, v be a positive integer, and Q be a v-complete subset of X. Assume that $c_{ii} = k > v$ for all $i \in Q$. Furthermore, let*

$$w = \max_{i \in Q} w_i$$

with r cells in Q having $w_i = w$.

If

$$L_0 = \sum_{i \in Q} w_i,$$

$$L_1 = v(L_0 - 1) + 1,$$

$$L_2 = k(m - 1) + v(r - 1) + 1,$$

$$k = qv + v_0, 0 \leq v_0 < v,$$

then

$$S_0(P) \geq L_1, \quad for\ L_1 - L_2 \geq (v - v_0)(w - 1),$$

$$S_0(P) \geq L_2, \quad for\ L_2 - L_1 \geq v_0(w - 1),$$

and

$$S_0(P) \geq L_3 = L_2 + (L_0 - r - q(w - 1))(v - v_0),$$

in all other cases.

Here, we consider another lower bound presented in [20] in the presence of co-site constraints.

Result 4. *(Theorem 2 in [20]): Let $P = (X, W, C)$ be a CAP and Q be a 1-complete subset of X. Let $i \in Q$. Assume $c_{ii} = k > v' > 1$ and there exists a subset of Q, R such that*

$$i \notin R$$

and

$$c_{ii} \geq v', \quad for\ all\ j \in R.$$

Furthermore, let $w_R = \sum_{j \in R} w_j$.

If $k - 2v' + 1 \leq 0$

$$S_0(P) \geq (w_i - 1)k + 1 + w_R,$$

else

$$S_0(P) \geq (w_i - 1)k + 1 + max(w_R - (w_i - 1)(k - 2v' + 1), 0).$$

3.4 Genetic Algorithm for Channel Assignment

While solving an optimization problem using a genetic algorithm, the parameter set of the optimization problem must be coded as a finite-length string or chromosome over some finite alphabet Q [47]. A collection of M (finite) such strings or chromosomes is called a population. A simple genetic algorithm is composed of three basic operators: i) reproduction or selection, ii) crossover, and iii) mutation [47]. A genetic algorithm (GA) starts with an initial population (randomly generated) and in each iteration, a new (hopefully improved) population of the same size is generated from the current population by applying the above-mentioned operators on the strings of the current population. The newly generated population is then used to generate the next (possibly improved) population and so on.

Let S_b be the best string (with respect to the fitness value) of the population generated up to iteration t. In the elitist model of genetic algorithm (EGA), if S_b or any string better than S_b is not in the population generated in iteration $(t + 1)$, we include S_b in the $(t + 1)$-th population [47]. We describe below this technique for solving the channel assignment problem as given in [29] which ensures that the population is improved in successive iterations.

3.4.1 Problem Formulation

Let us assume that the channel assignment problem is represented by a CAP graph and the frequency separation constraints are given by the matrix $C = (c_{ij})$ as described in Section 3.1.2. Suppose the CAP graph has n nodes. A random order of these nodes is considered a string S or chromosome. For example, S = ((20), (21), (10), (11), (00)) is a string corresponding to the CAP graph of Figure 3.1(a). Let M be the population size (we have taken M as an even integer). Let *cp* be the crossover probability. We set *cp* to a high value, say 0.95, in the algorithm.

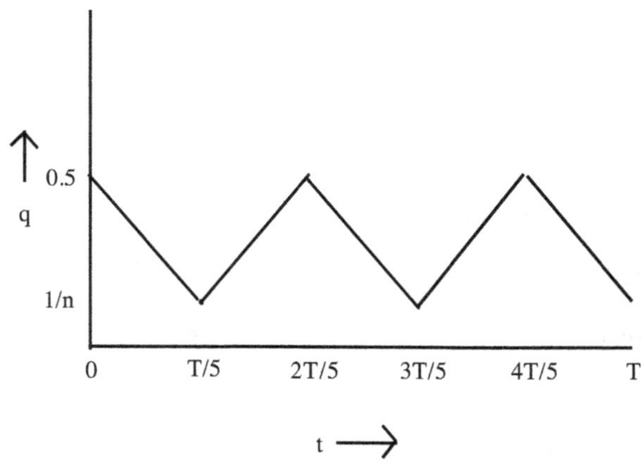

FIGURE 3.9: Variation of mutation probability (q) with number of iterations (t)

Let q be the mutation probability and T be the total number of iterations. Usually, T is a very large positive integer. We divide the total number of iterations into five equal intervals (assuming that T is a multiple of 5) as $[0, T/5], [T/5, 2T/5], [2T/5, 3T/5], [3T/5, 4T/5]$ and $[4T/5, T]$. We start with a mutation probability of q = 0.5 and then it is varied with the number of iterations as shown in Figure 3.9. The way in which the mutation probability is varied here is similar to that used in [48]. The reason for varying the mutation probability q in this fashion is that we are to increase the value of q to maintain the diversity of the population; similarly as the optimal string is approached, q is to be reduced. Figure 3.9 shows one such technique for varying q with the number of iterations.

The mutation probability at the t-th iteration is given by the following function.

```
Function m_probability(int t)

begin
    W ← 5(1 − 0.5n)/(nT)
    for i = 0 to 2 do
    begin
        V ← 2i + 1
        if (t ≤ VT/5) do
        begin
            U ← t − (V − 1)T/5
            return (0.5 + WU)
            break
        end;
        if (t ≤ (V + 1)T/5) do
        begin
            U ← (V + 1)T/5 − t
            return (0.5 + WU)
            break
        end;
    end
end.
```

The fitness function $Fit(S)$ used in the algorithm is described by the following function.

function $Fit(S)$ *// S is a string.//*

begin

$t[0] \leftarrow 0$

// $t[i]$ *is the frequency assigned to the i-th node*

$node_i$ *(*$0 \leq i \leq n-1$*) of S //*

for $i = 1$ *to* $n-1$ *do*

begin

Set $t[i]$ *to the smallest integer without violating the*

frequency separation requirements specified by

the matrix C *with all the previously assigned*

values $t[0], t[1], ..., t[i-1]$.

end

return $max(t[0], t[1], ..., t[n-1])$

end.

Next, we describe the *Algorithm GA* to resolve the channel assignment problem.

3.4.2 *Algorithm GA*

Step 1:

Set the iteration number $t \leftarrow 0$; Set $cp \leftarrow 0.95$; Set $M \leftarrow 20$.

Step 2: *(Initial population)*

begin

for $i = 0$ *to* $M-1$ *do*

begin

Generate a random order of the nodes in the CAP

graph and consider it as a string S_i.

end.

set $q_t \leftarrow \{S_0, S_1, \cdots, S_{M-1}\}$ *as the initial population.*

end

Example 2. *For the CAP graph shown in Figure 3.1(a), with* M = *4,* $\{S_0$, S_1, S_2, $S_3\}$ *is an initial population where* S_0 = S_1 = *((21), (20), (11), (00), (10)),* S_2 = *((00), (20), (11), (21), (10)) and* S_3 = *((11), (20), (21), (00), (10)).*

Note: The initial population may contain multiple copies of one or more strings. In particular, M copies of a single string may also constitute an initial population.

Step 3:

Compute $Fit(S_i)$ for each string S_i $(0 \leq i \leq M - 1)$ of q_t. Find the best string S_{best1} (i.e., the string with the least fitness value) and the worst string S_{worst1} (i.e., the string with the highest fitness value) of q_t. If S_{best1} or S_{worst1} is not unique, choose one arbitrarily.

Step 4: *(Selection or Reproduction)*

Perform the selection operation as described below on the strings of q_t to generate a mating pool q_{mat} of size M.

Selection operation

- *Calculate the probability* p_i *of selection of* S_i $(i = 0, 1, \cdots, M - 1)$

$$p_i = \frac{\frac{1}{Fit(S_i)}}{\sum_{i=0}^{M-1} \frac{1}{Fit(S_i)}}.$$

Note : With this choice of p_i*, the probability of selection of* S_i *will be higher if* $Fit(S_i)$ *is smaller. This is required because we are dealing with the minimization problem.*

- *Calculate the cumulative probability* q_i *for* S_i $(i = 0, 1, \cdots, M-1)$

$$q_i = \sum_{j=0}^{i} p_j.$$

- *Generate a random number* r_j *from [0, 1] for* $j = 0, 1, \cdots, M-1$. *Now, if* $r_j \le q_0$, *select* S_0*; otherwise select* S_i $(1 \le i \le M-1)$ *if* $q_{i-1} < r_j \le q_i$.

Note: $p_0 = q_0$ and $p_i = q_i - q_{i-1}$ for $1 \le i \le M-1$.

Remark 2. *With this implementation, if* p_i *is higher, i.e., the interval* $(q_{i-1}, q_i]$ *is of greater length, the probability that the random number generated falls within the interval* $(q_{i-1}, q_i]$ *will be higher. In other words, the probability of selection of* S_i *will be higher if* $Fit(S_i)$ *is smaller. In this way, from a current population of size* M, *a mating pool* q_{mat} *of the same size with potential strings is constructed. Since we are dealing with a minimization problem, we will say the string* S_i *has more potential than the string* S_j *if* $Fit(S_i) < Fit(S_j)$. *The strings of the mating pool then undergo crossover and mutation. This strategy for construction of the mating pool is close to that used in [49].*

Step 5: *(Crossover)*

Perform the crossover operation as described below on the strings of q_{mat} and obtain a population q_{temp1} of size M.

Crossover operation

Form $M/2$ *pairs by pairing the i-th and* $(M/2 + i)$*-th string from* q_{mat} $(i = 0, 1, \cdots, (M/2 - 1))$.

for each pair of strings do

begin

generate a random number R from [0, 1].

if ($R \leq cp$) *do*

//to produce two offsprings for the next generation.//

begin

generate two random numbers from $[0, n-1]$ *to define a matching section. Use this matching section to effect a cross through position-by-position exchange operation.*

end.

end.

Note : The above reordering operator known as partially matched crossover (PMX) [47] has the effect of both inversion and crossover.

Example 3. *Let us assume that the two random numbers generated are 2 and 3. Then, PMX on* S_1 *= ((00), (10), (11), (20), (21)) and* S_2 *= ((20), (00), (21), (10), (11)) will produce the two offsprings* S_1' *= ((00), (20), (21), (10), (11)) and* S_2' *= ((10), (00), (11), (20), (21)).*

Remark 3. *We have used the two-point crossover operation and PMX operator. Multi-point crossover and other forms of reordering operators such as order crossover (OX) and cycle crossover (CX) [47] can also be used.*

Step 6: *(Mutation)*

Perform the mutation operation as described below on the strings of q_{temp1} and obtain a population q_{temp2} of size M.

Mutation operation

Set $q \leftarrow m\_probability(t)$

for each string S_i *of* q_{temp1} $(0 \leq i \leq M-1)$ *do begin*

  for each node $node_j$ *(*$0 \leq j \leq n-1$*) of string* S_i *do*

  begin

    generate a random number from [0, 1], say m*.*

      if $(m \leq q)$ *do*

      begin

        exchange $node_j$ *of* S_i *with any other*

        randomly selected node $node_k$ *of* S_i

        $(0 \leq k \leq n-1,\ k \neq j)$

      end

  end

end.

Example 4. *If the node (00) is mutated by the node (21), after mutation the string* S_1 *= ((00), (10), (11), (20), (21)) is changed to the string* S_1' *= ((21), (10), (11), (20), (00)).*

Step 7:

Calculate $Fit(S_i)$ for each string S_i $(0 \leq i \leq M-1)$ of q_{temp2}. Find the best string S_{best2} and the worst string S_{worst2} of q_{temp2}. If S_{best2} or S_{worst2} is not unique, choose one arbitrarily.

Step 8: *(Elitism)*

Compare S_{best1} of q_t and S_{best2} of q_{temp2}. If $Fit(S_{best2}) > Fit(S_{best1})$, then replace S_{worst2} with S_{best1}. Rename q_{temp2} as q_t.

Step 9:

$t \leftarrow t + 1$; Go to step 3 if $t < T$.

Bhandari et al. [49] provided the proof that an EGA converges to the global optimal solution with any choice of initial population as the number of iterations goes to infinity. They also proved that no finite stopping time can guarantee the optimal solution. In practice, we must terminate the process after finitely many iterations with a high probability that the process has achieved the global optimal solution. In the present problem, the optimal string is not necessarily unique. There may be many strings which provide the optimal value. It was shown in [49] that if the number of strings having the optimal value is larger, the probability of fast convergence is higher. Practically, GA runs for a fixed number of iterations or terminates if no improvement is found at some stage within the fixed number of iterations.

3.4.3 GA on Special Cases of CAP

The *Algorithm GA*, presented in the previous section is applicable to any arbitrary CAP graph. We now consider the application of this algorithm to some special cases of the channel assignment problem. We assume that the cellular graph is of hexagonal structure and the channel interference extends only up to two cells from the call originating cell. First, we consider the case where every cell of the network has a demand of only one channel. In fact, we apply GA on a small subset of nodes to find its assignment and next we extend the assignment regularly to cover the whole network for faster convergence of GA.

We present below three frequency assignment schemes as given in [29] that depend on the relative values of s_1 and s_2.

3.4.3.1 Scheme 1: $s_1 = s_2$

Lemma 3. *In a hexagonal cellular network of two-band buffering with a demand of only one channel per cell, there exists a frequency assignment scheme which requires a bandwidth of* $6s_2$ *for* $s_1 = s_2$.

Proof: Consider the subgraph shown in Figure 3.10(a). The frequency separation constraint is specified by the matrix whose (i, j)-th element is $s_1 = s_2$, if cells i

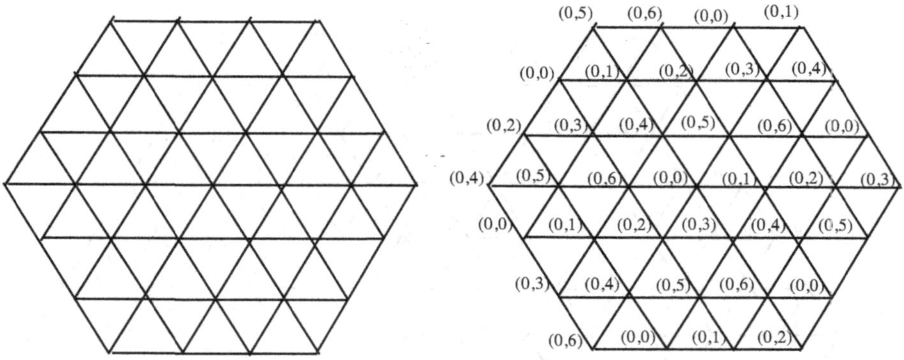

(a) (b)

FIGURE 3.10: (a) Subgraph of cellular graph and (b) frequency assignment when $s_1 = s_2$

and j are within Distance 2 and 0 otherwise. With this frequency separation matrix, we apply the *Algorithm GA* over the subgraph shown in the figure. The resulting assignment is shown in Figure 3.10(b). Observe that the frequency channels (0, 0), (0, 1), (0, 2), (0, 3), (0, 4), (0, 5) and (0, 6) are assigned repeatedly to the nodes over the subgraph in a regular and symmetrical way. The symmetry is clear from Figure 3.10(b). Using this symmetry, frequencies can be assigned to the nodes over the entire network. This assignment is shown in Figure 3.11 and is characterized as follows.

We first consider three directions x, y and z in the cellular graph in Figure 3.12. In each direction, we have definite types of lines and each type of line is identified by the sequence of frequencies assigned to the nodes lying on that line. A sequence is of Period b if each frequency is periodically repeated at Distance b on that line.

In x direction, there is only one type of line (*type x*) identified by the repetitive sequence of Period 7 as (0, 0) (0, 1) (0, 2) (0, 3) (0, 4) (0, 5) (0, 6) $\cdots$.

In y direction, there is only one type of line (*type y*) identified by the repetitive sequence of Period 7 as (0, 0) (0, 5) (0, 3) (0, 1) (0, 6) (0, 4) (0, 2) $\cdots$.

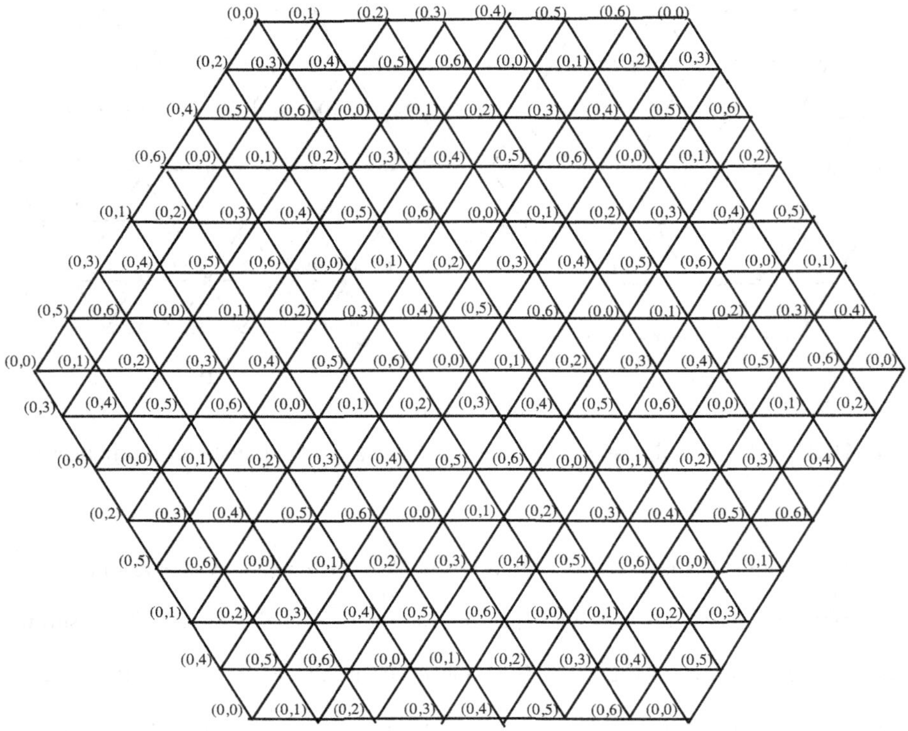

FIGURE 3.11: Frequency assignment of two-band buffering system for $s_1 = s_2$

In z direction, there is only one type of line (*type z*) identified by the repetitive sequence of Period 7 as (0, 0) (0, 4) (0, 1) (0, 5) (0, 2) (0, 6) (0, 3) $\cdots$.

It follows from Figure 3.11 that along any of the lines of *type x*, *type y* and *type z*, there is no conflict. Therefore, this assignment is conflict-free over the entire network.

Hence, by using the scheme described above, the frequencies can be assigned to the nodes of the entire network using a bandwidth of $6s_2$, when $s_1 = s_2$. □

We now generalize Lemma 3 and state the following theorem for the assignment of w channels to every node of the entire cellular network.

Theorem 3. *In a hexagonal cellular network with homogeneous demand w and two-band buffering, there exists a frequency assignment scheme for $s_1 = s_2$, which requires a bandwidth of:*

Case 1. $(w-1)7s_2 + 6s_2$ *when* $s_0 \leq 7s_2$*, and*

Case 2. $(w-1)s_0 + 6s_2$ *when* $s_0 \geq 7s_2$.

Proof: Follows from Lemma 3. □

Remark 4. *For the second case of Theorem 3, i.e., when $s_1 = s_2$ and $s_0 \geq 7s_2$, the required bandwidth is close to ws_0.*

Remark 5. *For $s_1 = s_2$, the above assignment requires a bandwidth equal to the corresponding lower bound presented in Theorem 1 for different relative values of s_0, s_1, and s_2. Hence this assignment is optimal.*

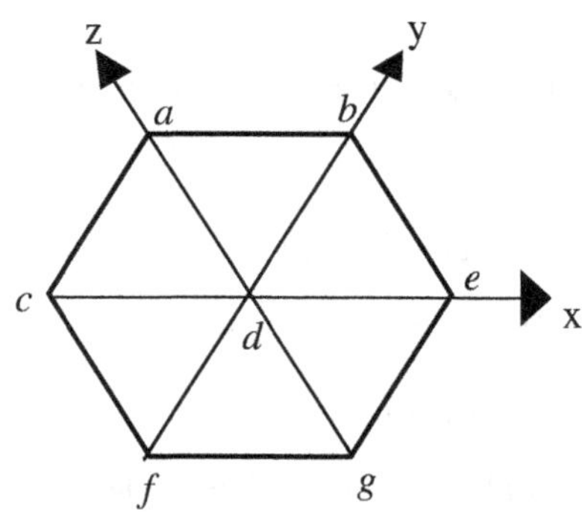

FIGURE 3.12: Three directions in hexagonal cellular graph

3.4.3.2 Scheme 2: For i) $s_2 < s_1 \leq 2s_2$ and ii) $s_1 \geq 2s_2$

Consider the nine-node subgraph with nodes a, b, c, d, e, f, g, h and i of the cellular graph as shown in Figure 3.13.

Each node of this subgraph represents a cell of the cellular network. Suppose

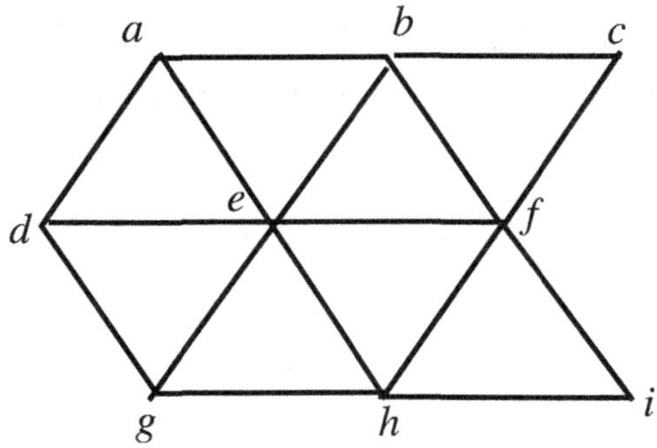

FIGURE 3.13: A nine-node subgraph of hexagonal cellular network

$\phi(\alpha)$ is the frequency assigned to the node α, $\alpha \in \{a, b, c, d, e, f, g, h, i\}$. We propose that only these nine frequencies be assigned to the nodes of the entire cellular graph in a regular and symmetrical fashion as shown in Figure 3.14. The idea behind this assignment is explained as follows.

We again consider the three directions x, y, and z in the cellular graph as shown in Figure 3.12.

In x direction, there are three different types of lines identified by the repetitive sequence of Period 3 as

type x_1*:* $\phi(a)\ \phi(b)\ \phi(c)\ \phi(a)\ \phi(b)\ \phi(c) \cdots$,

type x_2*:* $\phi(d)\ \phi(e)\ \phi(f)\ \phi(d)\ \phi(e)\ \phi(f) \cdots$, and

type x_3*:* $\phi(g)\ \phi(h)\ \phi(i)\ \phi(g)\ \phi(h)\ \phi(i) \cdots$.

In y direction, there is only one type of line (*type y*) identified by the repetitive sequence of Period 9 as $\phi(a)\ \phi(h)\ \phi(f)\ \phi(c)\ \phi(g)\ \phi(e)\ \phi(b)\ \phi(i)\ \phi(d) \cdots$.

In z direction, there is only one type of line (*type z*) identified by the repetitive sequence of Period 9 as $\phi(c)\ \phi(i)\ \phi(f)\ \phi(b)\ \phi(h)\ \phi(e)\ \phi(a)\ \phi(g)\ \phi(d) \cdots$.

Based on this assignment scheme, we construct a table of frequency separation constraints between every pair of these nine nodes, as shown in Table 3.6 (assuming that only one channel is assigned to each node), such that

1. Along each of the lines of *type* x_1, *type* x_2 and *type* x_3 in x direction, there is no conflict.

2. Along all the lines of *type y* in y direction, there is no conflict.

3. Along all the lines of *type z* in z direction, there is no conflict.

Therefore, these nine frequencies can be repeated over the entire network without conflict.

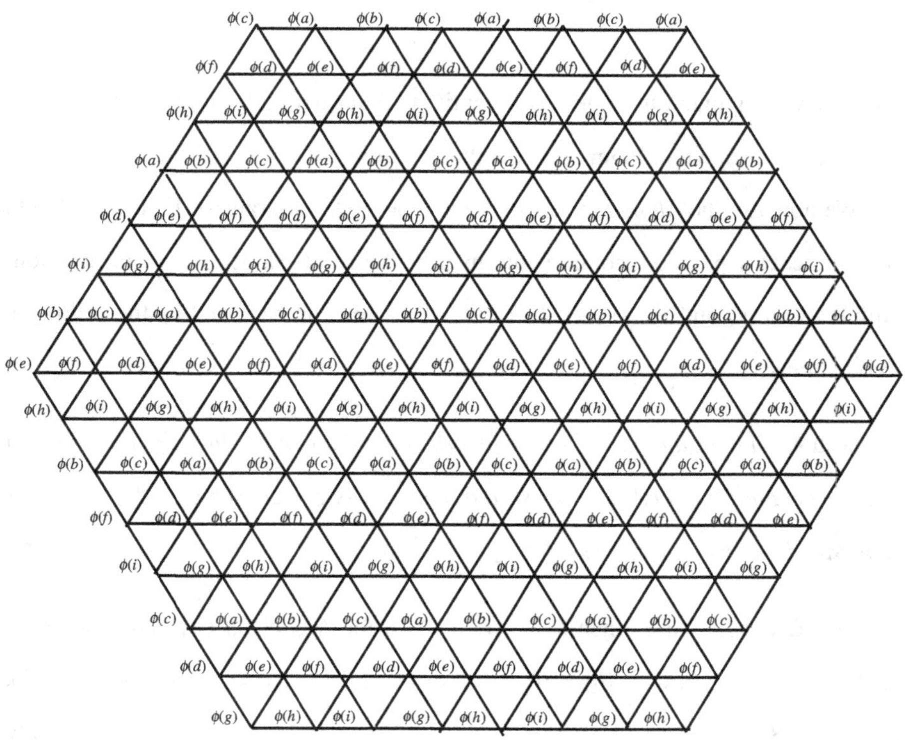

FIGURE 3.14: Frequency assignment of two-band buffering system using nine frequency channels

It is important to note that Table 3.6 has been so constructed that only nine frequency channels assigned to the nodes of Figure 3.13 would be sufficient for the

TABLE 3.6: Frequency separation matrix for Scheme 2

$Nodes \rightarrow$ $\downarrow$	a	b	c	d	e	f	g	h	i
a	0	s_1	s_1	s_1	s_1	s_2	s_1	s_1	s_2
b	s_1	0	s_1	s_2	s_1	s_1	s_2	s_1	s_1
c	s_1	s_1	0	s_1	s_2	s_1	s_1	s_2	s_1
d	s_1	s_2	s_1	0	s_1	s_1	s_1	s_2	s_1
e	s_1	s_1	s_2	s_1	0	s_1	s_1	s_1	s_2
f	s_2	s_1	s_1	s_1	s_1	0	s_2	s_1	s_1
g	s_1	s_2	s_1	s_1	s_1	s_2	0	s_1	s_1
h	s_1	s_1	s_2	s_2	s_1	s_1	s_1	0	s_1
i	s_2	s_1	s_1	s_1	s_2	s_1	s_1	s_1	0

frequency assignment to nodes of the entire cellular network following the assignment patterns shown in Figure 3.14, without conflict.

We now use the *Algorithm GA* over the nine-node subgraph of Figure 3.13 with the frequency separation requirements given by Table 3.6. Depending on the relationship between s_1 and s_2, the results of the algorithm can be stated by the following two lemmas.

Lemma 4. *In a hexagonal cellular network of two-band buffering with a demand of only one channel per cell, there exists a frequency assignment scheme which requires a bandwidth of* $8s_2$ *for* $s_2 < s_1 \leq 2s_2$.

Proof: For the case when $s_2 < s_1 \leq 2s_2$, the *Algorithm GA* gives the result such that $\phi(a)$ = (0, 0), $\phi(b)$ = (0, 6), $\phi(c)$ = (0, 3), $\phi(d)$ = (0, 5), $\phi(e)$ = (0, 2), $\phi(f)$ = (0, 8), $\phi(g)$ = (0, 7), $\phi(h)$ = (0, 4) and $\phi(i)$ = (0, 1). Note that, the inequality $2s_2 \geq s_1$ has been used here because the frequencies (0, 6) and (0, 8) are assigned to nodes that are Distance 1 apart from each other. In this case, the maximum frequency assigned is (0, 8). This assignment is shown in Figure 3.15 and obtained by replacing the values of $\phi(a)$, $\phi(b)$, $\phi(c)$, $\phi(d)$, $\phi(e)$, $\phi(f)$, $\phi(g)$, $\phi(h)$, and $\phi(i)$ by (0, 0), (0, 6), (0, 3), (0, 5), (0, 2), (0, 8), (0, 7), (0, 4), and (0, 1), respectively, in Figure 3.14.

Hence, by using the frequency assignment scheme described above, the frequencies can be assigned to the nodes of the entire network using a bandwidth of $8s_2$ when $2s_2 \geq s_1 > s_2$. □

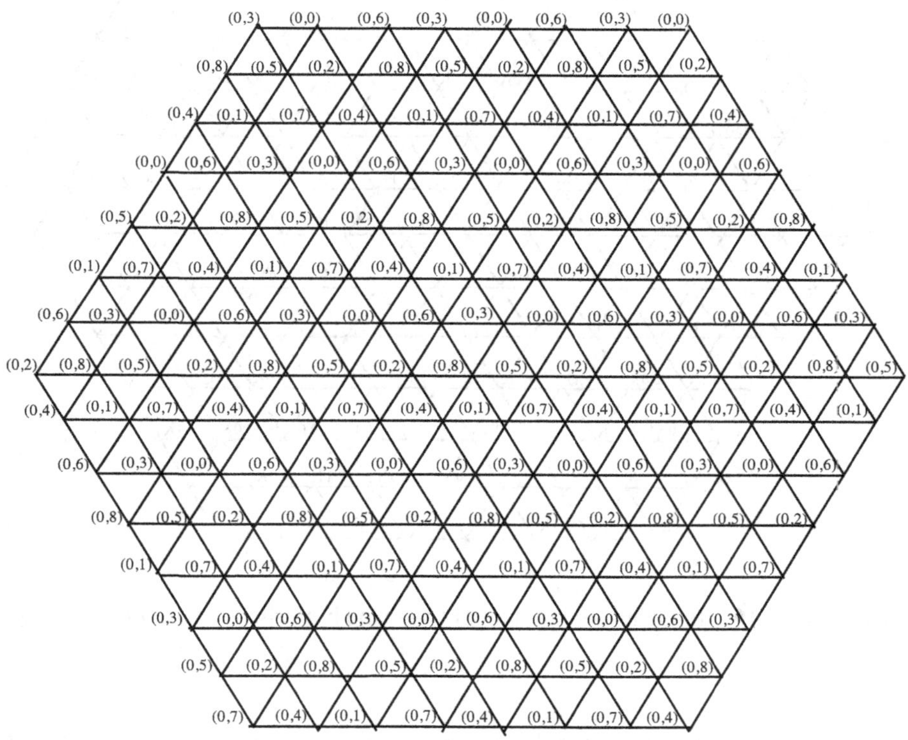

FIGURE 3.15: Frequency assignment of two-band buffering system for $s_2 < s_1 \leq 2s_2$

Lemma 5. *In a hexagonal cellular network of two-band buffering with a demand of only one channel per cell, there exists a frequency assignment scheme which requires a bandwidth of* $4s_1$ *for* $s_1 \geq 2s_2$.

Proof: Exactly the same as in Lemma 4. This assignment is shown in Figure 3.16. □

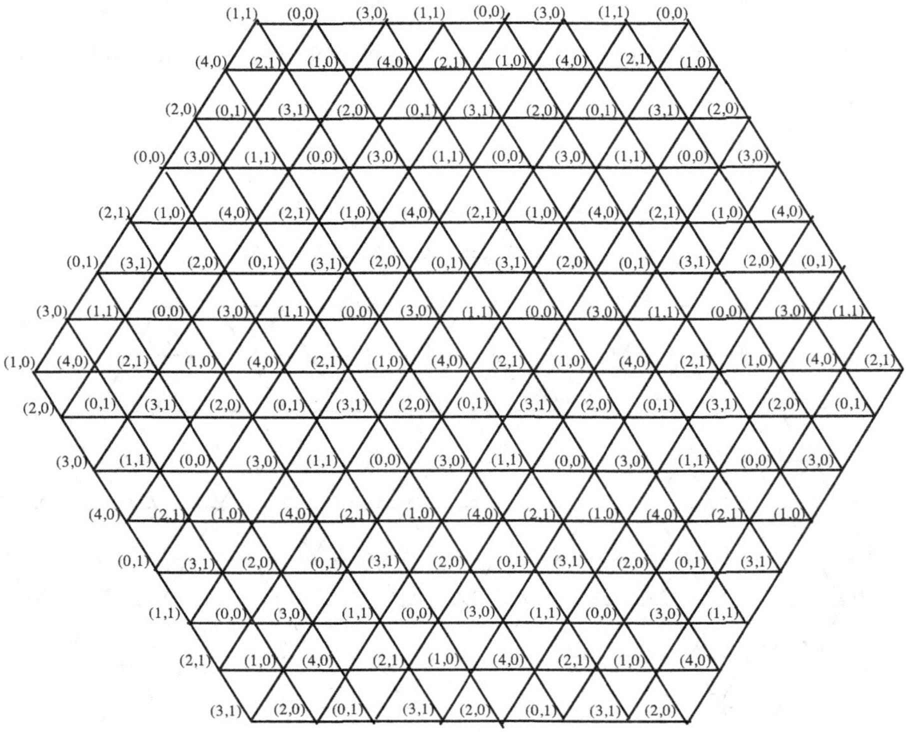

FIGURE 3.16: Frequency assignment of two-band buffering system for $s_1 \geq 2s_2$

Next we consider the case of homogeneous demand with w channels. The next two theorems simply extend the results of Lemmas 4 and 5, respectively.

Theorem 4. *In a hexagonal cellular network with homogeneous demand w and 2-band buffering, there exists a frequency assignment scheme for $s_2 < s_1 \leq 2s_2$, which requires a bandwidth of:*

Case 1. $(w-1)9s_2 + 8s_2$ *when* $s_0 \leq 9s_2$*, and*

Case 2. $(w-1)s_0 + 8s_2$ *when* $s_0 \geq 9s_2$.

Proof: Follows from Lemma 4. For $w = 2$, the assignments are illustrated in Figures 3.17(a) and (b) for $s_0 \leq 9s_2$ and $s_0 \geq 9s_2$, respectively. □

Remark 6. *For the second case of Theorem 4, i.e., when $s_2 < s_1 \leq 2s_2$ and $s_0 \geq 9s_2$, the required bandwidth is close to ws_0.*

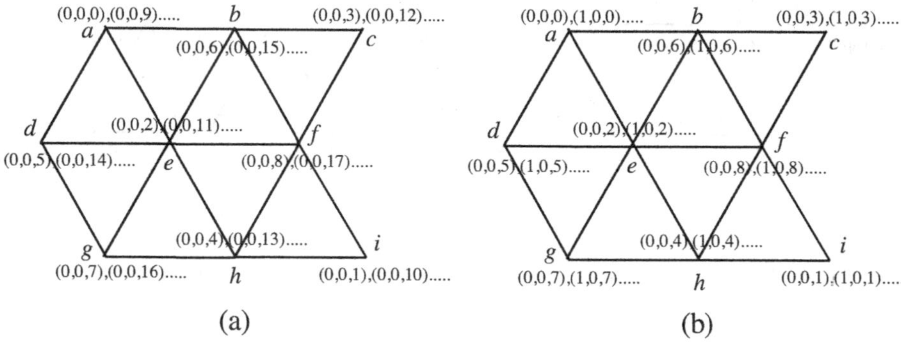

FIGURE 3.17: Frequency assignments to nine-node subgraph for $s_2 < s_1 \leq 2s_2$

Theorem 5. *In a hexagonal cellular structure with homogeneous demand w and two-band buffering, there exists a frequency assignment scheme for $s_1 \geq 2s_2$, which requires a bandwidth of:*

Case 1. $(w-1)(4s_1+s_2)+4s_1$ *when* $s_0 \leq (4s_1+s_2)$, *and*

Case 2. $(w-1)s_0+4s_1$ *when* $s_0 \geq (4s_1+s_2)$.

Proof: Follows from Lemma 5. For $w = 2$, the corresponding assignments are shown in Figures 3.18(a) and (b) for $s_0 \leq (4s_1+s_2)$ and $s_0 \geq (4s_1+s_2)$, respectively. □

Remark 7. *For the second case of Theorem 5, i.e., when $s_1 \geq 2s_2$ and $s_0 \geq (4s_1+s_2)$, the required bandwidth is close to ws_0.*

3.4.3.3 Scheme 3: For $s_1 \geq s_2$

Consider again the nine-node subgraph of Figure 3.13. In this scheme as well, we propose to use only nine frequencies assigned to these nine nodes in a regular and symmetric fashion for the whole cellular graph shown in Figure 3.19. However,

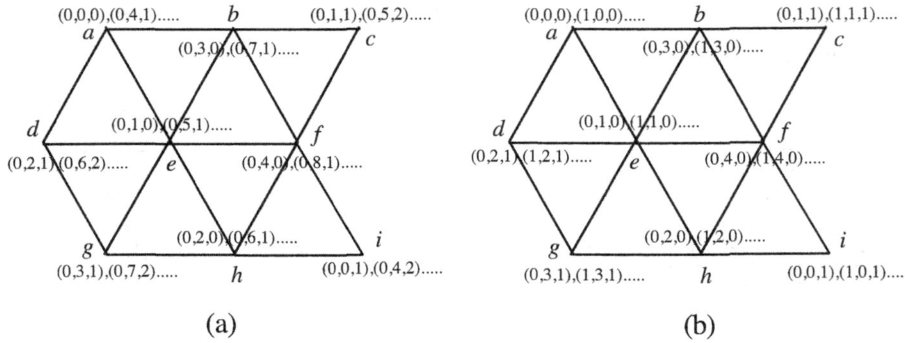

FIGURE 3.18: Frequency assignments to nine-node subgraph for $s_1 \geq 2s_2$

we would use a different frequency separation matrix as given in Table 3.7. The difference between Tables 3.6 and 3.7 is in the submatrix corresponding to the rows for a, b, c and columns for g, h and i. Because of this change, the repetition pattern to be used here will be slightly different from that of Scheme 2 as discussed below.

We again consider the three directions x, y, and z in the cellular graph in Figure 3.12.

In x direction, there are three different types of lines (*types* x_1, x_2 *and* x_3) as in Scheme 2 above.

In y direction, there are three different types of lines identified by the repetitive sequence of Period 3 as:

type y_1: $\phi(b)\ \phi(g)\ \phi(e)\ \phi(b)\ \phi(g)\ \phi(e) \cdots$,

type y_2: $\phi(a)\ \phi(i)\ \phi(d)\ \phi(a)\ \phi(i)\ \phi(d) \cdots$, and

type y_3: $\phi(c)\ \phi(h)\ \phi(f)\ \phi(c)\ \phi(h)\ \phi(f) \cdots$.

In z direction, there are also three different types of lines identified by the repetitive sequence of Period 3 as:

type z_1: $\phi(g)\ \phi(d)\ \phi(c)\ \phi(g)\ \phi(d)\ \phi(c) \cdots$,

type z_2: $\phi(b)\ \phi(i)\ \phi(f)\ \phi(b)\ \phi(i)\ \phi(f) \cdots$, and

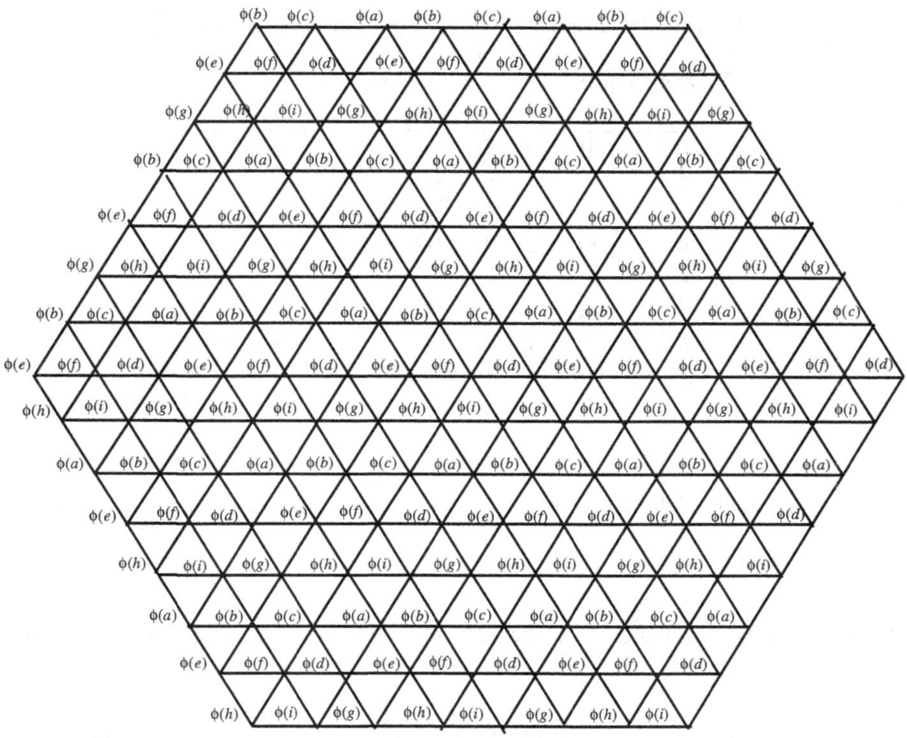

FIGURE 3.19: Frequency assignment of two-band buffering system using nine frequency channels

type z_3: $\phi(h)\ \phi(e)\ \phi(a)\ \phi(h)\ \phi(e)\ \phi(a) \cdots$.

It follows from Figure 3.19 and the frequency separation matrix given by Table 3.7 that

1. Along each of the lines of *type* x_1, *type* x_2 and *type* x_3 in x direction, there is no conflict.

2. Along each of the lines of *type* y_1, *type* y_2 and *type* y_3 in y direction, there is no conflict.

TABLE 3.7: Frequency separation matrix for Scheme 3

$Nodes \rightarrow$ $\downarrow$	a	b	c	d	e	f	g	h	i
a	0	s_1	s_1	s_1	s_1	s_2	s_2	s_1	s_1
b	s_1	0	s_1	s_2	s_1	s_1	s_1	s_2	s_1
c	s_1	s_1	0	s_1	s_2	s_1	s_1	s_1	s_2
d	s_1	s_2	s_1	0	s_1	s_1	s_1	s_2	s_1
e	s_1	s_1	s_2	s_1	0	s_1	s_1	s_1	s_2
f	s_2	s_1	s_1	s_1	s_1	0	s_2	s_1	s_1
g	s_2	s_1	s_1	s_1	s_1	s_2	0	s_1	s_1
h	s_1	s_2	s_1	s_2	s_1	s_1	s_1	0	s_1
i	s_1	s_1	s_2	s_1	s_2	s_1	s_1	s_1	0

3. Along each of the lines of *type* z_1, *type* z_2 and *type* z_3 in z direction, there is no conflict.

Therefore, these frequencies can be repeated over the entire network without any conflict.

It is important to note that Table 3.7 has been so constructed that only nine frequency channels assigned to the nodes of Figure 3.13 would be sufficient for the frequency assignment to nodes of the entire cellular network following the assignment patterns as shown in Figure 3.19 without any conflict.

We now use the *Algorithm GA* over the nine-node subgraph of Figure 3.13 with the frequency separation requirements given by Table 3.7. The results of the algorithm can be stated by the following lemma.

Lemma 6. *In a hexagonal cellular network of two-band buffering with a demand of only one channel per cell, there exists a frequency assignment scheme which requires a bandwidth of* $(2s_1 + 6s_2)$ *for all values of* s_1 *and* s_2 *where* $s_1 \geq s_2$.

Proof: For this case, the *Algorithm GA* gives the result $\phi(a)$ = (0, 0), $\phi(b)$ = (2, 4), $\phi(c)$ = (1, 3), $\phi(d)$ = (2, 5), $\phi(e)$ = (1, 4), $\phi(f)$ = (0, 2), $\phi(g)$ = (0, 1), $\phi(h)$ = (2, 6) and $\phi(i)$ = (1, 2). Note that the maximum frequency assigned is (2, 6). This assignment is shown in Figure 3.20 which is obtained by replacing the values of $\phi(a)$, $\phi(b)$, $\phi(c)$, $\phi(d)$, $\phi(e)$, $\phi(f)$, $\phi(g)$, $\phi(h)$, and $\phi(i)$ by (0, 0), (2, 4), (1, 3), (2, 5), (1, 4), (0, 2), (0, 1), (2, 6), and (1, 2), respectively, in Figure 3.19.

By using the scheme described above, assignment to the nodes of the entire network can be done using a bandwidth of $(2s_1 + 6s_2)$ for all cases of $s_1 \geq s_2$. □

We now state the following result for assigning w frequency channels to each of the nodes of the entire cellular network by simply extending the results of Lemma 6.

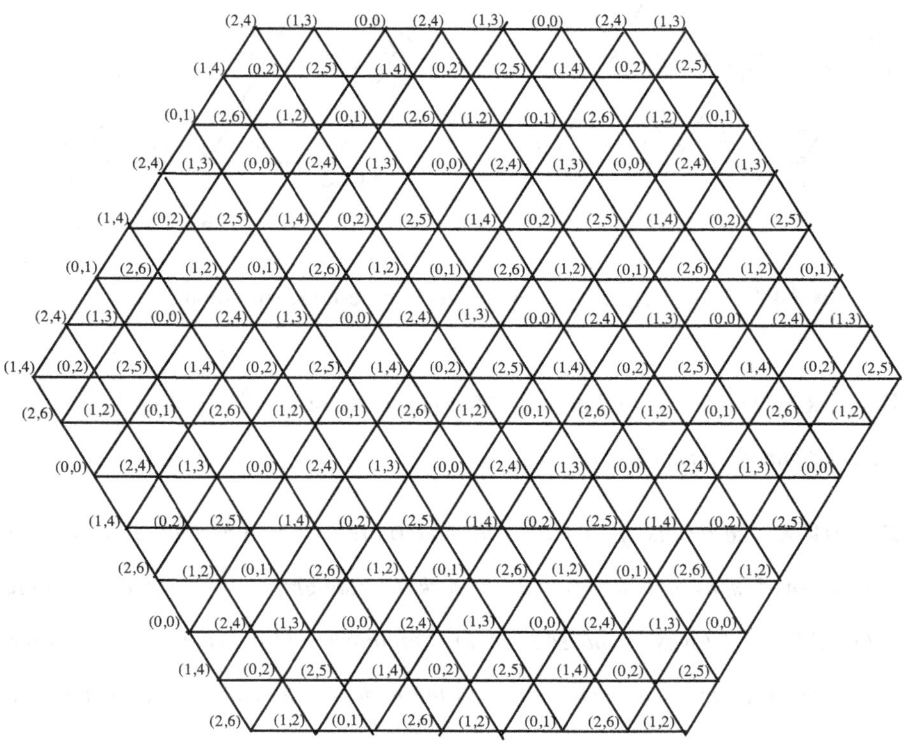

FIGURE 3.20: Frequency assignment of two-band buffering system for $s_1 \geq 3s_2$

Theorem 6. *In a hexagonal cellular network of two-band buffering with a homogeneous demand w, there exists a frequency assignment scheme for all values of s_1 and*

s_2 *where* $s_1 \geq s_2$, *which requires a bandwidth of:*

Case 1. $(w-1)(3s_1+6s_2)+(2s_1+6s_2)$, *for* $s_0 \leq (3s_1+6s_2)$, *and*

Case 2. $(w-1)s_0+(2s_1+6s_2)$, *for* $s_0 \geq (3s_1+6s_2)$.

Proof: Follows from Lemma 6. For $w = 2$, the corresponding assignments are illustrated in Figures 3.21(a) and (b) for $s_0 \leq (3s_1+6s_2)$ and $s_0 \geq (3s_1+6s_2)$, respectively. □

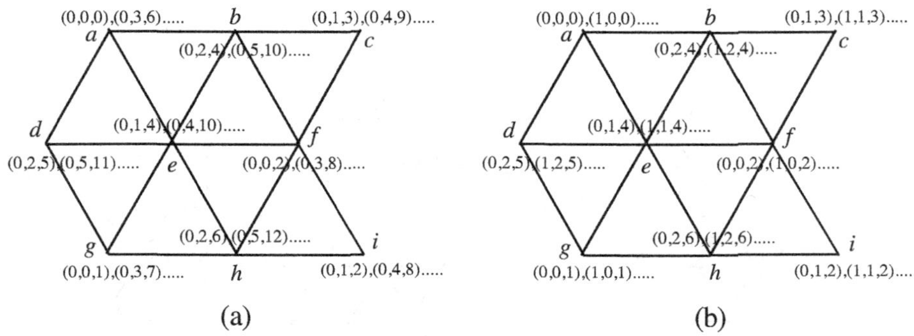

FIGURE 3.21: Frequency assignments to nine-node subgraph for $s_1 \geq s_2$

Remark 8. *For the second case of Theorem 6, i.e., when* $s_0 \geq (3s_1+6s_2)$, *the required bandwidth is close to* ws_0.

Remark 9. *The results of Lemmas 4 through 6 are based on assigning frequencies to a nine-node subgraph of the entire network. This subgraph with the corresponding Tables 3.6 and 3.7 was so chosen to ensure that only nine frequency channels would be sufficient for frequency assignment to the nodes of the entire cellular network. Different choices of the subgraph or the frequency separation matrix would give different assignment patterns. Besides that, we get the results of Lemmas 4 through 6 by applying genetic algorithm. Therefore, it is likely that we would not always get the best solution. Since the results of Theorems 4 through 6 are based on these lemmas, it is not possible to claim that Theorems 4 through 6 give minimal bandwidth for as-*

signment. However, for different relative values of s_1 and s_2, the required bandwidth by the above assignment schemes is either less than or equal to that reported in [1].

3.4.4 Overall Bandwidth Requirement

A frequency assignment scheme was described in [1] for hexagonal cellular network with a demand of only one channel per cell and two-band buffering, requiring a bandwidth of $(2s_1 + 6s_2)$ for all cases of $s_1 \geq s_2$. Let I be the factor of improvement in bandwidth achieved by the above schemes over that in [1]. Summarizing all the results of Lemmas 3 through 6, we now consider the following cases :

When $s_1 = s_2$.

$$I = \frac{2s_1 + 6s_2}{6s_2} = 4/3,$$

leading to an improvement by 25%.

When $s_2 < s_1 \leq 2s_2$.

$$I = \frac{2s_1 + 6s_2}{8s_2}.$$

Note that, I lies within the interval (1, 5/4], leading to an improvement of 20% at best.

When $2s_2 \leq s_1 \leq 3s_2$.

$$I = \frac{2s_1 + 6s_2}{4s_1}.$$

Here I lies within the interval [1, 5/4], leading to an improvement of 20% at best.

When $s_1 \geq 3s_2$.

$$I = \frac{2s_1 + 6s_2}{2s_1 + 6s_2} = 1.$$

Considering all the cases, we conclude that the above schemes result in a 25% improvement in bandwidth at best over that in [1].

For homogeneous demands of w channels per cell, we summarize below the results of Theorems 3 through 6 on bandwidth requirements for different relative values of s_0, s_1 and s_2.

1. When $s_1 = s_2$ (Theorem 3)
   (a) $(w-1)7s_2 + 6s_2$ for $s_0 \leq 7s_2$ (Case 1 of Theorem 3)
   (b) $(w-1)s_0 + 6s_2$ for $s_0 \geq 7s_2$ (Case 2 of Theorem 3)
2. When $s_2 < s_1 \leq 2s_2$ (Theorem 4)
   (a) $(w-1)9s_2 + 8s_2$ for $s_0 \leq 9s_2$ (Case 1 of Theorem 4)
   (b) $(w-1)s_0 + 8s_2$ for $s_0 \geq 9s_2$ (Case 2 of Theorem 4)
3. When $s_1 \geq 2s_2$ (Theorem 5 or 6, whichever gives the minimum bandwidth)
   $Min[(w-1)(4s_1+s_2)+4s_1$ for $s_0 \leq (4s_1+s_2)$ (Case 1 of Theorem 5), $(w-1)s_0 + 4s_1$ for $s_0 \geq (4s_1+s_2)$ (Case 2 of Theorem 5), $(w-1)(3s_1+6s_2) + (2s_1+6s_2)$ for $s_0 \leq (3s_1+6s_2)$ (Case 1 of Theorem 6), and $(w-1)s_0 + (2s_1+6s_2)$ for $s_0 \geq (3s_1+6s_2)$ (Case 2 of Theorem 6)]

Example 5. *Let* $s_0 = 8$, $s_1 = 1$, *and* $s_2 = 1$. *In this case the bandwidth requirement will be given by Case 2 of Theorem 3, which corresponds to item 1 (b) above.*

3.5 Coalesced CAP

We now present an elegant technique [51] for solving the channel assignment problem (CAP) for 2G cellular mobile networks, where channel allocation is made on a quasi-fixed basis and all sessions are connection-oriented. It first maps a given CAP $\mathbf{P}$ to a modified coalesced CAP $\mathbf{P}'$ on a smaller subset of cells of the network, which appreciably reduces the search space. This helps to solve $\mathbf{P}'$ by applying approximate algorithms very efficiently, reducing the computing time drastically. This solution to $\mathbf{P}'$ is then used to solve the original problem $\mathbf{P}$ by using a modified version of the FAR (forced assignment with rearrangement) operation reported in [17]. This technique has been tested on well-known benchmark problems and produced optimal solutions for all cases with improved computation times. For instance, it needs only about 10 and 20 seconds (on an unloaded DEC Alpha station 200 4/233) to get an optimal assignment for the two most difficult benchmark Problems 2 and 6, respectively, with zero call blocking, in contrast to 60 and 72 seconds, respectively, on an unloaded Sun Ultra 60 workstation reported in [51]. Moreover, as a by-product of this approach, there remain, many unused or redundant channels that may be used for accommodating small perturbations in demands dynamically.

3.5.1 Construction of Coalesced CAP

For a given CAP P, initially the first category of algorithms is applied to find a solution assuming a single demand per cell. This solution is now used to construct the coalesced CAP P'. Here follows the algorithm to generate a coalesced CAP P' from the given CAP P.

Algorithm *Construct_Coalesced_CAP*

Step 1: Define the CAP $P^* = (X^*, W^*, C^*)$ from the given CAP $P = (X, W, C)$, such that $X^* = X$, $C^* = C$, but $W^* = (w_i^*) = (1)$, i.e., $w_i^* = 1 \; \forall i$ $(0 \leq i \leq n-1)$. Note that, P^* is nothing more than P with the homogeneous single demand per cell.

Step 2: Find an admissible frequency assignment Φ^* for P^* by applying a suitable algorithm of the first category (e.g., the algorithm GA in [29]). Let a_0, a_1, $\cdots$, a_{z-1} be the z $(z \leq n)$ different channels assigned by Φ^*.

TABLE 3.8: Homogeneous single channel assignment for Problem 9

Cell nos.	0	1	2	3	4	5	6	7	8	9	10	11	12	13	14	15	16	17	18	19	20	21	22	23	24
Φ^*	5	6	3	5	1	2	1	0	6	4	2	2	7	0	3	6	2	5	1	4	0	7	6	1	0

Example 6. *To demonstrate this step, let us consider a practical assignment problem from Helsinki, Finland [32, 26, 24], to be referred later as Problem 9. The example CAP $P = (X, W, C)$, has been formulated on a 25-cell system of non-hexagonal structure, whose frequency separation matrix C and demand vector W are shown in Tables 3.3 and 3.4, respectively. The entry corresponding to the i-th row and j-th column of Table 3.3, i.e., c_{ij}, represents the minimum frequency separation requirement between a call in cell i and a call in cell j $(0 \leq i, j \leq 24)$. The column-i of the row D_3 in Table 3.4 indicates the channel demand w_i from cell i. Next, the problem $P^* = (X^*, W^*, C^*)$ is derived from P (Problem 9 defined above), where $X^* = X$, $C^* = C$, but $W^* = (w_i^*) = (1)$. Table 3.8 shows an admissible solution Φ^* for P^*, where column-i indicates the channel assigned to cell i. Note that Φ^* needs $z = 8$ channels namely $a_0 = 0$, $a_1 = 1$, $\cdots$, $a_7 = 7$ respectively.*

Step 3: *Construct the coalesced CAP $P' = (X', W', C')$ from Φ^* using the following four sub-steps :*

TABLE 3.9: Matrix $c'(a_i, a_j)$ for Problem 9

$c'(a_i, a_j)$	a_0	a_1	a_2	a_3	a_4	a_5	a_6	a_7
a_0	2	1	1	1	1	1	1	1
a_1	1	2	1	1	1	1	1	1
a_2	1	1	2	1	1	1	1	1
a_3	1	1	1	2	1	1	1	1
a_4	1	1	1	1	2	1	1	1
a_5	1	1	1	1	1	2	1	1
a_6	1	1	1	1	1	1	2	1
a_7	1	1	1	1	1	1	1	2

Step 3.1: Let Y(i) be the set of all cells where channel a_i has been assigned in Φ^*. Since every cell demands only one channel, $Y(i) \cap Y(j) = NULL$ for $i \neq j$. On the other hand, since Φ^* is *admissible*, $X^* = \cup_{i=0}^{i=z-1} Y(i)$.

Example 7. *For the problem P^* of Example 6, the admissible assignment Φ^* is shown in Table 3.8. Note that, channel* 0 *has been assigned to the cells 7, 13, 20, and 24. Hence, $Y(0) = \{7, 13, 20, 24\}$. Similarly, $Y(1) = \{4, 6, 18, 23\}$, $Y(2) = \{5, 10, 11, 16\}$, $Y(3) = \{2, 14\}$, $Y(4) = \{9, 19\}$, $Y(5) = \{0, 3, 17\}$, $Y(6) = \{1, 8, 15, 22\}$, and $Y(7) = \{12, 21\}$.*

Step 3.2: Compute the maximum entry in C (the frequency separation matrix of P) for every pair of cells u and v, $u \in Y(i)$, $v \in Y(j)$ and denote it by $c'(a_i, a_j)$. That is,

$$c'(a_i, a_j) = c'(Y(i), Y(j)) = max_{u \in Y(i), v \in Y(j)} \{c_{uv}\},$$

where $0 \leq i, j \leq z - 1$.

Example 8. *For the problem of Example 6, from Table 3.3, we get $c'(Y(0), Y(1)) = c'(a_0, a_1) = 1$, i.e., 1 is the maximum among all c_{ij}'s where $i \in Y(0) = \{7, 13, 20, 24\}$ and $j \in Y(1) = \{4, 6, 18, 23\}$. All $c'(a_i, a_j)$'s ($0 \leq i, j \leq 7$) are shown in Table 3.9.*

Step 3.3: Find the maximum weight among all cells in $Y(i)$ and denote it by $M(a_i)$. That is,

$$M(a_i) = M(Y(i)) = max_{j \in Y(i)}\{w_j\}$$

where $0 \leq i \leq z - 1$.

Example 9. *For the problem of Example 6, the demand of each cell is given in Table 3.4. From this table,* $M(Y(0)) = M(a_0) = 7$*, i.e., 7 is the maximum demand among the cells in* $Y(0) = \{7, 13, 20, 24\}$*. Similarly,* $M(a_1) = 9$, $M(a_2) = 9$, $M(a_3) = 9$, $M(a_4) = 8$, $M(a_5) = 10$, $M(a_6) = 11$, *and* $M(a_7) = 10$.

Step 3.4: Represent all the cells in $Y(i)$ by a single node $N(Y(i))$ of a weighted graph G' where the weight of a node $N(Y(i))$ is $M(Y(i))$. Connect nodes $N(Y(i))$ and $N(Y(j))$ by an edge with weight $c'(Y(i), Y(j))$ if $c'(Y(i), Y(j)) > 0$, and terminate.

This graph G' is termed the *coalesced CAP graph*. The corresponding coalesced CAP $P' = (X', W', C')$ is represented by the following components:

1. A set $X' = (N(Y(i)))$ $(0 \leq i \leq z - 1)$ of z distinct nodes where node $N(Y(i))$ represents the set $Y(i)$ in Φ^* of P^*.

2. A demand vector $W' = (M(Y(i)))$ where $M(Y(i))$ represents the weight of the node $N(Y(i))$ in G', $0 \leq i \leq z - 1$.

3. A frequency separation matrix $C' = (c'(Y(i), Y(j)))$ where $c'(Y(i), Y(j))$ represents the weight of the edge between nodes $N(Y(i))$ and $N(Y(j))$ in G', $0 \leq i, j \leq z - 1$.

4. A frequency assignment matrix $\Phi' = (\phi'_{ij})$, where ϕ'_{ij} represents the frequency assigned to call j in the node $N(Y(i))$ $(0 \leq i \leq z - 1, 0 \leq j \leq M(Y(i)) - 1)$.

5. A set of frequency separation constraints specified by the frequency separation matrix:
   $|\phi'_{ik} - \phi'_{jl}| \geq c'(Y(i), Y(j))$ for all i, j, k, l (except when both $i = j$ and $k = l$). □

Once P' is constructed as above, the objective is to find an assignment Φ' for this P' with a given bandwidth B.

Example 10. *The coalesced CAP $P' = (X', W', C')$ for the problem $P = (X, W, C)$ of Example 6, is given by $X' = (N(a_i))$, $0 \leq i \leq 7$, W'=(7, 9, 9, 9, 8, 10, 11, 10), and C' is given in Table 3.9.*

Lemma 7. *The given CAP P and the coalesced CAP P' are equivalent if $z = n$. For $z < n$, the total search space of P' is always less than that of P.*

Proof: Clearly, if $z = n$, problems P and P' have the same number of nodes having the same weights, and $C = C'$. Hence, there is no reduction in search space after transferring P to P'.

As explained in Section 3.1.3, the number of nodes and their demands in a CAP graph determines the total search space for a given CAP. Let the sum of demands on all cells for problems P and P' be $\mathcal{T}$ and $\mathcal{T}'$, respectively. Hence, the CAP graphs for P and P' will have $\mathcal{T}$ and $\mathcal{T}'$ nodes, respectively. If $z < n$, $\mathcal{T}'$ must be less than $\mathcal{T}$ because the sum of the demands of all the cells in $Y(i)$ has been replaced by the maximum of those to contribute to $\mathcal{T}'$. Hence, the proof. □

3.5.2 Technique for Solving Original CAP from Coalesced CAP

Transforming the original CAP $P = (X, W, C)$ to the coalesced CAP $P' = (X', W', C')$, we apply a suitable algorithm of the second category of CAP to find a frequency assignment Φ' of P' with a view to minimizing the number of blocked calls BL_{total}. The assignment Φ' may or may not be admissible, depending on the available bandwidth B. Hence, to derive the required assignments for P, we consider the following two cases.

Case 1: Assignment Φ' is admissible

An admissible frequency assignment for P can be derived by using the Φ' by means of the following theorem.

Theorem 7. *Given the problem $P = (X, W, C)$ and the bandwidth B, if the frequency assignment Φ' for P' is admissible, an admissible frequency assignment for P can be derived from Φ'.*

Proof: To get an assignment of P from Φ', all the cells in $Y(i)$ $(0 \leq i \leq z-1)$, are assigned the same set of channels assigned to $N(Y(i))$ in Φ'. This assignment must satisfy the interference constraints because in P', $c'(Y(i), Y(j))$ is the maximum among all c_{ij}'s in C where $i \in Y(i)$ and $j \in Y(j)$. This assignment must also satisfy the demand vector $W = (w_i)$, since in P', $M(Y(i))$ is the maximum among all w_i's in W where $i \in Y(i)$. □

When it is admissible, Φ' not only satisfies all the requirements of P, but also provides some redundant channels. If cell i has been assigned w'_i channels while the requirement was w_i, and $w'_i > w_i$, then $r_i = (w'_i - w_i)$ number of channels remains unused or redundant in cell i. This set of redundant channels is represented by the vector $R = (r_i)$. The *total number of redundant channels* R_{total} of the system is

$$R_{total} = \sum_{i=0}^{i=n-1} r_i.$$

Example 11. *For the problem of Example 6, the derived problem P' has been described in Example 10 above. One solution to P' has been obtained by the algorithm in [15] (a GA-based algorithm for the second category of CAP) as shown in Table 3.10, where entries in row T_a indicate the total number of channels assigned to each node and those in row T_r indicate the total number of channels actually required* *for each node (in all subsequent tables T_a and T_r will indicate the same meanings noted here). In Table 3.10, the rows T_a and T_r are identical with the demand vector W' of P'. In other words, the frequency assignment of Table 3.10 is admissible. The complete assignment following Theorem 7 has been shown in Table 3.11. Note that the same set of channels assigned to node $N(Y(0))$ has also been assigned to all the cells in $Y(0) = \{7, 13, 20, 24\}$. It is easy to verify that this assignment is also admissible for P. In addition, this assignment keeps five redundant channels in cell 3. Similarly the redundant channels for other cells can be computed and we get $R = (0,0,0,5,0,5,4,0,7,0,1,0,0,0,2,5,5,5,4,1,1,6,6,2,2)$ and $R_{total} = 61$.*

TABLE 3.10: Channel assignment for coalesced CAP P' of Problem 9

$Nodes \rightarrow$	a_0	a_1	a_2	a_3	a_4	a_5	a_6	a_7
	21	0	4	1	14	2	3	12
	25	5	7	6	26	9	13	20
	27	11	10	8	29	16	17	22
	40	34	31	15	32	19	23	30
	44	45	47	18	39	24	36	35
	58	48	51	28	55	33	38	37
	71	60	63	43	62	53	42	41
		65	66	57	72	61	50	46
		68	70	67		64	52	49
						69	56	54
							59	
$T_a \rightarrow$	7	9	9	9	8	10	11	10
$T_r \rightarrow$	7	9	9	9	8	10	11	10

Case 2: Assignment Φ' is not admissible

In this case, the given bandwidth B is not enough to satisfy all the requirements for P'. Let us assume that Φ' satisfies the demand vector $W'' = (w''_i)$ instead of W' where $w''_i < w'_i$ for some or all i. From Φ', if we assign all the cells in $Y(i)$ $(0 \leq i \leq z-1)$ the same set of channels assigned to $N(Y(i))$, there will be some blocked calls in some cells, as well as some redundant channels in other cells. We denote the blocked calls and redundant channels produced by this assignment as $BL = (b_i)$ and $R = (r_j)$, respectively, where $b_i = w_i - w''_i$, if $w''_i < w_i$ and 0 otherwise, and $r_j = (w''_j - w_j)$, if $w''_j > w_j$ and 0 otherwise. We then try to assign the blocked calls in BL by appropriately using these redundant channels in R and other available free channels by an approach similar to the forced assignment with rearrangement (FAR) operation in [17]. For the sake of completeness, we briefly describe below the essential features of the FAR operation reported in [17].

TABLE 3.11: Complete channel assignment for Problem 9

$Cells \rightarrow$	0	1	2	3	4	5	6	7	8	9	10	11	12
	2	3	1	2	0	4	0	21	3	14	4	4	12
	9	13	6	9	5	7	5	25	13	26	7	7	20
	16	17	8	16	11	10	11	27	17	29	10	10	22
	19	23	15	19	34	31	34	40	23	32	31	31	30
	24	36	18	24	45	47	45	44	36	39	47	47	35
	33	38	28	33	48	51	48	58	38	55	51	51	37
	53	42	43	53	60	63	60	71	42	62	63	63	41
	61	50	57	61	65	66	65		50	72	66	66	46
	64	52	67	64	68	70	68		52		70	70	49
	69	56		69					56				54
		59							59				
$T_a \rightarrow$	10	11	9	10	9	9	9	7	11	8	9	9	10
$T_r \rightarrow$	10	11	9	5	9	4	5	7	4	8	8	9	10

$Cells \rightarrow$	13	14	15	16	17	18	19	20	21	22	23	24
	21	1	3	4	2	0	14	21	12	3	0	21
	25	6	13	7	9	5	26	25	20	13	5	25
	27	8	17	10	16	11	29	27	22	17	11	27
	40	15	23	31	19	34	32	40	30	23	34	40
	44	18	36	47	24	45	39	44	35	36	45	44
	58	28	38	51	33	48	55	58	37	38	48	58
	71	43	42	63	53	60	62	71	41	42	60	71
		57	50	66	61	65	72		46	50	65	
		67	52	70	64	68			49	52	68	
			56		69				54	56		
			59							59		
$T_a \rightarrow$	7	9	11	9	10	9	8	7	10	11	9	7
$T_r \rightarrow$	7	7	6	4	5	5	7	6	4	5	7	5

Essence of FAR operation [17]: Let b_i be an unassigned requirement and Q denote the set of already assigned frequencies. Suppose for b_i there is no frequency available to be assigned without conflict among the already assigned frequencies of Q. FAR attempts to assign a frequency in L (where L is the given list of available frequencies) to satisfy the requirement b_i with minimum change or perturbation on the present assignment Q. The essence of FAR is to identify a minimal subset $S(b_i)$ of Q, each requirement of which can be simultaneously reassigned with an alternative feasible frequency, so that b_i can be assigned a frequency without conflicting to the present assignment of Q. Let $B(b_i, f_i)$ denote the subset of requirements in Q which are conflicting if we assign frequency f_i to requirement b_i. In other words, f_i becomes a feasible frequency for b_i, if the frequency assignments for $B(b_i, f_i)$ are undone. To identify one $S(b_i)$, we examine a sequence of f_i's such that each time a $B(b_i, f_i)$ is generated, we undo the corresponding portion of frequency assignment in Q and try to assign an alternative feasible frequency to each requirement of $B(b_i, f_i)$ by the unforced assignment (UA) operation. The UA operation finds the lowest frequency in L feasible for the present assignments in Q. If the frequency assignment of $B(b_i, f_i)$ is successfully made, $B(b_i, f_i)$ becomes $S(b_i)$. In case such a frequency reassignment cannot be made for some $b_j \in B(b_i, f_i)$, one proceeds to identify $B(b_j, f_j)$ and attempts to assign an alternative feasible frequency to each $b_k \in B(b_j, f_j)$. Such $B(b_j, f_j)$'s are blockers at the second depth level. Generalizing this, the FAR operation is encoded to render the so-called v-th breadth level and w-th depth level (Bv-Dw) procedures. In Bv-Dw, we consider blockers only within the cardinality of v (i.e., $|B(t, f_i)| \leq v$) and limit the number of successive downward searches to w. The complexity of FAR prohibits the direct implementation of this general Bv-Dw procedure.

In the algorithm described below, ($B1$-$D1$), ($B2$-$D1$) and ($B1$-$D2$) are implemented by incorporating the concept of redundant channels as described above. We call this the modified FAR (MFAR) operation. This modification is based on the notion of a free channel to be assigned to an unassigned requirement, say, t in BL. We consider a channel free and suitable to be assigned to t even if it conflicts with the requirements of other cells containing redundant channels. However, when we choose such a channel for assignment to t, we may need to undo some of the assignments in the neighboring cells, and adjust the assignments in other cells to maintain the degree of perturbation (number of changes in the existing assignments) as low as possible, following the techniques similar to FAR operation [17].

We now provide a formal description of the algorithm (*Derive-Assignment*(P, P')) to derive the solution to $P = (X, W, C)$ from the solution to P'.

Algorithm *Derive-Assignment*(P, P')

Step 1: Assign all the cells in $Y(i)$ $(0 \leq i \leq z - 1)$ the same set of channels assigned to $N(Y(i))$ in Φ' of P'.

Remark 10. *By similar arguments as in the proof of Theorem 7, this assignment satisfies the interference constraints specified by C.*

Step 2: Apply MFAR operation to minimize the blocked calls in $BL = (b_i)$ using the redundant channels in $R = (r_j)$ appropriately. □

Example 12. *To illustrate this step, we consider the Philadelphia benchmark Problem 5 (Table 3.2). We first construct* P' *for this problem as follows. A solution* Φ^* *of* P^* *is shown in Figure 3.22 where the label* $[\alpha]$ *associated with a node indicates that a frequency* α *is assigned to that node. Only seven channels* $0, 1, \cdots, 6$ *have been assigned repeatedly in the scheme of Figure 3.22, where* $Y(0) = \{2, 6, 11, 19\}$, $Y(1) = \{9, 14\}$, $Y(2) = \{10, 12, 15\}$, $Y(3) = \{1, 5, 17, 18\}$, $Y(4) = \{4, 8, 13\}$, $Y(5) = \{0, 16\}$, *and* $Y(6) = \{3, 7, 20\}$. *Corresponding frequency separations* $c'(a_i, a_j)$ $(0 \leq i, j \leq 6)$ *are shown in Table 3.12. We find* $M(a_0) = 45$, $M(a_1) = 40$, $M(a_2) = 40$, $M(a_3) = 30$, $M(a_4) = 30$, $M(a_5) = 15$, *and* $M(a_6) = 25$. *Therefore,* $P' = (X', W', C')$ *is given by* $X' = (N(Y(i)))$, $0 \leq i \leq 6$, *W′=(45, 40, 40, 30, 30, 15, 25) and* C' *is as given in Table 3.12.*

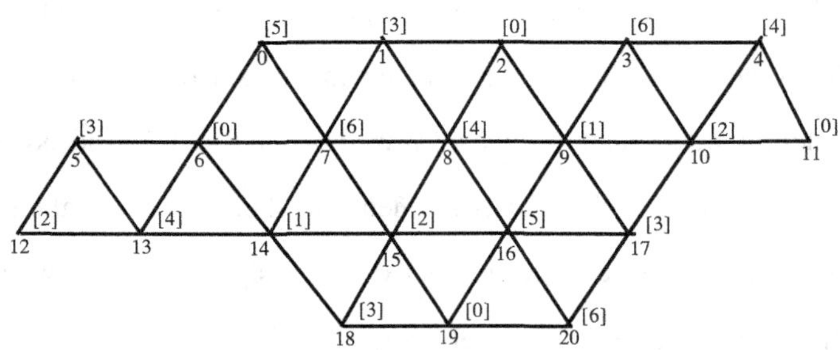

FIGURE 3.22: Single channel assignment of benchmark Problem 5

Now to get Φ', *we partition* P' *into several subnetworks with homogeneous weights, following the critical block approach reported in [28]. However, during the multiple weight assignment, we consider the channels not exceeding the given bandwidth B only. One solution to* P' *as obtained by this approach has been shown in Table 3.13. It follows that all the requirements for* P' *as given by W′=(45, 40, 40, 30, 30, 15, 25) are not satisfied; this assignment results in four blocked calls in node* $N(Y(5))$. *In other words, the derived assignment is not admissible.*

TABLE 3.12: Frequency separation matrix for P^* of Problem 5

$c'(a_i, a_j)$	a_0	a_1	a_2	a_3	a_4	a_5	a_6
a_0	7	1	1	1	1	1	1
a_1	1	7	1	1	1	1	1
a_2	1	1	7	1	1	1	1
a_3	1	1	1	7	1	1	1
a_4	1	1	1	1	7	1	1
a_5	1	1	1	1	1	7	1
a_6	1	1	1	1	1	1	7

We now apply step 1 of the algorithm Derive-Assignment(P, P') *to find the assignment for P from the solution to* P' *in Table 3.13. The derived assignment for P is shown in Table 3.14. Note that the same set of channels assigned to node* $N(Y(0))$ *has also been assigned to all the cells in* $Y(0) = \{2, 6, 11, 19\}$. *This assignment leaves four blocked calls* $(15 - 11 = 4)$ *in cell 16 and at the same time it also produces many redundant channels in other cells (e.g., cell 2 has* 40 *redundant channels).*

We now apply step 2 of the algorithm Derive-Assignment(P, P')*, to get the assignment as shown in Table 3.15. Note in the assignment of Table 3.14 the four blocked calls in cell 16. However, MFAR operation finds that channels 0, 5, 10, and 15 can be assigned to cell 16 if the assignment of channels 0, 5, 10, and 15 from cells 2 and 19 is undone. Note that there are more than four redundant channels in both cells 2 and 19. The assignments of channels 0, 5, 10 and 15 in cell 16 have been underlined and those of cells 2 and 19 have been marked by asterisks* $(*)$.

TABLE 3.13: Derived channel assignment for P' of Problem 5

$Nodes \rightarrow$	a_0	a_1	a_2	a_3	a_4	a_5	a_6
	0	1	2	3	4	159	96
	5	6	7	8	9	164	103
	10	11	12	13	14	169	108
	15	16	17	18	19	174	113
	20	21	22	23	24	184	118
	25	26	27	28	29	189	123
	30	31	32	33	34	194	128
	35	36	37	38	39	199	133
	40	41	42	43	44	209	138
	45	46	47	48	49	214	143
	50	51	52	53	54	219	148
	55	56	57	58	59		153
	60	61	62	63	64		158
	65	66	67	68	69		163
	70	71	72	73	74		168
	75	76	77	78	79		173
	80	81	82	83	84		178
	85	86	87	88	89		183
	90	91	92	93	94		188
	95	101	97	98	99		193
	100	106	102	154	104		198
	105	111	107	177	109		203
	110	116	112	182	114		208
	115	121	117	187	119		213
	120	126	122	192	124		218
	125	131	127	197	129		
	130	136	132	202	134		
	135	141	137	207	139		
	140	146	142	212	144		
	145	151	147	217	149		
	150	156	152				
	155	161	157				
	160	166	162				
	165	171	167				
	170	176	172				
	175	181	179				
	180	186	201				
	185	191	206				
	190	196	211				
	195	204	216				
	200						
	205						
	210						
	215						
	220						
$T_a \rightarrow$	45	40	40	30	30	11	25
$T_r \rightarrow$	45	40	40	30	30	15	25

TABLE 3.14: Derived channel assignment for Problem 5

Cells →	0	1	2	3	4	5	6	7	8	9	10	11	12	13	14	15	16	17	18	19	20
	159	3	0	96	4	3	0	96	4	1	2	0	2	4	1	2	159	3	3	0	96
	164	8	5	103	9	8	5	103	9	6	7	5	7	9	6	7	164	8	8	5	103
	169	13	10	108	14	13	10	108	14	11	12	10	12	14	11	12	169	13	13	10	108
	174	18	15	113	19	18	15	113	19	16	17	15	17	19	16	17	174	18	18	15	113
	184	23	20	118	24	23	20	118	24	21	22	20	22	24	21	22	184	23	23	20	118
	189	28	25	123	29	28	25	123	29	26	27	25	27	29	26	27	189	28	28	25	123
	194	33	30	128	34	33	30	128	34	31	32	30	32	34	31	32	194	33	33	30	128
	199	38	35	133	39	38	35	133	39	36	37	35	37	39	36	37	199	38	38	35	133
	209	43	40	138	44	43	40	138	44	41	42	40	42	44	41	42	209	43	43	40	138
	214	48	45	143	49	48	45	143	49	46	47	45	47	49	46	47	214	48	48	45	143
	219	53	50	148	54	53	50	148	54	51	52	50	52	54	51	52	219	53	53	50	148
		58	55	153	59	58	55	153	59	56	57	55	57	59	56	57		58	58	55	153
		63	60	158	64	63	60	158	64	61	62	60	62	64	61	62		63	63	60	158
		68	65	163	69	68	65	163	69	66	67	65	67	69	66	67		68	68	65	163
		73	70	168	74	73	70	168	74	71	72	70	72	74	71	72		73	73	70	168
		78	75	173	79	78	75	173	79	76	77	75	77	79	76	77		78	78	75	173
		83	80	178	84	83	80	178	84	81	82	80	82	84	81	82		83	83	80	178
		88	85	183	89	88	85	183	89	86	87	85	87	89	86	87		88	88	85	183
		93	90	188	94	93	90	188	94	91	92	90	92	94	91	92		93	93	90	188
		98	95	193	99	98	95	193	99	101	97	95	97	99	101	97		98	98	95	193
		154	100	198	104	154	100	198	104	106	102	100	102	104	106	102		154	154	100	198
		177	105	203	109	177	105	203	109	111	107	105	107	109	111	107		177	177	105	203
		182	110	208	114	182	110	208	114	116	112	110	112	114	116	112		182	182	110	208
		187	115	213	119	187	115	213	119	121	117	115	117	119	121	117		187	187	115	213
		192	120	218	124	192	120	218	124	126	122	120	122	124	126	122		192	192	120	218
		197	125		129	197	125		129	131	127	125	127	129	131	127		197	197	125	
		202	130		134	202	130		134	136	132	130	132	134	136	132		202	202	130	
		207	135		139	207	135		139	141	137	135	137	139	141	137		207	207	135	
		212	140		144	212	140		144	146	142	140	142	144	146	142		212	212	140	
		217	145		149	217	145		149	151	147	145	147	149	151	147		217	217	145	
			150				150			156	152	150	152		156	152				150	
			155				155			161	157	155	157		161	157				155	
			160				160			166	162	160	162		166	162				160	
			165				165			171	167	165	167		171	167				165	
			170				170			176	172	170	172		176	172				170	
			175				175			181	179	175	179		181	179				175	
			180				180			186	201	180	201		186	201				180	
			185				185			191	206	185	206		191	206				185	
			190				190			196	211	190	211		196	211				190	
			195				195			204	216	195	216		204	216				195	
			200				200					200								200	
			205				205					205								205	
			210				210					210								210	
			215				215					215								215	
			220				220					220								220	
T_a →	11	30	45	25	30	30	45	25	30	40	40	45	40	30	40	40	11	30	30	45	25
T_r →	5	5	5	8	12	25	30	25	30	40	40	45	20	30	25	15	15	30	20	20	25

TABLE 3.15: Complete channel assignment for Problem 5

Cells →	0	1	2	3	4	5	6	7	8	9	10	11	12	13	14	15	16	17	18	19	20
	159	3	*0	96	4	3	0	96	4	1	2	0	2	4	1	2	0	3	3	*0	96
	164	8	*5	103	9	8	5	103	9	6	7	5	7	9	6	7	5	8	8	*5	103
	169	13	*10	108	14	13	10	108	14	11	12	10	12	14	11	12	10	13	13	*10	108
	174	18	*15	113	19	18	15	113	19	16	17	15	17	19	16	17	15	18	18	*15	113
	184	23	20	118	24	23	20	118	24	21	22	20	22	24	21	22	159	23	23	20	118
	189	28	25	123	29	28	25	123	29	26	27	25	27	29	26	27	164	28	28	25	123
	194	33	30	128	34	33	30	128	34	31	32	30	32	34	31	32	169	33	33	30	128
	199	38	35	133	39	38	35	133	39	36	37	35	37	39	36	37	174	38	38	35	133
	209	43	40	138	44	43	40	138	44	41	42	40	42	44	41	42	184	43	43	40	138
	214	48	45	143	49	48	45	143	49	46	47	45	47	49	46	47	189	48	48	45	143
	219	53	50	148	54	53	50	148	54	51	52	50	52	54	51	52	194	53	53	50	148
		58	55	153	59	58	55	153	59	56	57	55	57	59	56	57	199	58	58	55	153
		63	60	158	64	63	60	158	64	61	62	60	62	64	61	62	209	63	63	60	158
		68	65	163	69	68	65	163	69	66	67	65	67	69	66	67	214	68	68	65	163
		73	70	168	74	73	70	168	74	71	72	70	72	74	71	72	219	73	73	70	168
		78	75	173	79	78	75	173	79	76	77	75	77	79	76	77		78	78	75	173
		83	80	178	84	83	80	178	84	81	82	80	82	84	81	82		83	83	80	178
		88	85	183	89	88	85	183	89	86	87	85	87	89	86	87		88	88	85	183
		93	90	188	94	93	90	188	94	91	92	90	92	94	91	92		93	93	90	188
		98	95	193	99	98	95	193	99	101	97	95	97	99	101	97		98	98	95	193
		154	100	198	104	154	100	198	104	106	102	100	102	104	106	102		154	154	100	198
		177	105	203	109	177	105	203	109	111	107	105	107	109	111	107		177	177	105	203
		182	110	208	114	182	110	208	114	116	112	110	112	114	116	112		182	182	110	208
		187	115	213	119	187	115	213	119	121	117	115	117	119	121	117		187	187	115	213
		192	120	218	124	192	120	218	124	126	122	120	122	124	126	122		192	192	120	218
		197	125		129	197	125		129	131	127	125	127	129	131	127		197	197	125	
		202	130		134	202	130		134	136	132	130	132	134	136	132		202	202	130	
		207	135		139	207	135		139	141	137	135	137	139	141	137		207	207	135	
		212	140		144	212	140		144	146	142	140	142	144	146	142		212	212	140	
		217	145		149	217	145		149	151	147	145	147	149	151	147		217	217	145	
			150				150			156	152	150	152		156	152				150	
			155				155			161	157	155	157		161	157				155	
			160				160			166	162	160	162		166	162				160	
			165				165			171	167	165	167		171	167				165	
			170				170			176	172	170	172		176	172				170	
			175				175			181	179	175	179		181	179				175	
			180				180			186	201	180	201		186	201				180	
			185				185			191	206	185	206		191	206				185	
			190				190			196	211	190	211		196	211				190	
			195				195			204	216	195	216		204	216				195	
			200				200					200								200	
			205				205					205								205	
			210				210					210								210	
			215				215					215								215	
			220				220					220								220	
T_a →	11	30	41	25	30	30	45	25	30	40	40	45	40	30	40	40	15	30	30	41	25
T_r →	5	5	5	8	12	25	30	25	30	40	40	45	20	30	25	15	15	30	20	20	25

3.6 Fast Near-Optimal Channel Assignment

We now present a novel method [50] for solving channel assignment problems (CAPs) in hexagonal cellular networks with non-homogeneous demands in a two-band buffering system (where channel interference does not extend beyond two cells). The CAP with non-homogeneous demand is first partitioned into a sequence of smaller subproblems, each of which has a homogeneous demand from a subset of the nodes of the original network. The solution to such a subproblem constitutes an assignment phase in which multiple homogeneous demands are assigned to the nodes corresponding to the subproblem, satisfying all the frequency separation constraints. The whole assignment process for the original network consists of a succession of multiple homogeneous assignments for all the subproblems. Based on this concept, we present a polynomial time approximation algorithm given in [50] for solving the channel assignment problem for cellular networks having non-homogeneous demands. The assignment algorithm, when executed on well-known benchmark instances, yields an assignment which is always within about 6% more than the optimal bandwidth, but requires very little execution time (less than 5 milliseconds on a HPxw8400 workstation). This algorithm is suitable for real-life situations where fast channel assignment is of primary importance and a marginal deviation (6%) from the optimal bandwidth is tolerable.

3.6.1 Assignment Algorithm

The ultimate objective of this study is to devise a greedy algorithm that can produce a very fast assignment of the required channels (in milliseconds), while requiring a bandwidth as close as possible to the optimal bandwidth. This is ag long-standing challenge in the study of the frequency assignment problem.

The essence of this approach is to partition the given channel assignment problem into a sequence of subproblems, each consisting of a specific subgraph with a uniform demand to be satisfied at each node. To satisfy a uniform demand from each node of the subgraph, we apply a greedy algorithm to first find a single frequency assignment to each node, and then multiple frequencies are assigned on a round-by-round basis (one frequency at each round) using the *increment* in the starting frequency for the next consecutive round. The concept of *increment* is such that if frequency $\phi(v)$ can be assigned at node v without any conflict with already assigned frequencies, the set of multiple frequencies $\phi(v)+k\times increment$ for $k = 1, 2, 3, \cdots$, can also be assigned round by round at node v without conflict. Since s_0 is the minimum frequency separation requirement between two calls in the same cell, the value of *increment* will be at least s_0. Thus, the multiple homogeneous demand can be satisfied easily using the single frequency assignment and the value of *increment*. The solution to the original problem will constitute a series of assignment phases. The starting frequency of a particular phase of assignment will be the frequency immediately next to the highest frequency assigned during the previous phase of assignment. When we move from one assignment phase, say phase i, to the next phase $i + 1$, we must ensure at the start that every node in the subgraph of phase $i + 1$ is assigned a frequency that does not cause any interference with the frequencies assigned at the last round to the nodes in phase i. It is sufficient to satisfy the last round frequencies assigned to the nodes in phase i while assigning frequencies at phase $i + 1$ because these are the only frequencies which may conflict with the frequencies of phase $i+1$. This is because even the highest frequency assigned in the second last round in phase i and the least frequency in the first round of phase $i + 1$ are separated by at least s_0 since the $increment$ is always greater than or equal to s_0. Since $s_0 \geq s_1 \geq s_2$, no frequencies assigned in the second last round of phase i can conflict with the frequencies in the first round of phase $i + 1$. This prevents conflict in the frequency assignment across subsequent phases of assignment.

We now describe the assignment algorithm, illustrating each step with the example of the benchmark Problem 6 defined on the hexagonal cellular network shown in Figure 3.2. For the purpose of illustration, the demand vector D_2 has been explicitly shown in Figure 3.2 where the label $[a]$ associated with a node denotes the demand of that node. The frequency separation constraints for this benchmark problem is given by $s_0 = 5$, $s_1 = 2$ and $s_2 = 1$. The frequency separation matrix $C = (c_{ij})$ can be derived by using the s_0, s_1 and s_2 values. In C, $c_{ii} = 5\ \forall i$ and c_{ij} is equal to 2, 1 and 0 if nodes i and j are distance 1, 2 and ≥ 3 apart, respectively from each other in the cellular graph of the figure.

As a first step of the assignment algorithm, we rearrange the nodes of the network according to the descending order of their demands. The resulting order may reveal several nodes with the same demand. We further rearrange these nodes according to the descending order of their S_i values, where S_i is the sum of the demands from adjacent nodes of node i. Let this ordered set of nodes be termed S.

We now consider the subgraph of the cellular graph containing only the highest demand nodes and look into its assignment by using a greedy algorithm. This greedy algorithm will start with assigning the frequency $origin$ to the first highest demand node in the ordered set S. It takes the next node from the ordered set S and tries to assign the minimum frequency to that node from the ordered sequence $< origin, origin + 1, origin + 2, \cdots >$, satisfying the frequency separation constraints with the previously assigned node. This process is repeated for all the remaining nodes of the set S taken in order. On completion of this process, we find the minimum value of *increment* such that the frequency channels in the interval $[origin, origin + increment)$ are sufficient for a single frequency assignment to each highest demand node. The multiple frequencies can be assigned to these nodes using the value of *increment* and the single frequency assignment.

We next consider the assignment of the whole network with the frequencies explicitly in the interval $[origin, origin + increment)$, satisfying all the frequency separation criteria. The way the value of *increment* will be computed, the frequencies in the interval $[origin, origin + increment)$ will definitely cover the assignment of all the highest demand nodes, but may or may not be able to cover the other nodes of the entire network. Let V be the set of nodes covered with frequencies in $[origin, origin + increment)$ in this phase of assignment and γ_{max} be their highest demand, and also let V' be the set of nodes left out and γ'_{max} be their highest demand. Let $\gamma_{max} - \gamma'_{max} = w$. We then repeat the assignment of the nodes in V for w rounds, avoiding any interference, so as to satisfy w demands of each of the nodes in V, and reducing their highest demand to γ'_{max}. For those nodes in V whose actual demand is less than the value of w, the assignment will be repeated for rounds equal to their respective actual demands. This will ensure that the highest demand node in V' will now move to the set of highest demand nodes of the entire network. As a result, in the next phase of assignment, this node must be covered by the algorithm. We will now consider the residual demand vector after reducing w or the actual demand corresponding to the nodes in V and start the second phase of assignment by setting the value of $origin$ equal to the frequency immediately next to the highest frequency channel used in this first phase of assignment. The entire process will be repeated until the residual demand vector becomes a vector of 0. We now present the assignment algorithm for a particular phase of assignment. Let $Q = (q_0, q_1, \cdots, q_{r-1})$ be the set of r $(0 \leq r \leq n)$ nodes whose residual demands are non-zero and to which frequencies have already been assigned in the previous phase of assignment. Let $\phi(Q)$ be the set of last round frequencies assigned to the nodes in Q in the previous phase of assignment, say $\phi(Q) = (\phi(q_0), \phi(q_1), \cdots, \phi(q_{r-1}))$. Let *origin* be the starting frequency of current phase of assignment. Non-trivial steps of the algorithm have been explained by explicitly mentioning the input and output of each step for ease of understanding.

Algorithm Assign_Frequency

Step 1 (initialization):

Set Q and $\phi(Q)$ to empty set and *origin* to zero.

Step 2 (sort the nodes based on demand):

Input: The current demand vector $W = (w_i)$ and the current node ordering $X = (x_i)$ where w_i represents the demand from node x_i $(0 \leq i \leq n-1)$.

Step 2.1: Sort the nodes in X according to the descending order of their demands in W.

Step 2.2: If multiple nodes have the same demand, sort them according to the descending order of their sum of the demands from adjacent nodes.

Step 2.3: If multiple nodes have the same sum of the demands from adjacent nodes, sort them according to the ascending order of their node IDs (i.e., x_i values).

Output: The sorted node ordering X' and the sorted demand vector W'.

Example 13. *The first two rows of Table 3.16 show the initial node ordering X and the initial demand vector W for Problem 6. The sorted demand vector W' and the sorted node ordering X' are shown in the last two rows of Table 3.16. Node 11 appears first in the sorted node-ordering in X' because node 11 has the highest demand in W. In X, nodes 9 and 10 have the same demand 40. The set of nodes adjacent to nodes 9 and 10 are {2,3,8,10,16,17} and {3,4,9,11,17}, respectively (see Figure 3.2). Now the sum of the demands from adjacent nodes of 9 and 10 are 128 $(5+8+30+40+15+30)$ and 135 $(8+12+40+45+30)$, respectively (see Figure 3.2). Thus, node 10 appears before 9 in the sorted node ordering in X'.*

Step 3 (compute increment):

Input: The sorted node ordering X' obtained in Step 2 above.

TABLE 3.16: Sorted node ordering according to ascending order of demands

X	0	1	2	3	4	5	6	7	8	9	10	11	12	13	14	15	16	17	18	19	20
W	5	5	5	8	12	25	30	25	30	40	40	45	20	30	25	15	15	30	20	20	25
X'	11	10	9	17	6	8	13	14	7	5	20	19	18	12	16	15	4	3	2	1	0
W'	45	40	40	30	30	30	30	25	25	25	25	20	20	20	15	15	12	8	5	5	5

Step 3.1: Create a subset Y of X' to hold all the nodes in X' whose demands are equal to the maximum demand in W by maintaining the order of their appearance in X'.

Remark 11. *Since X' is the sorted node ordering according to the descending order of their demands in W, the first m nodes in X' will actually be stored in Y if there are m nodes having maximum demand in W.*

Step 3.2: Execute algorithm **Find_Increment** as given in Figure 3.23 to find the value of *increment* by satisfying the following three constraints.

$$|\phi(y_i) - \phi(q_k)| \geq c_{y_i q_k} \forall k \in [0, r-1] \tag{3.2}$$

$$|\phi(y_i) - \phi(y_j)| \geq c_{y_i y_j} \forall j \in [0, i-1] \tag{3.3}$$

$$increment - |\phi(y_i) - \phi(y_j)| \geq c_{y_i y_j} \forall j \in [0, i-1]. \tag{3.4}$$

Output: The value of *increment* for which the frequencies in $[origin, origin + increment)$ are sufficient to assign a frequency to each node in Y.

Remark 12. *Basically,* **Find_Increment** *assigns frequencies to the nodes in Y as per the given ordering $(y_0, y_1, \cdots, y_{m-1})$ starting with frequency $origin$ and returns the value of increment for which the frequencies in $[origin, origin + increment)$ are sufficient to assign a frequency to each node in Y. Here, constraint* (3.2) *ensures that $\phi(y_i)$, the frequency assigned at node y_i, satisfies the frequency separation requirement as specified by C with the frequencies $\phi(Q)$ assigned at the nodes of Q in the last round of the previous phase of assignment. Constraint* (3.3) *ensures that*

$\phi(y_i)$ satisfies the frequency separation requirement as specified by C with the previously assigned frequencies in this current phase at y_0, y_1, $\cdots$, y_{i-1}. Constraint (3.4) *ensures that if $\phi(y_i)$ can be assigned at node y_i without any conflict with already assigned frequencies, the set of frequencies $\phi(y_i) + k \times increment$ for $k = 1, 2, 3, \cdots$, can also be assigned at node y_i without any conflict. Note that for finding $\phi(y_0)$, we need to satisfy only Constraint* (3.2)*. Constraints* (3.3) *and* (3.4) *are active for all other $\phi(y_i)$'s where $0 < i \leq m - 1$.*

Algorithm Find_increment
$Y = (y_0, y_1, \cdots, y_{m-1})$
$Q = (q_0, q_1, \cdots, q_{r-1})$
$\phi(Q) = (\phi(q_0), \phi(q_1), \cdots, \phi(q_{r-1}))$
$origin$

```
begin
Set increment = s0;
    For i = 0 to m − 1 do
    begin
        Find φ(yi) as the least frequency in [origin, origin + increment) which
        satisfies constraints (3.2), (3.3) and (3.4) as mentioned under Step 3.2.
        If such φ(yi) does not exist, set increment = increment + 1.
    end
Return increment.
end
```

FIGURE 3.23: Algorithm for finding increment

Example 14. *Consider X' as shown in Table 3.16. The maximum demand in W is* 45*. Note that there is only one node in X' (node* 11*) having the maximum demand in W. Hence, $m = 1$ and Y includes only the first node (node* 11*) in X'. The increment value obtained from* **Find_Increment** (Y, Q, $\phi(Q)$, *origin*) *is $s_0 = 5$ which is the initial value. In the subsequent phase of assignment, the demand vector will be modified and as a result, there may be multiple maximum demand nodes for which the increment value may be higher.*

Step 4 (assign a single frequency):

Algorithm Find_assignment
$X' = (x'_0, x'_1, \cdots, x'_{t-1})$
$Q = (q_0, q_1, \cdots, q_{r-1})$
$\phi(Q) = (\phi(q_0), \phi(q_1), \cdots, \phi(q_{r-1}))$
$origin, increment$

```
begin
Set both V and V' to ∅;
    For i = 0 to t − 1 do
    begin
        Find φ(x'_i) as the least frequency in [origin, origin + increment)
        which satisfies constraints (3.2), (3.3) and (3.4).
        If such φ(x'_i) exists, put x'_i and φ(x'_i) into the sets V and φ(V),
        respectively, otherwise, put x'_i into the set V'.
    end
Return V, φ(V) and V'.
end
```

FIGURE 3.24: Algorithm for finding single assignment

Input: The X' and the value of *increment* as obtained in Step 2 and Step 3, respectively.

Assign frequencies in $[origin, origin + increment)$ to as many nodes as possible in X' by applying the algorithm **Find_Assignment** (X', Q, $\phi(Q)$, $origin, increment$) as given in Figure 3.24. Set V to the set of nodes which are assigned a frequency by this algorithm and V' to those not assigned by the algorithm. Set $\phi(V)$ to set of frequencies assigned to the nodes in V.

Output: V, $\phi(V)$ and V'.

Remark 13. *Algorithm* **Find_assignment** *basically assigns frequencies in* $[origin, origin$ $increment)$ *to the nodes in* X' *as per the given ordering* $(x'_0, x'_1, \cdots, x'_{t-1})$ *starting with frequency* $origin$. *Since the frequencies in* $[origin, origin + increment)$ *only are allowed to be assigned to the nodes in* X' *by the* **Find_Assignment** *algorithm, depending on the value of increment, these frequencies may or may not be sufficient for assigning a frequency to each node in* X'. *As a result, some nodes may remain unassigned. Here V is the set of nodes assigned a frequency and* V' *are those not assigned* *by the algorithm* **Find_Assignment**. *Note that the value of increment is computed in such a way that the set V must include the first m nodes in X' having maximum demand. In other words, all the maximum demand nodes must be assigned a frequency by the* **Find_Assignment** *algorithm. Here, $\phi(V)$ represents the first round frequency assignment to V where each node has been assigned a frequency.*

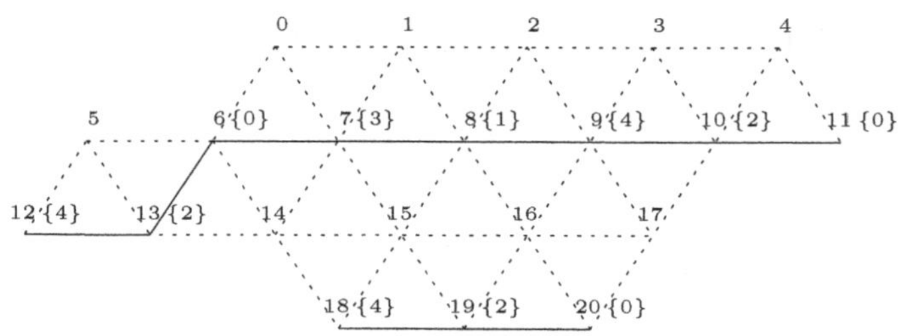

FIGURE 3.25: Single channel assignment to benchmark cellular network

Example 15. *Consider X' as shown in Table 3.16 and let $origin = 0$. Note that the value of $increment$ as obtained in Step 3 above is 5. The sets V, $\phi(V)$ and V' obtained by algorithm* **Find_Assignment** *are $V = \{11, 10, 9, 6, 8, 13, 7, 20, 19, 18, 12\}$, $\phi(V) = \{0, 2, 4, 0, 1, 2, 3, 0, 2, 4, 4\}$, and $V' = \{17, 14, 5, 16, 15, 4, 3, 2, 1, 0\}$ respectively. The frequency assignment $\phi(V)$ obtained by algorithm* **Find_Assignment** *using the frequencies in $[0, 5)$ has been shown in Figure 3.25, where the label $\{a\}$ associated with a node denotes that frequency a has been assigned to that node during the single frequency assignment of the nodes in V.*

Step 5 (assign multiple frequencies):

Input: The demand vector W; sets V, V' and $\phi(V)$ as computed in Step 4 and the value of *increment* as computed in Step 3.

Step 5.1: Compute γ_{max} and γ'_{max} to act as the maximum demands among the nodes in V and V', respectively.

Step 5.2: Compute the number of rounds d_v that the frequencies will be assigned

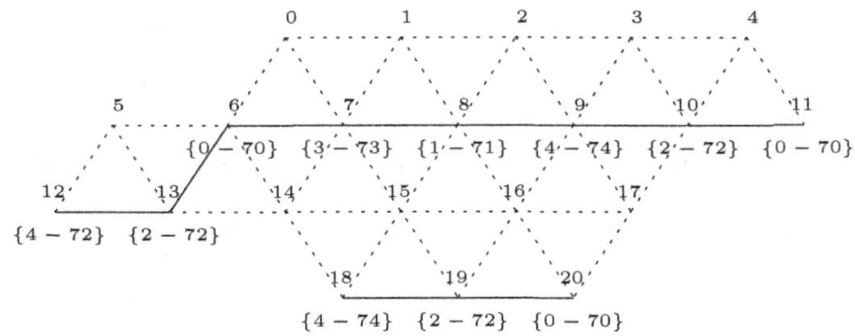

FIGURE 3.26: Set of assignment paths in first phase

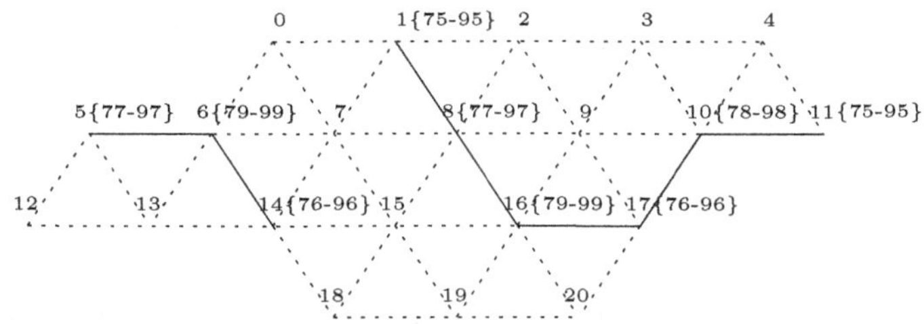

FIGURE 3.27: Set of assignment paths in second phase

to every node $v \in V$ as: $d_v = (\gamma_{max} - \gamma'_{max})$ if $w_v \geq (\gamma_{max} - \gamma'_{max})$, otherwise, $d_v = w_v$. Denote $D = (d_v)$ as the round vector where $v \in V$.

Step 5.3: Assign d_v frequencies to the node $v \in V$ using the value of *increment* and the first round frequency $\phi(v)$ already assigned to v as follows: $\phi(v) + j \times increment$, where $j \in \{0, 1, \cdots (d_v - 1)\}$.

Output: The set of d_v frequencies assigned to each node $v \in V$.

Example 16. *Consider the sets V and V′ obtained in Step 4 above and the current demand vector W shown in Table 3.16. Here* γ_{max}*, the maximum demand among the nodes in V* = {*11, 10, 9, 6, 8, 13, 7, 20, 19, 18, 12*}*, and* γ'_{max}*, the maximum demand among the nodes in V′* = {*17, 14, 5, 16, 15, 4, 3, 2, 1, 0*}*, are* 45 *and* 30*, respectively. The round vector as computed by Step 5 above is D* = {*15, 15, 15, 15, 15, 15, 15, 15, 15, 15, 15*} *as* $w_v \geq (\gamma_{max} - \gamma'_{max}) = (45 - 30) = 15$ *for all* $v \in V$*. Using* $\phi(V)$ *and the value of increment, the multiple frequency assignment to the nodes in V is shown in Figure 3.26, where the label* $\{a - b\}$ *associated with a node denotes that* *frequency* $a, a+increment, a+2\times increment, \cdots, b = a+(d_v-1)\times increment$ *has been assigned to that node where* $increment = 5$ *and* $d_v = 15\ \forall\ v \in V$. *Similar to Figure 3.26, the label* $\{a - b\}$ *associated with a node in Figures 3.27 through 3.31 denotes that frequency* $a, a+increment, a+2\times increment, \cdots, b = a + (d_v - 1) \times increment$ *has been assigned to that node. From Figure 3.26, the highest frequency assigned in this multiple frequency assignment is* 74.

Step 6 (update):

Step 6.1: Modify the current demand vector W as follows: $w_v = w_v$ if $v \in V'$; $w_v = w_v - (\gamma_{max} - \gamma'_{max})$ if $v \in V$ and $w_v \geq (\gamma_{max} - \gamma'_{max})$; and $w_v = 0$ if $v \in V$ and $w_v < (\gamma_{max} - \gamma'_{max})$. Denote this modified demand vector as the residual demand vector R.

Step 6.2: Reset Q to hold the subset of the nodes in V for which the residual demand is non-zero.

Step 6.3: Reset $\phi(Q)$ to hold the last round frequencies assigned at the nodes in Q at Step 5.

Step 6.3: Reset *origin* to the immediate next frequency of the maximum frequency in $\phi(Q)$.

Step 6.4: Consider the residual demand vector R and the sorted node ordering X' as the current demand vector W and current node ordering X, respectively.

Step 7: Repeat Step 2 - Step 6 until the residual demand vector becomes 0.

Remark 14. *Note that the next phase of assignment will be started with the frequency immediately next to the highest frequency used in the previous phase of assignment. The reason to hold in* $\phi(Q)$ *the last round frequencies of the nodes in* Q *is that these are the potential frequencies that may conflict with the frequencies that will be assigned in the first round of the next phase of assignment. Therefore it is required to assign frequencies in the first round of the next phase of assignment satisfying the last round frequencies of the previous phase of assignment held in* $\phi(Q)$.

Example 17. *The residual demand vector R, Q, $\phi(Q)$ and $origin$ as obtained by Step 6 are R ={30, 25, 25, 30, 15, 15, 15, 25, 10, 25, 10, 5, 5, 5, 15, 15, 12, 8, 5, 5, 5}, Q ={11, 10, 9, 6, 8, 13, 7, 20, 19, 18, 12}, $\phi(Q)$ ={70, 72, 74, 70, 71, 72, 73, 70, 72, 74, 74} and $origin = 75$, respectively. The next phase of assignment obtained by repeating Step 2 through Step 6 with $origin = 75$ is shown in Figure 3.27. The set W ={30, 25, 25, 30, 15, 15, 15, 25, 10, 25, 10, 5, 5, 5, 15, 15, 12, 8, 5, 5, 5} is then sorted as W′ ={30, 30, 25, 25, 25, 25, 15, 15, 15, 15, 15, 12, 10, 10, 8, 5, 5, 5, 5, 5, 5}, which produces X′ = {17, 11, 10, 9, 14, 5, 16, 6, 8, 15, 13, 4, 7, 20, 3, 2, 18, 19, 12, 1, 0}. The set Y is computed as = $\{11, 17\}$ where both the nodes 11 and 17 have demand 30. The increment is computed as 5. Therefore, the available frequency channels are 75, 76, 77, 78 and 79, which are assigned to the set X′. Now V, $\phi(V)$ and V′ are computed as V = {17, 11, 10, 14, 5, 16, 6, 8, 1}, $\phi(V)$ = {76, 75, 78, 76, 77, 79, 79, 77, 75}, and V′ = {9, 15, 13, 4, 7, 20, 3, 2, 18, 19, 12, 0}, respectively. Note that the set of nodes in V in the second phase again forms a network structure of a* set of sparse paths, *as shown in Figure 3.27. The number of rounds that the frequencies in $\phi(V)$ will be repeated is given by D = {5, 5, 5, 5, 5, 5, 5, 5, 5} as $\gamma_{max} - \gamma'_{max} = 30 - 25 = 5$ and demands of all the nodes in V are greater than or equal to 5. Now the residual demand vector is given by R = {25, 25, 20, 25, 20, 20, 10, 10, 10, 15, 15, 12, 10, 10, 8, 5, 5, 5, 5, 0, 5}. Since demands of all the nodes in V are greater than or equal to 5, the set Q and set V are same. The last round frequencies assigned at the nodes in V is given by $\phi(Q)$ = {96, 95, 98, 96, 97, 99, 99, 97, 95}. The origin will now become 100 as 99 is the highest frequency assigned at $\phi(Q)$. Following the same procedure, subsequent phases of assignment are shown in Figures 3.28 through 3.31. In the sixth phase of the assignment, the residual demand vector becomes 0 and the process terminates. The highest frequency assigned in Figure 3.31 is 267 which corresponds to the final bandwidth requirement of 268.*

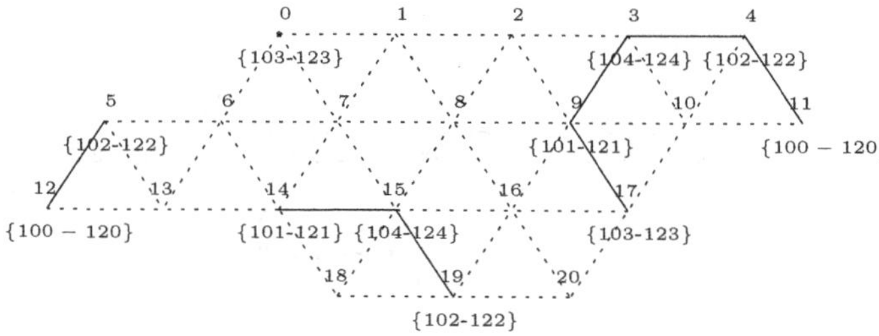

FIGURE 3.28: Set of assignment paths in third phase

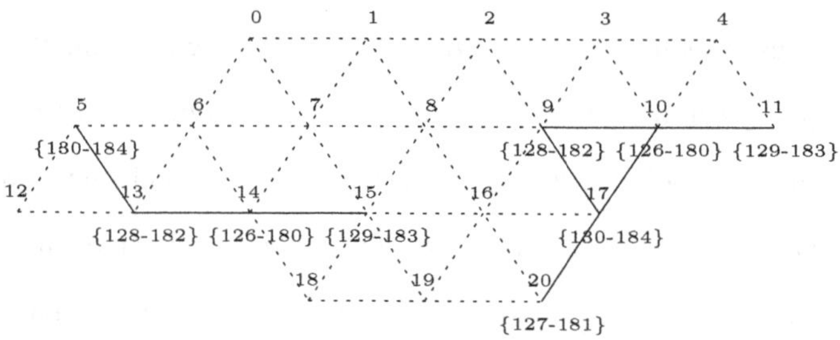

FIGURE 3.29: Path augmented with triangle in fourth phase

3.6.2 Convergence Behavior of Algorithm

Theorem 8. *The above assignment algorithm is bound to terminate after n phases of assignments, where n is the number of nodes in the entire network.*

Proof: The algorithm **Assign_Frequency** will terminate when the residual demand vector becomes 0. Note that Step 3 finds the value of *increment* such that all the highest demand nodes in the current demand vector W can be assigned

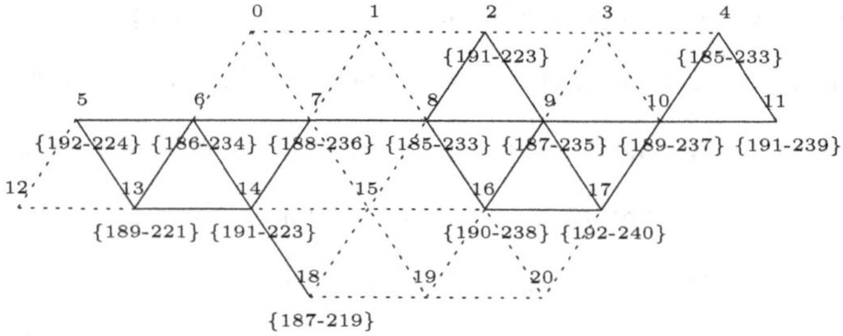

FIGURE 3.30: Path augmented with quadrilaterals in fifth phase

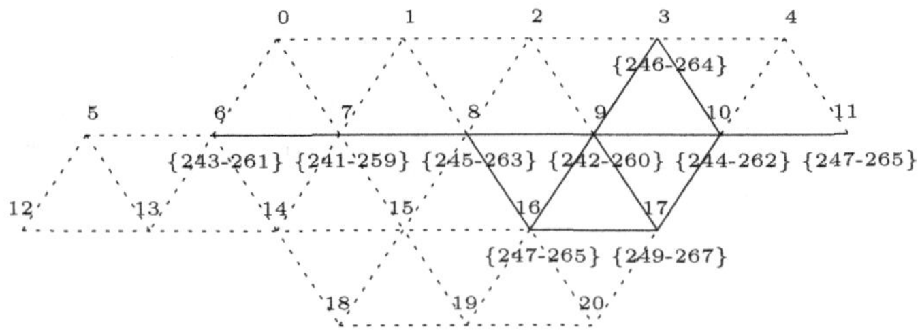

FIGURE 3.31: Assignment of nodes in the last phase

with frequencies in $[origin, origin + increment)$. After computing this *increment* value in Step 4, given the whole set of nodes of the network in order, we find the assignment of frequencies to as many nodes as possible with the frequencies in $[origin, origin+increment)$. V is the set of nodes assigned a frequency and V' are those not assigned in Step 4. Note that the value of *increment* is computed in such a way that the set V must include all the highest demand nodes in W. Next, we have derived the round vector $D = (d_v)$ where $d_v = (\gamma_{max} - \gamma'_{max})$ if $w_v \geq (\gamma_{max} - \gamma'_{max})$ and $d_v = w_v$ otherwise. The algorithm then repeats the assignment of the nodes in V for d_v rounds and the residual demand vector after reducing d_v demands corresponding to the nodes in V is computed. However, the demands corresponding to the nodes in V' remain unchanged. Thus, after satisfying d_v demands of the nodes in V, the highest demand of the nodes of V will reduce to γ'_{max}. This will ensure that the highest demand node in V' will now move to the set of highest demand nodes in the residual demand vector. As a result, in the next phase of assignment, this node must be covered by the assignment algorithm. Note that for the first phase of assignment, the set V must include all the highest demand nodes of the entire network. Therefore, in the worst case, if V' includes the second highest demand of the entire network, the highest demand in the residual demand vector will become the second highest demand of the entire network. Proceeding similarly, after n phases, the residual demand vector must be 0, if all the demands in the original demand vector are distinct and at each phase k the frequency separation matrix forces to include the $(k+1)$-th highest demand node in V', where n is the number of nodes in the entire network. Therefore, in the worst case, the algorithm must terminate after n phases of assignment.

3.6.3 Worst-Case Time Complexity of Assignment Algorithm

Step 1 of the algorithm **Assign_Frequency** needs $O(1)$ time. Step 2.1 involves sorting the nodes of W in descending order of their demands, which requires $O(n\log(n))$ time. For multiple nodes having the same demand, Step 2.2 involves re-sorting these nodes according to the descending order of their S_i values, where S_i is the sum of the demands from adjacent nodes of node i (a node may be adjacent to at most six nodes in a hexagonal cellular network). Thus, computing the S_i values of all nodes requires $O(n)$ time. Therefore, in the worst case when all demands are equal, finding the S_i values of all nodes and then sorting the demand vector involves $O(n + n\log(n)) = O(n\log(n))$ time. In Step 3, to find the value of $increment$, we effectively test every frequency in the range $[origin, origin + increment)$ if it can be assigned to a node satisfying the frequency separation requirement with all the previously assigned nodes. Thus, for testing a given frequency on all nodes requires $O(n^2)$ time in the worst case. Considering all such frequencies in the above range, the total time complexity for this step will be $increment \times O(n^2)$. To find the maximum value of $increment$, we first note that for a single frequency assignment on all nodes, the bandwidth is $(2s_1 + 6s_2)$ for all values of s_1 and s_2 where $s_1 \geq s_2$ [29]. Hence, for multiple frequency assignment, the maximum value of the $increment$ can be $max(s_0, 3s_1 + 6s_2)$, where the term $(3s_1 + 6s_2)$ arises in the following situation: a node u assigned with the frequency $origin$ is at a distance one from some node v which has been assigned with the frequency $(2s_1 + 6s_2)$. Thus the second frequency that can be assigned on node u is $max(s_0, 3s_1 + 6s_2)$. Therefore, the worst case complexity of Step 3 is $max(s_0, 3s_1 + 6s_2) \times O(n^2) = O(I \times n^2)$, where $I = max(s_0, 3s_1 + 6s_2)$. Given a set of nodes in order, Step 4 constitutes finding assignment of frequencies to as many nodes as possible with the frequencies in $[origin, origin + increment)$ using algorithm **Find_Assignment**. Therefore, the complexity of algorithm **Find_Assignment** is again $O(I \times n^2)$. Step 5 involves assigning multiple frequencies to the nodes selected at Step 4 using the single frequency assignment to them and the value of *increment*. In the worst case, if all nodes have been selected and w_{max} channels have been assigned to each of them, its complexity becomes $O(w_{max} \times n)$, where w_{max} is the maximum demand in W. Step 6 requires $O(n)$ time. Therefore, the overall time complexity of a single phase of assignment is $O(I \times n^2 + w_{max} \times n)$ and that for all phases becomes $number\_of\_phases \times O(I \times n^2 + w_{max} \times n)$. From Theorem 8, the algorithm terminates after n phases of assignment. Hence we get the following result:

Theorem 9. *The worst case time complexity of the algorithm* **Assign_Frequency** *is* $O(I \times n^3 + w_{max} \times n^2)$ *where* $I = max(s_0, 3s_1 + 6s_2)$.

3.6.4 Worst-Case Deviation of Bandwidth from Optimality

Consider the demand vector W. Let there exist h distinct demands $\alpha_h, \alpha_{h-1}, \cdots, \alpha_1$ in W such that $\alpha_h > \alpha_{h-1} > \cdots > \alpha_1$. Suppose there are $\beta_h, \beta_{h-1}, \cdots, \beta_1$ nodes having $\alpha_h, \alpha_{h-1}, \cdots, \alpha_1$ demands, respectively. Hence $\sum_{i=1}^{h} \beta_i = n$ where n is the total number of nodes. The first round of the first phase of algorithm **Assign_Frequency** finds a single frequency assignment to all β_h nodes having α_h demands. The value of $increment_1$ (the required increment in the first phase) will be computed by the algorithm **Find_increment** such that the frequencies in the range $[origin, origin + increment_1)$ are sufficient for a single frequency assignment to all these β_h nodes. Note that these β_h nodes may induce an arbitrary subgraph of the hexagonal cellular network. The algorithm **Find_Assignment** finds single frequency assignment of the whole network using the frequencies explicitly in the interval $[origin, origin + increment_1)$ by considering an arbitrary ordering that may be produced by Step 2 of the algorithm **Assign_Frequency**. Based on the way the value of $increment_1$ is computed, the frequencies in the interval $[origin, origin + increment_1)$ will not only cover the assignment of all the β_h nodes having α_h demand but also may cover some of the other nodes of the network. Let V be the set of nodes covered with frequencies in $[origin, origin+increment_1)$ and V' be the set of nodes left out. Let there exist at least one among the β_{h-1} nodes of demand α_{h-1} which is left out. In such a case, the maximum demand in V' will be α_{h-1}. Hence the number of rounds in the first phase will be $\alpha_h - \alpha_{h-1} = y_1$ and the bandwidth requirement of this first phase will be $(2s_1 + 6s_2) + increment_1 \times y_1$. In general, in the i-th phase, all the $\beta_h + \beta_{h-1} + \cdots + \beta_{h-(i-1)}$ nodes having demands $\alpha_h, \alpha_{h-1}, \cdots, \alpha_{h-(i-1)}$ must be covered and accordingly $increment_i$, the increment in the i-th phase, will be computed and the number of rounds in this phase will be $\alpha_{h-(i-1)} - \alpha_{h-(i-2)} = y_i$. There can be at most h phases and finally the bandwidth requirement becomes $(2s_1 + 6s_2) + \sum_{i=1}^{h}(increment_i \times y_i)$. We note that $s_0 \leq increment_i \leq max(s_0, 3s_1 + 6s_2)\ \forall i$. Hence, $(2s_1 + 6s_2) + s_0(\sum_{i=1}^{h} y_i) \leq (2s_1+6s_2)+\sum_{i=1}^{h}(increment_i \times y_i) \leq (2s_1+6s_2)+max(s_0, 3s_1+6s_2)(\sum_{i=1}^{h} y_i)$. Note that $\sum_{i=1}^{h} y_i = \alpha_h - 1$ and hence $(2s_1 + 6s_2) + s_0(\alpha_h - 1) \leq (2s_1 + 6s_2) + \sum_{i=1}^{h}(increment_i \times y_i) \leq (2s_1+6s_2)+max(s_0, 3s_1+6s_2)(\alpha_h - 1)$. The worst case situation occurs when all nodes have equal demands ($h = 1$) and the greedy algorithm produces the worst possible ordering for finding a single frequency assignment to each node of the entire network.

The lower bound on bandwidth requirement of the cellular network having homogeneous demands of w channels per node can be found from the results reported in [29]. From Theorems 1 and 2 in [29], it can be verified [50] that the worst case ratio of the bandwidth required by the above greedy algorithm to the optimal bandwidth is always less than or equal to $5/3$. However, where 1) $s_0 \geq (s_1 + 5s_2)$ with $s_2 \leq s_1 \leq 2s_2$, and 2) $s_0 \geq (2s_1 + 3s_2)$ with $s_1 \geq 2s_2$, this ratio becomes $10/7$. The benchmark Problems 3, 4, 7, 8, 10 and 11 all fall under one of these two cases. It should be noted that the above analysis is based on a conservative estimate of the increment (taking it always to be the maximum) and thus the above ratio is a very loose upper bound on the deviation of the required bandwidth from optimality. However, computing the average case approximation ratio appears to be very difficult and is left as an open problem.

3.7 Conclusion

We have discussed lower bounds on the required bandwidth for assigning w channels to each node of a hexagonal cellular network with two-band buffering for different relative values of s_0, s_1, and s_2 [29]. Next we described an algorithm based on GA for solving the channel assignment problem in its general form, then demonstrated how this technique can be effectively applied to the special cases of hexagonal cellular networks with homogeneous demands and two-band buffering. We have also shown that this methodology can easily be applied to the cases of non-homogeneous demands.

We explained an elegant technique for solving the channel assignment problem (CAP) which is applicable even to a cellular network of non-hexagonal structure. The assignment is performed for a given bandwidth B with a view to minimizing call blocking. This technique can achieve the optimal solutions for all the well-known benchmark problems when the given bandwidth B is equal to the lower bound of the corresponding problem. The required computation time has also been improved over the critical block approach in [28]. How the unused or redundant channels pro-

TABLE 3.17: Derived channel assignment for Problem 2

Cells →	0	1	2	3	4	5	6	7	8	9	10	11	12	13	14	15	16	17	18	19	20
	230	272	228	129	2	3	1	127	0	3	126	0	0	126	4	2	128	1	226	126	4
	236	279	234	134	7	8	6	132	5	8	131	5	5	131	9	7	133	6	232	131	9
	242	286	240	139	12	13	11	137	10	13	136	10	10	136	14	12	138	11	238	136	14
	248	293	246	144	17	18	16	142	15	18	141	15	15	141	19	17	143	16	244	141	19
	254	300	252	149	22	23	21	147	20	23	146	20	20	146	24	22	148	21	250	146	24
	260	307	258	154	27	28	26	152	25	28	151	25	25	151	29	27	153	26	256	151	29
	266	314	264	159	32	33	31	157	30	33	156	30	30	156	34	32	158	31	262	156	34
	393	321	389	164	37	38	36	162	35	38	161	35	35	161	39	37	163	36	268	161	39
		328				43	41	167	40	43	166	40	40	166	44	42	168		275	166	
		335				48	46	172	45	48	171	45	45	171	49	47	173		282	171	
		342				53	51	177	50	53	176	50	50	176	54	52	178			176	
		349				58	56	182	55	58	181	55	55	181	59	57	183			181	
		356				63	61	187	60	63	186	60	60	186	64	62	188			186	
		363				68	66	192	65	68		65	65	191	69	67	193				
		370				73	71	197	70	73		70	70	196	74	72	198				
		377					76	202	75	78			75		79	77	203				
		384					81	207	80	83			80		84	82	208				
		391					86	212	85	88			85		89	87	213				
		399						217	90	93			90		94	92	218				
		407						222	95	98			95		99	97	223				
		415						227	100	103			100		104	102	392				
		423						233	105	108			105		109	107	400				
		431						239	110	113			110		114	112	408				
		439						245	115	118			115		119	117	416				
		447						251	120	123			120		124	122	424				
								257	125	270			128		129	229	432				
								263	130	277			133		134	235	440				
								269	135	284			138		139	241	448				
								276	140				143		144	247					
								283	145				148		149	253					
								290	150				153		154	259					
								297	155						159	265					
								304	160						164	271					
								311	165						169	278					
								318	170						174	285					
								325	175						179	292					
								332	180							299					
								339	185							306					
								346	190							313					
								353	195							320					
								360	200							327					
								367	205							334					
								374	210							341					
								381	215							348					
								388	220							355					
								396	225							362					
								404	231							369					
								412	237							376					
								420	243							383					
								428	249							390					
								436	255							398					
								444	261							406					
									267							414					
									274							422					
									281							430					
									288							438					
									295							446					
									302												
									309												
									316												
									323												
									330												
									337												
									344												
									351												
									358												
									365												
									372												
									379												
									386												
									394												
									402												
									410												
									418												
									426												
									434												
									442												
Demands →	8	25	8	8	8	15	18	52	77	28	13	15	31	15	36	57	28	8	10	13	8

TABLE 3.18: Derived channel assignment for Problem 6

Cells →	0	1	2	3	4	5	6	7	8	9	10	11	12	13	14	15	16	17	18	19	20
	103	75	191	104	102	77	0	3	1	4	2	0	4	2	76	104	79	76	4	2	0
	108	80	199	109	107	82	5	8	6	9	7	5	9	7	81	109	84	81	9	7	5
	113	85	207	114	112	87	10	13	11	14	12	10	14	12	86	114	89	86	14	12	10
	118	90	215	119	117	92	15	18	16	19	17	15	19	17	91	119	94	91	19	17	15
	123	95	223	124	122	97	20	23	21	24	22	20	24	22	96	124	99	96	24	22	20
				246	185	102	25	28	26	29	27	25	29	27	101	129	190	103	29	27	25
				255	193	107	30	33	31	34	32	30	34	32	106	135	198	108	34	32	30
				264	201	112	35	38	36	39	37	35	39	37	111	141	206	113	39	37	35
					209	117	40	43	41	44	42	40	44	42	116	147	214	118	44	42	40
					217	122	45	48	46	49	47	45	49	47	121	153	222	123	49	47	45
					225	130	50	53	51	54	52	50	54	52	126	159	230	130	54	52	50
					233	136	55	58	56	59	57	55	59	57	132	165	238	136	59	57	55
						142	60	63	61	64	62	60	64	62	138	171	247	142	64	62	60
						148	65	68	66	69	67	65	69	67	144	177	256	148	69	67	65
						154	70	73	71	74	72	70	74	72	150	183	265	154	74	72	70
						160	79	188	77	101	78	75	100	128	156			160	187	102	127
						166	84	196	82	106	83	80	105	134	162			166	195	107	133
						172	89	204	87	111	88	85	110	140	168			172	203	112	139
						178	94	212	92	116	93	90	115	146	174			178	211	117	145
						184	99	220	97	121	98	95	120	152	180			184	219	122	151
						192	186	228	185	128	126	100		158	191			192			157
						200	194	236	193	134	132	105		164	199			200			163
						208	202	241	201	140	138	110		170	207			208			169
						216	210	250	209	146	144	115		176	215			216			175
						224	218	259	217	152	150	120		182	223			224			181
							226		225	158	156	129		189				232			
							234		233	164	162	135		197				240			
							243		245	170	168	141		205				249			
							252		254	176	174	147		213				258			
							261		263	182	180	153		221				267			
										187	189	159									
										195	197	165									
										203	205	171									
										211	213	177									
										219	221	183									
										227	229	191									
										235	237	199									
										242	244	207									
										251	253	215									
										260	262	223									
												231									
												239									
												247									
												256									
												265									
Demands →	5	5	5	8	12	25	30	25	30	40	40	45	20	30	25	15	15	30	20	20	25

vided by this approach can be utilized for accommodating small changes in demands dynamically is a topic for further study.

We also described a polynomial time algorithm for near-optimal channel assignment in hexagonal cellular networks with non-homogeneous demands in a two-band buffering system. The CAP for non-homogeneous demands is divided into a number of smaller subproblems, each of which has a homogeneous demand from a subset of the nodes of the original network. A simple assignment scheme gives priority assignment to the higher demand nodes to solve each subproblem. We illustrated this algorithm with the help of the Philadelphia benchmark Problem 6, which is regarded as the most difficult one in the literature. This algorithm yields an assignment which never takes more than 6% of the optimal bandwidth and requires very little execution time (less than 5 milliseconds on a HPxw8400 workstation) for most difficult benchmark problems. Thus, this algorithm is suitable for real-life situations where fast channel assignment is demanded, although a marginal deviation (say, about 6%) from the optimal bandwidth may be tolerated.

3.8 Exercises

1. Describe the mathematical model of the channel assignment problem in a cellular mobile network.

2. Represent the channel assignment problem in a cellular mobile network in terms of the classical vertex coloring problem.

3. Derive a lower bound on the bandwidth requirement for assigning channels in a cellular mobile network with a single demand on each of the constituent nodes of the network.

4. Derive a lower bound on the bandwidth requirement of a given seven-node hexagonal subgraph (which consists of one central node surrounded by six peripheral nodes at six corners of a hexagon) of a cellular network having non-homogeneous demands from different nodes.

5. Describe the *frequency exhaustive* strategy and the *requirement exhaustive* strategy for solving the CAP.

6. Describe a method by which the CAP with non-homogeneous demand can be partitioned into a sequence of smaller subproblems where each subproblem has a homogeneous demand from a subset of nodes of the network.

7. Suppose *Algorithm A* can solve the CAP with the homogeneous demand of a single channel. Describe an algorithm to solve the CAP with non-homogeneous demands by using *Algorithm A*.

8. Describe a polynomial time algorithm that would solve the channel assignment problem in a cellular mobile network with a β- approximation bandwidth, i.e., the bandwidth required by the assignment algorithm would be within β times the optimal bandwidth for the problem.

9. Devise a polynomial time approximation algorithm for assigning channels to meet the non-homogeneous demands of the nodes in a network that will be more efficient than those proposed in this chapter. The worst case execution time of your proposed algorithm for all the benchmark problems needs to be less than that determined by the existing algorithms and the approximation factor by which the required worst case bandwidth will be more than the optimal bandwidth needs to be less than 5/3.

Bibliography

[1] A. Sen, T. Roxborough, B. P. Sinha, "On an optimal algorithm for channel assignment in cellular network," *Proc. of IEEE International Conf. Communications,* Vancouver, Canada, pp. 1147–1151, June 1999.

[2] A. Sen, T. Roxborough and S. Medidi, "Upper and lower bounds of a class of channel assignment problems in cellular networks," *Proc. of IEEE INFOCOM '98*, April 1998.

[3] W. K. Hale, "Frequency assignment: theory and application," *Proc. IEEE,* Vol. 68, pp. 1497–1514, 1980.

[4] M. Duque-Anton, D. Kunz and B. Ruber, "Channel assignment for cellular radio using simulated annealing," *IEEE Trans. Vehicular Technology,* Vol. 42, No. 1, pp. 14–21, Feb. 1993.

[5] R. Mathar and J. Mattfeldt, "Channel assignment in cellular radio networks," *IEEE Trans. Vehicular Technology,* Vol. 42, No. 4, pp. 647–656, Nov. 1993.

[6] D. Kunz, "Channel assignment for cellular radio using neural networks," *IEEE Trans. Vehicular Technology,* Vol. 40, No. 1, pp. 188–193, Feb. 1991.

[7] D. Tcha, Y. Chung and T. Choi, "A new lower bound for the frequency assignment problem," *IEEE/ACM Trans. Networking,* Vol. 5, No. 1, Feb. 1997.

[8] K. N. Sivarajan, R. J. McEliece and J. W. Ketchum, "Channel assignment in cellular radio," *Proc. of 39th IEEE Vehicular Technology Conf.,* pp. 846–850, May 1989.

[9] W. Wang and C. K. Rushforth, "An adaptive local-search algorithm for the channel assignment problem," *IEEE Trans. Vehicular Technology,* Vol. 45, No. 3, pp. 459–466, Aug. 1996.

[10] S. Khanna and K. Kumaran, "On wireless spectrum estimation and generalized graph coloring," *Proc. of IEEE INFOCOM,* April 1998.

[11] R. A. Leese, "A unified approach to the assignment of radio channels on a rectangular hexagonal grid," *IEEE Trans. Vehicular Technology,* Vol. 46, No. 4, pp. 968–980, Nov. 1997.

[12] L. Narayanan and S. Shende, "Static frequency assignment in cellular networks," *Proc. of 4th International Colloquium on Structural Information and Communication Complexity,* 1997.

[13] A. Sen, T. Roxborough and S. Medidi, "Upper and lower bounds of a class of channel assignment problems in cellular networks," *Proc. of IEEE INFOCOM,* April 1998.

[14] S. Kim and S. Kim, "A two-phase algorithm for frequency assignment in cellular mobile systems," *IEEE Trans. Vehicular Technology,* Vol. 43, No. 3, pp. 542–548, Aug. 1994.

[15] D. Beckmann and U. Killat, "A new strategy for the application of genetic algorithms to the channel assignment problem," *IEEE Trans. Vehicular Technology,* Vol. 48, No. 4, pp. 1261–1269, July 1999.

[16] C. Y. Ngo and V. O. K. Li, "Fixed channel assignment in cellular radio networks using a modified genetic algorithm," *IEEE Trans. Vehicular Technology,* Vol. 47, No. 1, pp. 163–172, Feb. 1998.

[17] D.-W. Tcha, J.-H. Kwon, T.-J. Choi, and S.-H. Oh, "Perturbation-minimizing frequency assignment in a changing TDMA/FDMA cellular environment," *IEEE Trans. Vehicular Technology,* Vol. 49, No. 2, pp. 390–396, March 2000.

[18] W. K. Lio and G. G. Coghill, "Channel assignment through evolutionary optimization," *IEEE Trans. Vehicular Technology,* Vol. 45, No. 1, pp. 91–96, Feb. 1996.

[19] J. S. Kim, S. H. Park, P. W. Dowd, and N. M. Nasrabadi, "Cellular radio channel assignment using a modified Hopfield network," *IEEE Trans. Vehicular Technology,* Vol. 46, pp. 957–967, Nov. 1997.

[20] C. W. Sung, and W. S. Wong, "Sequential packing algorithm for channel assignment under co-channel and adjacent-channel interference constraint," *IEEE Trans. Vehicular Technology,* Vol. 46, pp. 676–686, Aug. 1997.

[21] J. S. Kim, S. H. Park, P. W. Dowd, and N. M. Nasrabadi, "Channel assignment in cellular radio using genetic algorithm," *Wireless Personal Commun.,* Vol. 3, No. 3, pp. 273–286, Aug. 1996.

[22] T.-M. Ko, "A frequency selective insertion strategy for fixed channel assignment," *in Proc. of 5th IEEE Int. Symp., Personal, Indoor and Mobile Radio Commun.,* pp. 311–314, Sept. 1994.

[23] N. Funabiki and Y. Takefuji, "A neural network parallel algorithm for channel assignment in cellular radio network," *IEEE Trans. Vehicular Technology,* Vol. 41, pp. 430–437, Nov. 1992.

[24] R. Battiti, A. Bertossi, and D. Cavallaro, "A randomized saturation degree heuristic for channel assignment in cellular radio networks," *IEEE Trans. Vehicular Technology,* Vol. 50, No. 2, pp. 364–374, Mar. 2001.

[25] K. Smith and M. Palaniswami, "Static and dynamic channel assignment using neural network," *IEEE J. Select. Areas Commun.*, Vol. 15, pp. 238–249, Feb. 1997.

[26] G. Chakraborty, "An efficient heuristic algorithm for channel assignment problem in cellular radio networks," *IEEE Trans. Vehicular Technology,* Vol. 50, No. 6, pp. 1528–1539, Nov. 2001.

[27] http://www.specbench.org/osg/cpu95/results/cint95(cfp95).html

[28] S. C. Ghosh, B. P. Sinha and N. Das, "A new approach to efficient channel assignment for hexagonal cellular networks," *Int. J. Foundations Computer Sci.*, Vol. 14, No. 3, pp. 439–463, June 2003.

[29] S. C. Ghosh, B. P. Sinha and N. Das, "Channel assignment using genetic algorithm based on geometric symmetry," *IEEE Trans. Vehicular Technology*, Vol. 52, No. 4, pp. 860–875, July 2003.

[30] A. Gamst, "A new lower bound for the channel assignment problem," *IEEE Trans. Vehicular Technology,* Vol. 49, No. 4, pp. 1265–1272, July. 2000.

[31] R. Montemanni, J. N. J. Moon, and D. H. Smith, "An improved tabu search algorithm for the fixed-spectrum frequency assignment problem," *IEEE Trans. Vehicular Technology,* Vol. 52, No. 3, pp. 891–901, May 2003.

[32] D. Kunz, "Channel assignment for cellular radio using neural networks," *IEEE Trans. Vehicular Technology,* Vol. 40, No. 1, pp. 188–193, Feb. 1991.

[33] W. Wang and C. K. Rushforth, "Local search for channel assignment in cellular mobile networks," in *DIMACS Series in Discrete Mathematics and Theoretical Computer Science*, Vol. 35, pp. 689–709, 1997.

[34] I. Katzela and M. Naghshineh, "Channel assignment schemes for cellular mobile telecommunication systemsa comprehensive survey," *IEEE Personal Commun.*, vol. 3, pp. 10–31, 1996.

[35] M. Zhang and T.-S. Yum, "The non-uniform compact pattern allocation algorithm for cellular mobile systems," *IEEE Trans. Vehicular Technology*, Vol. 40, No. 2, pp 387–391, 1991.

[36] S.-H. Oh and D.-W. Tcha, "Prioritized channel assignment in cellular radio networks," *IEEE Trans. Commun.*, Vol. 40, pp. 1259–1269, 1992.

[37] P. Venkata Krishna, N. Ch and S.N. Iyengar, "Optimal channel allocation algorithm with efficient channel reservation in cellular networks," *Int. J. Commun. Networks Distributed Systems*, Vol. 1, No. 1, pp. 33–51, 2008.

[38] G. Vidyarthi, A. Ngom, I. Stojmenovic, "A hybrid channel assignment approach using an efficient evolutionary strategy in wireless mobile networks," *IEEE Trans. Vehicular Technology* Vol. 54, No. 5, pp. 1887–1895, 2005.

[39] G. Kendall, M. Mohamad, "Channel assignment optimization using a hyper-heuristic," *Proc. of IEEE Conference on Cybernetics and Intelligent Systems*, Singapore, December 2004.

[40] A. Boukerche, K.E. Khatib, T. Huang, "A performance evaluation of distributed dynamic channel allocation protocols for mobile networks," *Wireless Commun. and Mobile Computing*, Vol. 7, No. 1, pp. 69–80, 2007.

[41] M.R. Garey, D.S. Johnson, "The complexity of near-optimal graphcoloring," *Journal of the ACM*, Vol. 23, pp. 43–49, 1976.

[42] A. Gamst, "Some lower bound for a class of frequency assignment problems," *IEEE Trans. Vehicular Technology*, Vol. 35, No. 1, pp. 8–14, 1986.

[43] A. Gamst, W. Rave, "On frequency assignment in mobile automatic telephonesystems," *Proceedings of IEEE GLOBECOM*, 1982.

[44] S.C. Ghosh, B.P. Sinha, N. Das, "More on lower bounds for channel assignment problem," *Proceedings of 6th International Conference on High Performance Computing in Asia Pacific Region*, Bangalore, India, Vol. II, PP. 522–527, December 2002.

[45] D.H. Smith, S. Hurley, S.M. Allen, "A new lower bound for the channel assignment problem," *IEEE Trans. Vehicular Technology,* Vol. 49, No. 4, pp. 1265–1272, 2000.

[46] R. Montemanni, D. Smith, S. Allen, "Lower bounds for fixed-spectrum frequency assignment," *Annals of Operations Research,* Vol. 107, No. 1, pp. 237–250, 2001.

[47] D. E. Goldberg, *Genetic Algorithm: Search, Optimization and Machine Learning*, Addison Wesley, 1989.

[48] C. A. Murthy and N. Chowdhury, "In search of optimal clusters using genetic algorithms," *Pattern Recognition Letters,* vol. 17, pp. 825–832, 1996.

[49] D. Bhandari, C. A. Murthy and S. K. Pal, "Genetic algorithm with elitist model and its convergence," *Int. J. Pattern Recognition Artificial Intelligence,* Vol. 10, 731–747, 1996.

[50] G. K. Audhya, K. Sinha, K. Mandal, R. Dattagupta, S. C. Ghosh and B. P. Sinha, "A new approach to fast near-optimal channel assignment in cellular mobile networks", *IEEE Trans. Mobile Computing*, Vol. 12, No. 9, pp. 1814–1827, September 2013.

[51] S. C. Ghosh, B. P. Sinha and N. Das, "Coalesced CAP: An improved technique for frequency assignment in cellular networks," *IEEE Trans. Vehicular Technology*, Vol. 55, No. 2, pp. 640–653, March 2006.

[52] G. K. Audhya, K. Sinha, S. C. Ghosh and B. P. Sinha, "A survey on the channel assignment problem in wireless networks," *Wireless Commununication and Mobile Computing*, Vol. 11, Issue 5, pp. 583–609, May 2011.

Chapter 4

Localization of Nodes in Mobile Networks

In this chapter, we discuss location identification algorithms for computing the location of a node in an ad hoc network. We start with a small percentage of nodes known as reference nodes which are assumed to know their locations with high precision, either through GPS or some other means. Starting from these reference nodes, locations of other nodes are successively computed. A beacon node is a reference node or a node whose location has been computed with the help of location information of its neighbors. We assume errors exist in the advertised locations of all beacon nodes, including the reference nodes. $\pm\epsilon$ denotes the maximum possible error in the advertised coordinates of a beacon node along either axis of a two-dimensional co-ordinate system. $\pm\delta$ is the maximum possible error along either axis in a measured range between a node and a beacon. We first present an intelligent beacon selection algorithm proposed by [11] such that the error in the computed location of a node is within a bound of $\pm[3\epsilon(1+\sqrt{2})+2\delta]$ to $\pm(3\epsilon+2\delta)$ along one axis and within $\pm[2\delta(1+\sqrt{2})+3\epsilon]$ to $\pm(3\epsilon+2\delta)$ along its orthogonal axis. Authors in [11] also derived a general expression for computing the bounds on an error that can be introduced in the computed location of a node and the conditions that must be met in selecting the beacon nodes to keep the error in the computed location of a node within these bounds.

4.1 System Model

An ad hoc network is modeled as a graph $G = (V, E)$ consisting of n nodes. V is the set of all nodes, $|V| = n$ and E is the set of edges in the graph G. The nodes may be stationary or mobile. All communication links are assumed to be bi-directional. A node v is said to be a *neighbor* of u if they are within each other's hearing zones. The neighborhood $N(i)$ of a node i consists of all nodes that are within its transmission range. A small percentage of the nodes are assumed to know their individual locations with high precision, either through GPS or some other means. These nodes serve as the *reference nodes* (RNs) in the network. Initially, nodes other than the RNs do not possess any knowledge of their locations. A node is defined as a *beacon node* if it is used in computing the unknown location of another node. A beacon node may be a reference node or a node whose location has been computed with the help of location information of its neighboring beacons.

Considering thermal noise at the receiver, NLOS errors and channel characteristics, the measured range between two nodes u and v can be expressed as,

$$r_{uv} = d_{uv} + \eta_{uv} + c\tau_{uv} \tag{4.1}$$

where d_{uv} represents the unknown Euclidian distance between u and v. η_{uv} completely models the combined additive effects of thermal receiver noise, signal bandwidth and signal-to-noise ratio. η_{uv} has been shown to be a zero-mean normal random variable and hence can be measured or pre-computed [1]. η_{uv} is assumed to be always additive. $c\tau_{uv}$ represents the NLOS distance error and is the dominant error contributor [2], c being the speed of light in air.

It will be assumed that beside the ranges measured for location computation, the advertised locations of the reference nodes (and beacon nodes in general) are erroneous within certain known bounds. The advertised co-ordinates of a reference node may have $\pm\epsilon$ error along both x-axis and y-axis. The error in a measured range can be erroneous by $\pm\delta$ along both axes.

4.2 Preliminaries

The above model assumes that errors may be present in the advertised coordinates of a beacon node (including reference nodes) and in a measured range between two nodes. We note that a number of existing location sensing technologies mention the accuracy of their location determination processes. This value can be used as the ϵ error in the advertised coordinates of the reference nodes. For example,

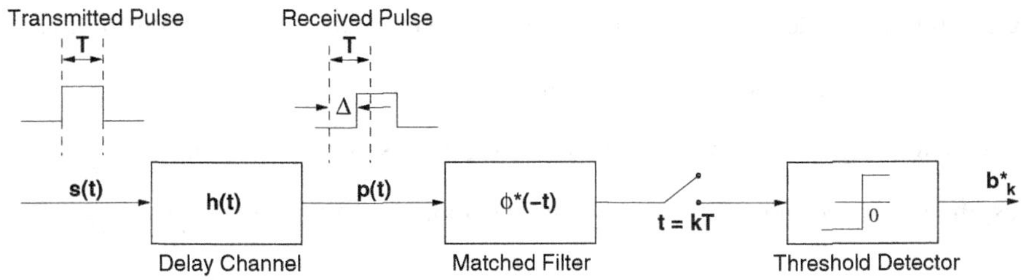

FIGURE 4.1: Delay channel and receiver design

GPS, by far the most widely used location sensing technology, has an accuracy of 1 to 5 meters in outdoor environments for 95 to 99% of the time [3]. Microsoft Radar [4] claims an accuracy of 3 to 4.3 meter, while PinPoint 3D-iD and WhereNet [3] have location accuracies of 1 to 3 meter. The Cricket system [5] has an accuracy of 4×4 feet regions nearly 100% of the time. Based on the application scenario, we can thus set an appropriate value for ϵ corresponding to the reference nodes.

To determine a $\pm\delta$ bound on range measurement errors, NLOS error is assumed to be the major contributor for errors in measured ranges. The error due to thermal noise (η_{uv} in Equation (4.1)) is generally much smaller than the NLOS error component and ignored in the subsequent discussions. In [6], experimental results show that for indoor locations where line of sight exists between the transmitter and the receiver, the zero mean Gaussian model for normalized distance error closely models the NLOS distance error. For the indoor obstructed line of sight (OLOS) case, while the Gaussian model is not suitable, a mixed distribution consisting of zero mean Gaussian and exponential distribution fits the distance error very well.

While it is not possible to know the extent of error introduced in the measured range due to NLOS propagation, it is possible to estimate the maximum NLOS error that can be tolerated so that a transmitted signal is correctly interpreted at the receiver. To analyze the amount of tolerable NLOS error, consider a *delay channel* and a receiver circuit as shown in Figure 4.1. In a delay channel, the output channel state at a given instant t is the same as the input channel state at instant $t - t'$. A receiver circuit consisting of a *matched filter* and a *threshold detector* will be assumed to be used for this purpose.

Let $s(t)$ represent the transmitted signal. Let b_n be the symbol during the nth slot. For simplicity, we assume a BPSK modulation scheme, implying $b_n \epsilon \{-1, +1\}$. If $\phi(t)$ represents the shaping pulse, then we have,

$$s(t) = \sum_{n=-\infty}^{\infty} b_n \cdot \phi(t - nT), \qquad b_n \epsilon \{-1, +1\}$$

. For a delay channel, the received signal, $p(t)$ is given by

$$p(t) = \begin{cases} 0 & \text{if } t \leq \Delta, \\ s(t) & \text{otherwise.} \end{cases}$$

At the receiver end, the signal output $p(t)$ of the channel is passed through a matched filter, $\phi^*(-t)$ and the output is sampled periodically at the symbol time period T. Matched filtering is a process for detecting a known piece of signal that is embedded in noise. Matched filter theory states that the maximum signal-to-noise ratio (SNR) at the output will occur when the filter has an impulse response that is the time reverse of the input signal [7, 8]. Let $R_n(t)$ represent the output of the matched filter passed through a threshold detector, with the threshold set to 0. Then from [8, 7],

$$\begin{aligned} R_n(t) &= \int_0^T p(t) \cdot \phi^*(-t) \cdot dt \\ &= \int_\Delta^T s(t) \cdot \phi^*(-t) \cdot dt \ . \end{aligned} \tag{4.2}$$

Suppose $b_n = 1$. The probability that the receiver is unable to correctly interpret a transmitted bit is given by

$$\text{Prob. of error, } P_e = P(R_n(t) < 0|_{b_n=1}). \tag{4.3}$$

Assuming $s(t) = 1$ in Equation (4.2), the normalized output of the matched filter is given as

$$R_n(t) = \frac{1}{T}\int_{\Delta}^{T} 1 \cdot 1 \cdot dt = \frac{T-\Delta}{T}. \tag{4.4}$$

The condition under which a transmitted symbol b_n is wrongly estimated at the receiver can be obtained by combining Equations 4.3 and 4.4:

$$\begin{aligned} P_e &= P(R_n(t) < 0|_{b_n=1}) \\ &\Rightarrow \frac{T-\Delta}{T} < 0 \\ &\Rightarrow \Delta > T \end{aligned} \tag{4.5}$$

From the above discussions, we get the following lemma:

Lemma 8. *For a delay channel, in order to estimate a transmitted signal correctly at the receiver in the presence of NLOS errors, the time shift in the received pulse introduced by the NLOS component must be less than the time period of the pulse.*

Lemma 8 provides a measure of $\pm\delta$ bound along both coordinate axes on the error in ranging. Typically a pulse width is of the order of 1 nanosecond for UWB-based networks, which is equivalent to 0.3 m of distance covered by a radio wave. Similarly for $IEEE802.11n$ standard (with an estimated data rate between 200 and

540 Mbps), the pulse width would be 5 to 1.85 ns, which translates to 1.5 to 0.555 meter for $c\tau_{uv}$ in Equation (4.1).

4.3 Estimation of Location Error

Assume that the location of a node u is obtained through triangulation, using its neighboring beacons whose locations are already known. The error in the estimated position of node u can arise from:

1. Uncertainties in the locations of the beacon nodes used for triangulating the location of node u,
2. Errors in ranging; NLOS errors are dominant.

4.3.1 Error due to Ambiguity in Node Coordinates

We first consider the case when the positions of the beacon nodes are not known accurately. Let us consider two beacon nodes B_1 and B_2 as in Figure 4.2, with corresponding estimated ranges as r_1 and r_2, respectively. Let T represent the computed position of node u. Let S be the foot of the normal drawn from T to the line joining B_1 and B_2 , with d_1 and d_2 being the distances of S from B_1 and B_2, respectively. Let $d = d_1 + d_2$. Then the coordinates of S are $(d_1, 0)$ and that of T are $(d_1, \sqrt{r_1^2 - d_1^2})$ with respect to the origin at $B_1(0,0)$ and considering the x-axis along B_1B_2.

If there is some error in the positional estimates of the beacon nodes, each of B_1 and B_2 can be taken anywhere within a circle of radius ϵ with B_1 and B_2 as centers as shown in Figure 4.3 (note that B_1 and B_2 are assumed to be the actual positions

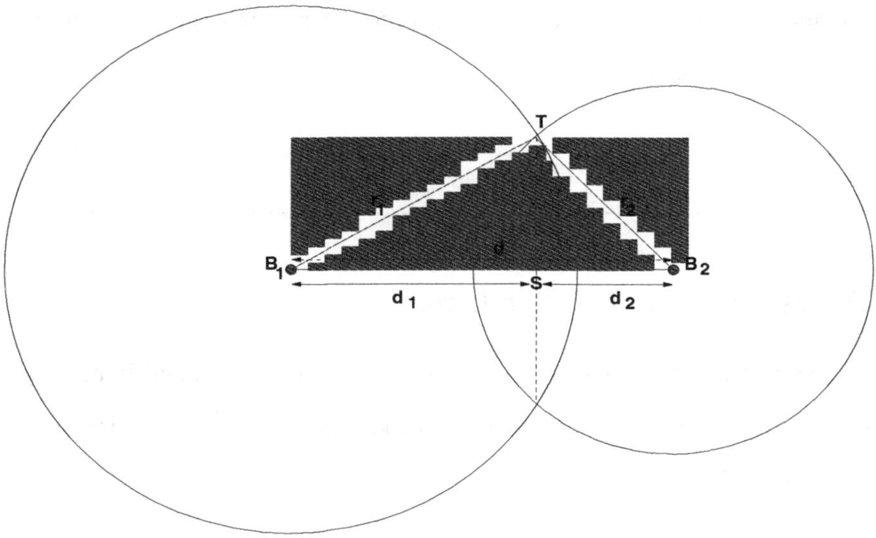

FIGURE 4.2: Location estimation using two beacon nodes

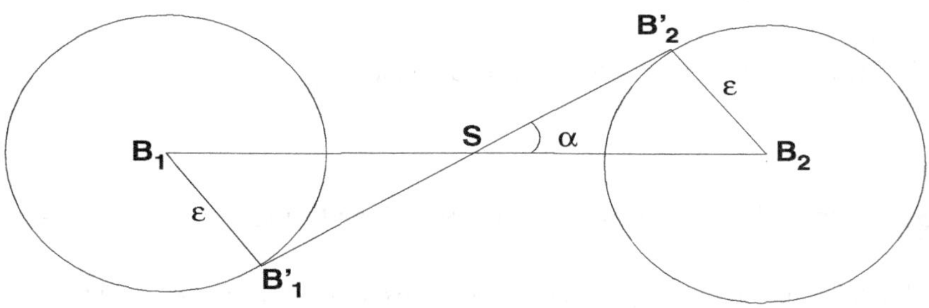

FIGURE 4.3: Location estimation when error in position of beacon nodes

of the beacon nodes) where $\pm\epsilon$ is the upper bound on the error in the coordinates of a beacon node. This means that the above coordinate values $(d_1, \sqrt{r_1^2 - d_1^2})$ of T are now referred to some new origin B_1' and some rotated x-axis $B_1'B_2'$ as shown in the figure, where B_1' and B_2' are points within the circles of radius ϵ around B_1 and B_2, respectively. We denote the coordinate values of T with respect to this new coordinate system as $(x', y') = (d_1, \sqrt{r_1^2 - d_1^2})$. The line $B_1'B_2'$ can make an angle of $\alpha \leq \frac{2\epsilon}{d}$ with B_1B_2.

We can convert these coordinate values to (x, y) with respect to B_1 and B_2 as:

$$\left.\begin{aligned} x = &\quad x' \cos\alpha - y' \sin\alpha \pm \epsilon \\ y = &\quad x' \sin\alpha + y' \cos\alpha \mp \epsilon \end{aligned}\right\} \tag{4.6}$$

Since, in most cases $\frac{2\epsilon}{d}$ would be a small quantity, we have from Equation (4.6),

$$(x, y) \approx (x' - y'\alpha \pm \epsilon,\ x'\alpha + y' \mp \epsilon).$$

The coordinates of T with respect to rotated x-axis formed by the line joining B_1 and B_2 are then

$$\begin{aligned} x &= d_1 - \sqrt{{r_1}^2 - {d_1}^2} \cdot \left(\frac{2\epsilon}{d}\right) \pm \epsilon \\ y &= d_1 \cdot \left(\frac{2\epsilon}{d}\right) + \sqrt{{r_1}^2 - {d_1}^2} \mp \epsilon. \end{aligned}$$

Thus, the error introduced in the computed position of node u due to a maximum ambiguity of $\pm\epsilon$ in the advertised locations of the beacon nodes is given by

$$\left.\begin{aligned} |x - x'| &= \sqrt{{r_1}^2 - {d_1}^2} \cdot \left(\frac{2\epsilon}{d}\right) + \epsilon \\ |y - y'| &= d_1 \cdot \left(\frac{2\epsilon}{d}\right) + \epsilon \qquad \leq 3\epsilon \end{aligned}\right\}. \tag{4.7}$$

From Equation (4.7) we conclude that if we start with an uncertainty of $\pm\epsilon$ in the positions of the individual beacon nodes, this uncertainty can contribute to a maximum error of $\pm 3\epsilon$ in the evaluated position of node u along the axis normal to the line joining $B_1 B_2$. However, no such bound can be given for the error in the coordinate value along the line $B_1 B_2$.

In order to calculate a bounded estimate of the error introduced in the computation of an unknown node's coordinates, we consider three beacon nodes instead of two. In Figure 4.4, the point T depicts one of the two possible positions of the node

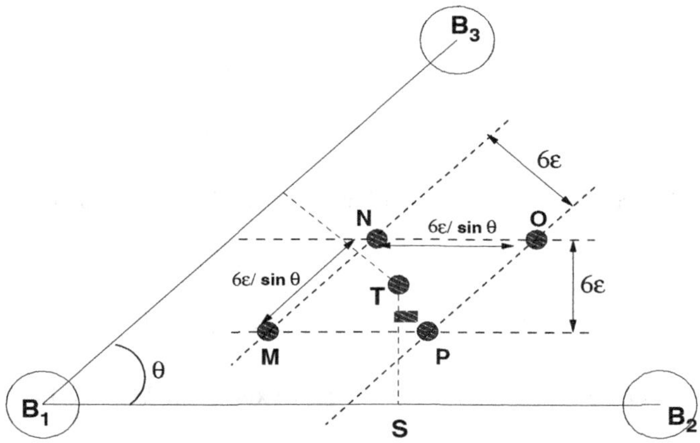

FIGURE 4.4: BS location error estimation when using three beacon nodes

u using B_1 and B_2 as beacon nodes and assuming no error either in the coordinates of the beacon nodes or in the measured ranges. Now from Equation (4.7)), an error of $\pm\epsilon$ in the positional estimates of the two beacon nodes, generates an error of $\pm 3\epsilon$ in the y-coordinate estimate of the unknown node. Hence, with respect to the line B_1B_2, we imagine a strip of width 6ϵ placed parallel to B_1B_2 such that the point T is at a distance of 3ϵ from each of the two bounding lines of the strip. Similarly, with respect to beacon node B_3, we place another strip of width 6ϵ parallel to the line B_1B_3 and having the point T at a distance of 3ϵ from each of the two bounding lines of the strip. The actual position of T can thus always be found within the rhombus $MNOP$, as shown in Figure 4.4. The length of the side of this rhombus is $\frac{6\epsilon}{\sin\theta}$, where θ is the angle between the lines B_1B_2 and B_1B_3. Thus, the side of this rhombus will be bounded to $6\sqrt{2}\epsilon$ or less if $45 \leq \theta \leq 90$ degrees, and within 12ϵ if $30 \leq \theta < 45$ degrees. We have the following lemma:

Lemma 9. *For an error of $\pm\epsilon$ in the x and y coordinates of each of the beacon nodes, the location of an unknown node can be bounded to inside a rhombus of side $\frac{6\epsilon}{\sin\theta}$ using three beacon nodes (say B_1, B_2 and B_3), where without any loss of generality, θ is the angle between the lines B_1B_2 and B_1B_3.*

4.3.2 Contribution by Range Estimation Error

Next we consider the case when there is only a maximum error of $\pm\delta$ in the individual range estimates (due to NLOS errors and Gaussian noise). Again, δ is typically assumed to be quite small compared to d, r_1 or r_2. Assume that the beacon node coordinates used are error-free. That is, we start with B_1 and B_2 as the actual coordinates of the beacon nodes but with a range error of δ, so that the measured ranges from B_1 and B_2 are now r_1' and r_2', respectively, where $r_1' = r_1 \pm \delta$, and $r_2' = r_2 \pm \delta$. Hence, the modified value for d_1 is given by

$$\begin{aligned} d_1' &= \frac{d_2 + {r_1'}^2 - {r_2'}^2}{2d} \\ &= \frac{d_2 + r_1^2 - r_2^2 \pm 2\delta r_1 \pm 2\delta r_2}{2d} && \text{(neglecting } O(\delta^2) \text{ terms).} \end{aligned}$$

Note that the distance d between the two beacon nodes B_1 and B_2 is assumed to be constant in the above expression for d_1'. Hence, the error resulting in the computed position of u along the x-axis (i.e., the line joining B_1 and B_2) is

$$\begin{aligned} |d_1' - d_1| &\leq \frac{2\delta(r_1 + r_2)}{2d} \\ &\leq \frac{2\delta \cdot 2d}{2d} && \text{since } r_1 \leq d \text{ and } r_2 \leq d \\ &= 2\delta. \end{aligned} \tag{4.8}$$

The modified value of $\sqrt{r_1^2 - d_1^2}$ is now

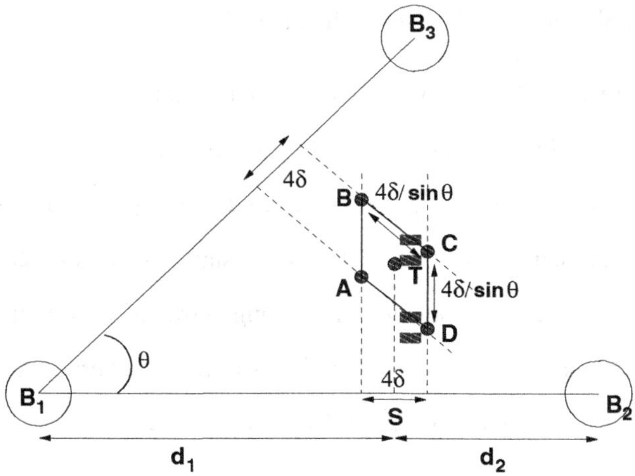

FIGURE 4.5: Contribution by ranging error when using three beacon nodes

$$\begin{aligned}\sqrt{{r'_1}^2 - {d'_1}^2} &= \sqrt{(r_1 \pm \delta)^2 - (d_1 \pm 2\delta)^2} \\ &= \sqrt{{r_1}^2 - {d_1}^2 \pm 2\delta r_1 \pm 4\delta d_1} \qquad \text{(neglecting } O(\delta^2) \text{ terms).}\end{aligned}$$

Thus, the error along the y-axis (along line TS in Figure 4.2) is

$$\left| \sqrt{{r'_1}^2 - {d'_1}^2} - \sqrt{{r_1}^2 - {d_1}^2} \right| \leq \frac{2\delta(r_1 + 2d_1)}{\sqrt{r_1^2 - d_1^2}}. \tag{4.9}$$

Thus, from Equations (4.8) and (4.9), the magnitude of the error in the coordinate value of u along the line B_1B_2 joining the two beacon nodes is upper bounded by 2δ, for a maximum error of $\pm\delta$ in each measured range. However, the error in the coordinate value normal to the line B_1B_2 caused by $\pm\delta$ error in each measured range is a function of the distances between the beacon nodes and between u and the beacon nodes and hence cannot be bounded.

In order to bound the error in a computed location caused by ranging errors, we again consider three beacon nodes instead of two, similar to the case of $\pm\epsilon$ error in the positional estimates of the beacon nodes. Figure 4.5 depicts how we use three beacon nodes to compute a bound on the estimates of the coordinates of the unknown node u, assuming an error of $\pm\delta$ in the measured ranges. With respect to B_1 and B_2, we consider a strip of width 4δ placed normal to B_1B_2, and at a distance d_1 from B_1. Point T will always lie within this strip. Similarly, with respect to another pair of beacon nodes, say B_1 and B_3, we imagine another strip of width 4δ placed normal to B_1B_3, over which T must also lie. Hence, the actual position of T can always be found within a rhombus (shown by solid lines in Figure 4.5) of side $\frac{4\delta}{\sin\theta}$, where θ is the angle between the lines B_1B_2 and B_1B_3. Thus, the side of this rhombus will be bounded to $4\sqrt{2}\delta$ or less if $45 \leq \theta \leq 90$ degrees, and within 8δ if $30 \leq \theta < 45$ degrees. We now have the following lemma:

Lemma 10. *For an error of $\pm\delta$ in the measured ranges with each of the beacon nodes, the location of an unknown node can be bounded inside a rhombus of side $\frac{4\delta}{\sin\theta}$ using three beacon nodes (say B_1, B_2 and B_3), where, without loss of generality, θ is the angle between the lines B_1B_2 and B_1B_3.*

4.3.3 Combined Error Estimation

Finally, we consider the maximum possible error of $\pm\epsilon$ in the advertised coordinates of the beacon nodes and also a possible error up to $\pm\delta$ in the measured range estimates.

In Sections 4.3.1 and 4.3.2, we derived a bound on the error that may be introduced in the computed location of an unknown node for each of the individual cases by considering three beacon nodes. In Figure 4.2, the point T represents the true location of the unknown node u. For the case of an error of $\pm\epsilon$ in the coordinates

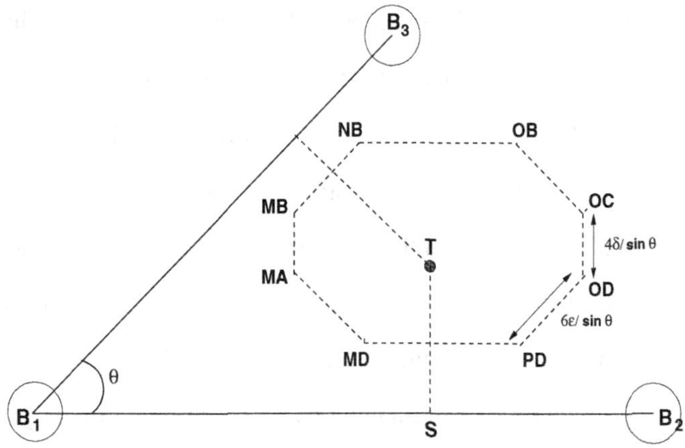

FIGURE 4.6: Minkowski's sum of the two errors

of each of the beacon nodes, we see that the true location of u must lie somewhere within a rhombus of side $\frac{6\epsilon}{\sin\theta}$ ($MNOP$ in Figure 4.4). Again, assuming the beacon node coordinates to be exact, the true location of u must lie somewhere within a rhombus of side $\frac{4\delta}{\sin\theta}$ for an error of $\pm\delta$ in each of the measured ranges with the beacon nodes.

Lemma 11. *If both the errors in the coordinates of the beacon nodes and the measured ranges occur simultaneously, a bounded estimate of the combined effect of the two errors can be obtained by considering the Minkowski's sum [9] of the two rhombi formed as stated in Lemmas 9 and 10.*

Proof. The combined effect of the two errors occurring simultaneously can be envisioned as follows.

Suppose we are trying to compute the coordinates of a node u. First we obtain the rhombus $MNOP$ (Figure 4.4) due to the error of $\pm\epsilon$ in the location of each of the beacon nodes. The rhombus would represent the location where node u must be found, assuming the measured ranges are error-free. In other words, any point inside and on the edges of the rhombus is potentially a location of node u. Let point X denote the location of u. If the measured ranges were also erroneous by $\pm\delta$, we can think of a rhombus $ABCD$ of side $\frac{4\delta}{\sin\theta}$ (Figure 4.5) placed such that X lies at the point of intersection of the two diagonals of $ABCD$ and the rhombus $ABCD$ is aligned to the Cartesian axes as in Figure 4.5. Since, point X can be anywhere on the rhombus $MNOP$, the region where node u is guaranteed to lie in the presence of both the errors is given by the boundary of the polygon formed by placing the rhombus $ABCD$ at all points on the rhombus $MNOP$, which is Minkowski's sum of the rhombuses $MNOP$ and $ABCD$. □

Figure 4.6 shows Minkowski's sum of the two rhombuses of sides $\frac{6\epsilon}{\sin\theta}$ and $\frac{4\delta}{\sin\theta}$, respectively. The labels of the vertices of the polygon of Figure 4.6 represent the respective vertices of the original rhombuses of Figures 4.4 and 4.5 from which they are formed. For example, the vertices MA and MB are formed from the vertex M of the rhombus $MNOP$ and the vertices A and B, respectively, of the rhombus $ABCD$.

Theorem 10. *Assuming an error of $\pm\epsilon$ in the coordinates of the beacon nodes and an error of $\pm\delta$ in the measured ranges, the error in the location of an unknown node due to the combined effect of these two errors is bounded by $\pm\left[\dfrac{3\epsilon(1+\cos\theta)}{\sin\theta}+2\delta\right]$ along the x-axis and $\pm\left[\dfrac{2\delta(1+\cos\theta)}{\sin\theta}+3\epsilon\right]$ along the y-axis.*

Proof. The projection of the edge PO on the x-axis is $6\epsilon\cot\theta$. Similarly, the projection of edge DA on the y-axis is $4\delta\cot\theta$. Now from Lemma 11, Minkowski's sum is formed by placing the polygon $ABCD$ on all the points along the edges of the polygon $MNOP$ such that the intersection of the two diagonals of $ABCD$ lies on the edges of $MNOP$. Hence the projection of $ABCD$ along the x-axis is 4δ and the projection of $MNOP$ along the y-axis is 6ϵ. Thus, the total span of the error along the x-axis due to the combined effect of the two errors is $\pm\left[\frac{6\epsilon}{\sin\theta}+6\epsilon\cot\theta+4\delta\right]$, and along the y-axis is $\pm\left[\frac{4\delta}{\sin\theta}+4\delta\cot\theta+6\epsilon\right]$. □

Corollary 1. *For $\frac{\pi}{4}\leq\theta\leq\frac{\pi}{2}$, the error induced on the estimated location of an* *unknown node i will be bounded within $\pm[3\epsilon(1+\sqrt{2})+2\delta]$ to $\pm(3\epsilon+2\delta)$ along the x-axis and within $\pm[2\delta(1+\sqrt{2})+3\epsilon]$ to $\pm(3\epsilon+2\delta)$ along the y-axis.*

4.4 Beacon Node Selection Algorithm

We would like to determine a set of beacon nodes from the neighbor set $N(i)$ of a node i that can be used to compute the location of i with an error bound given by Theorem 10. Let us define a *pivot beacon* as a beacon node with respect to which the angle θ is measured. Thus, in Figures 4.5 and 4.6, node B_1 is the pivot beacon.

Algorithm 4.4 described below first constructs all possible three-tuple combinations of the beacon nodes from the neighbor set of a node whose location is to be computed. For each such three-tuple, each node of the tuple is considered individually as the pivot beacon and the angle θ subtended at it by the other two beacon nodes is computed. In the procedure SELECTBEACONS, $TupleSet_i^j$ denotes the jth tuple constructed from the set $N(i)$ of a node i while $TupleSet_i^j(k)$ denotes the kth element of the jth tuple of node i. If for any pivot of a tuple, $\frac{\pi}{4} \leq \theta \leq \frac{\pi}{2}$, then the given tuple is copied to the output set and the process is repeated for the next tuple. The first node of an output three-tuple is assumed to be the pivot beacon for that tuple and the remaining two nodes are the other beacons.

Constructing all possible three-tuple combinations of the beacon nodes from the neighborhood set of a node i requires $O(m^3)$ time, where $m = |N(i)|$ (line 8 of the procedure SELECTBEACONS). Also, the statements inside the nested loop in the procedure may be executed $3\binom{m}{3}$ times in the worst case. Procedure SELECTBEACONS thus has a time complexity of $O(m^3)$.

If we start with an error bound of $\pm\epsilon$ along both axes for the advertised locations

```
procedure SELECTBEACONS(i, N(i), PosSet(i))
    /* The algorithm finds the set of 3-tuples */
    /* from the neighbor set of a i (beacons) */
    /* that satisfy the condition stated in corollary 1 */
    /* PosSet(i) = {p(j) : ∀j ∈ N(i), p(j) = co-ordinates of node j} */
    BSSet ← ∅;                                   ▷ Output set of chosen 3-tuples
    m ← |N(i)|;                                  ▷ Cardinality of N(i)
    TupleSet_i ← all possible 3-tuple combinations from set N(i);
    p ← |TupleSet_i|;                            ▷ Cardinality of TupleSet_i = (m choose 3)
    /* Let TupleSet_i^j ← j^th tuple of TupleSet_i */
    for j = 1 to p do                            ▷ Consider j^th tuple
        for k = 1 to 3 do
            /* Choose the pivot beacon */
            B_1 ← TupleSet_i^j(k);               ▷ pivot = k^th node of j^th tuple
            /* Let B_2 and B_3 be the other 2 nodes of TupleSet_i^j */
            L_12 ← line joining points p(B_1) and p(B_2)     ▷ for pivot and B_2
            L_13 ← line joining points p(B_1) and p(B_3)     ▷ for pivot and B_3
            theta ← angle between L_12 and L_13              ▷ theta = ∠B_2B_1B_3
            if theta ≥ π/4 and theta ≤ π/2 then
                /* Add the j^th 3-tuple to the set */
                /* of beacons that can be used to compute */
                /* location of node i, i.e., BSSet */
                BSSet ← BSSet ∪ TupleSet_i^j;
                break;                           ▷ go for next tuple
            end if
        end for
    end for
end procedure
```

of each of the reference nodes, as the location information propagates through the network in a hop-by-hop fashion, SELECTBEACONS allows the computation of the maximum possible error in the computed location of a node that is k hops away from a reference node, $k \geq 1$.

4.5 Location Region Identification

As mentioned in Chapter 1, while reasonably accurate radio propagation models exist for outdoor conditions, unfortunately there are no such unanimously accepted models for indoor environments. Hence, in the absence of a suitable model for predicting the location of a mobile terminal, it is possible that the node may be far away from the estimated point.

Rather than doing a location estimation, we might consider the location discovery problem in terms of finding the *region* where a node is *guaranteed* to be found. The objective of the location identification process then becomes minimizing the size of such a region. We assume again that a small percentage of the nodes in the network know their locations with a high degree of accuracy – possibly through GPS access or by some other means. We call them reference nodes. We describe below a distributed algorithm using computational geometric techniques [6] to compute the smallest *region of residence* where a node is guaranteed to be found, for every non-reference node in the network. A unique feature of this algorithm is that the location regions of the nodes in the network are improved through the exchange of location information between the neighbors in $O(nD)$ time, where n and D are the number of nodes and diameter of the network, respectively. Simulation results reported by the authors also demonstrate that this algorithm succeeds in finding reasonably small stable regions of residence for all non-reference nodes.

4.5.1 Preliminaries

The algorithm proposed by Sinha et al. [6] is based on the triangulation technique to compute the region where a node is guaranteed to be found.

Considering thermal noise at the receiver, NLOS errors and channel characteristics, the measured range between two nodes u and v is given by Equation (4.1). Let RR denote the set of all such measured ranges for all node pairs in the network, i.e., $RR = \{r_{ij} : r_{ij} \in E, \forall i, j \in V\}$. Also, $RR_i = \{r_{ij}, j \in N(i)\}$.

Definition 2. The *region of residence* R_i of a node i is defined to be the region where i is guaranteed to be found.

The region of residence of a reference node is assumed to be a point location of zero area. All other nodes have non-zero finite area regions of residence. Our objective is to find the minimum region of residence of a node i.

Lemma 12. *The range measurements obtained for a node u from a neighbor $v \in N(u)$ will always be greater than or equal to the Euclidian distance between u and v.*

Proof. Follows directly from Equation (4.1). □

Given two nodes u and v and a range measurement r_{uv} from v to u, the region of residence of u in the view of node v is the region formed by extending v's region of residence R_v in every direction by the measured range value, r_{uv}. We denote this by the operator $\oplus$, whose left operand is a region of residence and the right operand is a range value. Thus the region of residence of u in the view of node v is $R_{uv} = R_v \oplus r_{uv}$. We designate R_{uv} as the *viewed region of residence* of node u.

Example 18. Suppose node v has a triangular region of residence ΔABC, and the range of u measured by v is r_{uv} as shown in Figure 4.7. We draw a line $A'B'$ parallel to AB at a distance r_{uv} from AB and on the opposite side of the node C such that

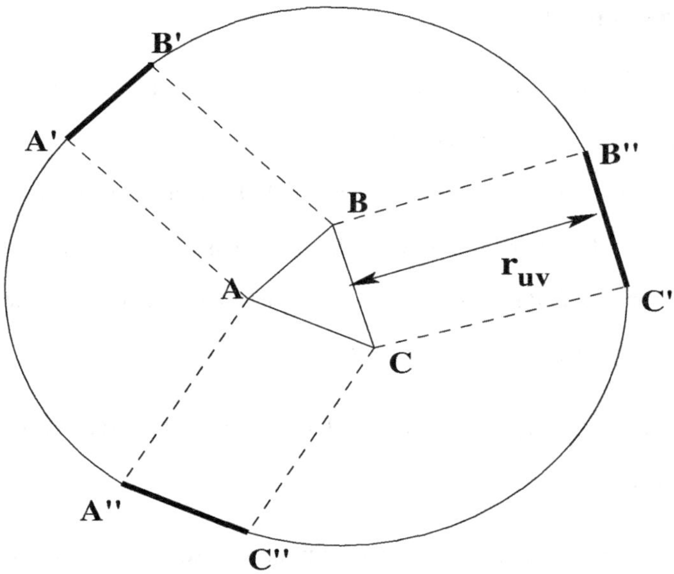

FIGURE 4.7: Construction of viewed region of residence of node

$AA'B'B$ forms a rectangle. Similarly, we draw a line $B''C'$ parallel to BC on the side opposite to that of A and distant r_{uv} from BC to form a rectangle $BB''C'C$, and also a line $C''A''$ parallel to CA on the side opposite to B and distant r_{uv} from CA so that $CC''A''A$ is a rectangle. Now from point A, draw a circular arc of radius r_{uv} to cut the lines $A'B'$ and $C''A''$ at A' and A'' respectively. Similarly, draw two other circular arcs of radius r_{uv}: i) from B to cut the lines $A'B'$ and $B''C'$ at B' and B'', respectively, and ii) from C to cut the lines $C'B''$ and $C''A''$ at C' and C'', respectively. The closed convex region $A'B'B''C'C''A''$ is the resulting R_{uv}.

Region R_{uv} can also be viewed as Minkowski's sum [9] of the region of residence R of the node v and a circle of radius r_{uv} centered at origin. We assume that the initial regions of residence of all nodes are bounded by straight line segments or by circular arcs. Hence, the region R_{uv} will also be bounded by straight line segments and/or circular arcs only. We now have the following result.

Lemma 13. *Node u is guaranteed to be found at some location inside R_{uv}.*

Proof. Follows from the construction of R_{uv}. □

Theorem 11. *The current minimum region of residence, $\mathcal{R}_u$ of a node u based on the information from its neighbors is the region formed by the intersection of the viewed regions of residence R_{ui}'s, $i \epsilon N(u)$, i.e., $\mathcal{R}_u = \bigcap_{i \epsilon N(u)} R_{ui}$.*

Proof. The proof follows from Lemma 13 as the common intersection region is the smallest region that satisfies Lemma 13 for all neighbors $i \epsilon N(u)$. □

Note that this current minimum region of residence may subsequently be refined (contracted in size) by improved viewing regions of residences from its neighbors.

Theorem 12. *The minimum region of residence of a node u, $\mathcal{R}_u$ (based on the information from its neighbors) cannot subsequently be made larger by an altered view of the region of residence, R_{ui} from any neighbor i.*

Proof. The proof follows directly from Theorem 11. □

To find the minimum region of residence, the algorithm presented in [6] proceeds in two steps:

1. Every node u in the network determines its current region of residence by ranging with each of its neighbors.

2. Once u has determined its current minimum region of residence, it attempts to improve the regions of residence of each neighbor, using its own region of residence and the range measurements obtained from respective neighbors.

Example 19 illustrates the working of the algorithm.

Example 19. Consider a node u with three neighbors i, j and k. Figure 4.8 demonstrates a probable situation where the triangles ΔABC and ΔDEF define the region of residence of nodes i and j, respectively. $PQRS$ is the region of residence of node

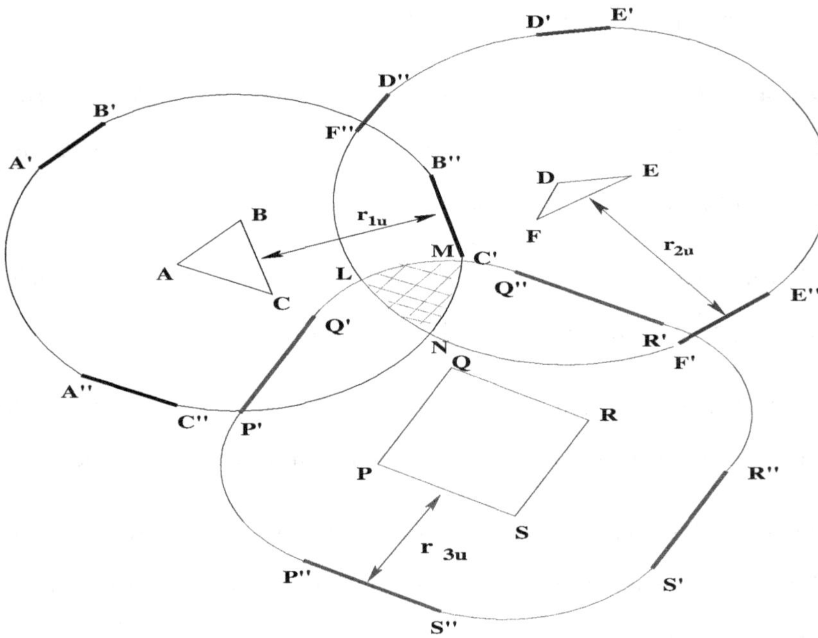

FIGURE 4.8: Computation of minimum region of residence

k. For simplicity, we assume the regions of residence as polygonal. Let r_{1u}, r_{2u} and r_{3u} be the range measurements that u obtains by ranging with i, j and k, respectively. According to the view of node i, u lies in the region R_{iu}, dictated by the shape $A'B'B''C'C''A''$ as demonstrated in Example 18. Similarly, $D'E'E''F'F''D''$ and $P'Q'Q''R'R''S'S''P''$ define the region of residence of u in the views of the nodes j and k, respectively. Following Theorem 11, the shaded region LMN defines the minimum region of residence of node u where u is guaranteed to be found.

Once the minimum region of residence of node u is found, node u then tries to refine the minimum region of residence of a neighbor v, $\forall v \in N(u)$, using $\mathcal{R}_u$ and the corresponding measured range r_{uv} from v. The new minimum region of residence of node v, $\mathcal{R}'_v$ is defined as the intersection of the viewed region of residence of v by u, R_{vu} and the current minimum region of residence of v, $\mathcal{R}_v$. In Figure 4.9, region LMN defines the minimum region of residence of node u and r_{3u} is the measured

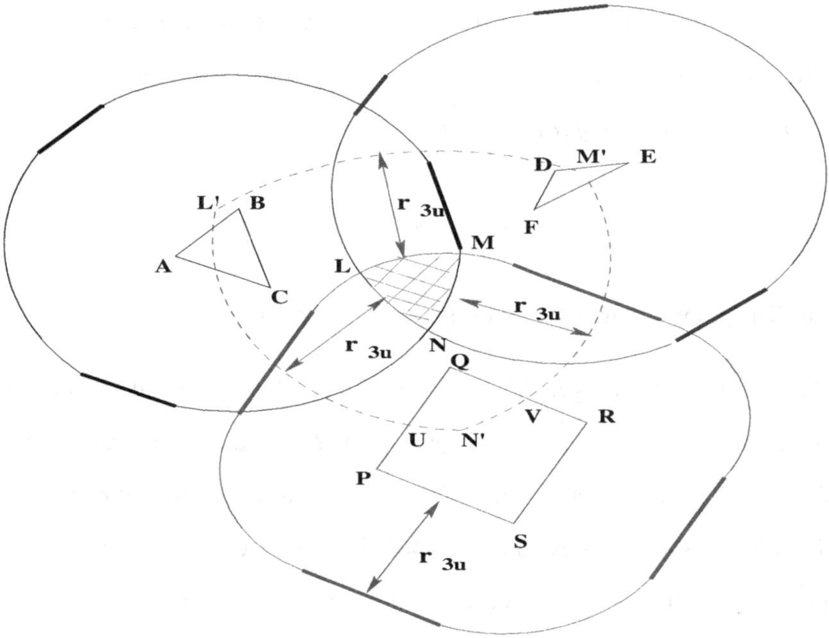

FIGURE 4.9: Improving minimum regions of residence of neighbors

range from k to u. The region defined by the dotted lines, $L'M'N'$ defines the viewed region of residence, R_{ku} of k by u. The region $UN'VQ$ is the intersection region of the current minimum region of residence of k, $PQRS$ and $L'M'N'$. Following Theorem 11, the new minimum region of residence of node k is the region $UN'VQ$. Node u tries to similarly improve the regions of nodes i and j using the measured ranges r_{1u} and r_{2u} and its minimum region of residence, $\mathcal{R}_u$.

A careful scrutiny of Figure 4.9 in the above example reveals that the part of the boundary of the region of residence of k which causes a computation (improvement) in the region of residence for u and the part of the boundary of the region of residence for k which is refined (improved) due to this computed part of the region of residence for u, are mutually disjoint. This observation holds even if the node k had an initial region of residence of a different shape.

Lemma 14. *Given a minimum region of residence of a node u, $\mathcal{R}_u$ and a measured* *range r_{uv} from a neighbor v, the improved minimum region of residence of v, $\mathcal{R}'_v$ is given by the intersection of the viewed region of residence of v by u, R_{vu} and the current minimum region of residence of v, $\mathcal{R}_v$.*

$$\mathcal{R}'_v = \mathcal{R}_v \cap R_{vu}. \tag{4.10}$$

Proof. The proof follows directly from Lemma 13 and Theorem 11. □

From Lemma 14, we get, $R'_v \subseteq R_v$. Note that R'_v is generated by introducing some extra arc and/or straight line segments on R_v due to the region computation initiated by the node u. Let us denote this set of new arcs and straight line segments by E^v_u. Elements of E^v_u are parallel to some boundary edge of R_u or a circular arc of a circle with radius r_{vu} centered at some vertex on the boundary of R_u.

Suppose there is a path in the network starting from a node u_0 to some node u_k, given by $u_0\ u_1\ u_2\ \cdots u_k$. If u_0 initiates its region computation by its neighbors and gets its region computed as R_{u_0}, then R_{u_0} may cause an improvement in the region of node u_1. This, in turn, may cause an improvement in the region of the node u_2, and so on, so that the process of region refinements may successively follow through the nodes $u_1, u_2, \cdots, u_k$. In particular, if we assume that $u_k = u_0$, i.e., $u_0\ u_1\ u_2\ \cdots u_k$ is a cycle, then it can be claimed that this process of successive region refinements will not be able to further refine the region R_{u_0} of u_0 after a finite number of steps. To establish this claim, we proceed as follows.

Let $E^{u_1}_{u_0}$ denote the set of newly introduced lines and/or arcs on the boundary of the minimum region of residence of u_1 due to the region computation of u_0 caused by all the immediate neighbors of u_0. The changed region R'_{u_1} of u_1 due to $E^{u_1}_{u_0}$ may cause a change in R_{u_2} by introducing some new lines and/or arcs which we denote by the set $E^{u_2}_{u_0,\ u_1}$. In general, we denote the set of newly added lines and/or arcs in the region of u_j, $1 \leq j \leq k$, by $E^{u_j}_{u_0,\ u_1\ \dots u_{j-1}}$. Because of the properties of Minkowski's sum of $R'_{u_{j-1}}$ and a circle of radius r_{u_{j-1},u_j} (range value between nodes u_{j-1} and u_j) with center at the origin, we see that for any j, $1 \leq j \leq k$, two possible cases may arise:

Case 1: A line segment (arc) in $E^{u_j}_{u_0,\, u_1\, ...u_{j-1}}$ is parallel to some line segment (arc) in $E^{u_{j-1}}_{u_0,\, u_1\, ...u_{j-2}}$ (for $j > 1$) or in R_{u_0} (for $j = 1$).

Case 2: An arc in $E^{u_j}_{u_0,\, u_1\, ...u_{j-1}}$ is

(a) not parallel to any arc in $E^{u_{j-1}}_{u_0,\, u_1\, ...u_{j-2}}$ (for $j > 1$) or in R_{u_0} (for $j = 1$),

(b) an arc of a circle with radius r_{u_{j-1},u_j}, having center at one point on the region $R'_{u_{j-1}}$ which is the point of intersection of two different arcs or two different line segments or an arc and a line segment, at least one of which must be in $E^{u_{j-1}}_{u_0,\, u_1\, ...u_{j-2}}$ (for $j > 1$) or in R_{u_0} (for $j = 1$).

This is illustrated in Figure 4.10 where the arcs α and β on $R'_{u_{j-1}}$ and R'_{u_j}, respectively, are parallel to each other, while the arc γ on R'_{u_j} is derived from the point Q on $R'_{u_{j-1}}$ (with Q as center and a radius equal to r_{u_{j-1},u_j}). We also see from Figure 4.10 that for every point on R'_{u_j}, there exists a unique point on $R'_{u_{j-1}}$ from which this point was derived. Thus, for the point T on R'_{u_j}, the corresponding point on $R'_{u_{j-1}}$ is Q, which is transitively derived from a point P on R_{u_0}.

Lemma 15. *Let T be any point on $E^{u_j}_{u_0,\, u_1\, ...u_{j-1}}$, and P be the corresponding point on R_{u_0} from which T was derived. The Euclidean distance PT is always greater than or equal to the maximum of all ($r_{u_{j-1},u_j}, \forall j, 1 \leq j \leq k$).*

Proof. We prove this by induction on j. The claim is trivially true for $j = 1$. Suppose the claim is true for $j = 1, 2, \cdots, j-1$. For $j \geq 1$, referring to Figure 4.10, the Euclidean distance PQ is then greater than or equal to the maximum of (r_{u_0,u_1}, r_{u_1,u_2},..., $r_{u_{j-2},u_{j-1}}$). If the point T is on an arc or line segment in $E^{u_j}_{u_0,\, u_1\, ...u_{j-1}}$ parallel to an arc or line segment in $E^{u_{j-1}}_{u_0,\, u_1\, ...u_{j-2}}$, the Euclidean distance PT =

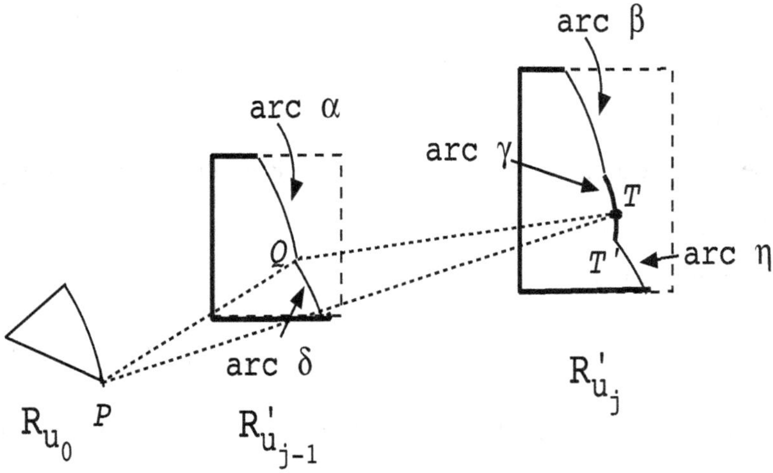

FIGURE 4.10: Refinements of regions of successive neighbor nodes

$PQ + QT$, from which the result follows as $QT = r_{u_{j-1},u_j}$. If, however, the point T is on an arc/line segment in $E^{u_j}_{u_0,\,u_1\,\ldots u_{j-1}}$, not parallel to any arc/line segment in $E^{u_{j-1}}_{u_0,\,u_1\,\ldots u_{j-2}}$, the corresponding point Q on $R'_{u_{j-1}}$ from which T is derived must be the point of intersection of two different arcs and/or line segments, as explained above. Without loss of generality, let Q be the point of intersection of two arcs α and δ in $E^{u_{j-1}}_{u_0,\,u_1\,\ldots u_{j-2}}$, as shown in Figure 4.10. The arc δ is mapped to the parallel arc η in $E^{u_j}_{u_0,\,u_1\,\ldots u_{j-1}}$. Let the arcs γ and η intersect at the point T'. Hence, the line QT' is normal to the tangent to the arc δ, and it follows that ΔPQT is an obtuse triangle with the obtuse angle at point Q. This implies that the distance PT is greater than of PQ or QT, and hence the lemma follows. □

Thus, we see that the part of the region boundary of a neighbor u_j of u_0, $\forall j, 1 \leq j \leq k$ which gets modified (refined) due to the region computation initiated by u_0 is always at a distance greater than or equal to r_{u_0,u_1} from the corresponding part of R_{u_0} which caused this refinement of R_{u_j}. Hence, we get the following important result which guarantees the termination of the successive refinement scheme.

Theorem 13. *If a node u initiates its region computation with the help of range readings from all of its neighbors, the computed region R_u of u may cause refinements of the successive neighbors through the whole network, but it will never be able to further refine R_u of u itself.*

Definition 3. The stable region of residence of a node u is the minimum region of residence of u which cannot be further improved upon using the current global set of range readings for all node pairs in the network. We denote such a region by $\mathcal{S}_u$.

Theorem 14. *A node u can compute its stable region of residence once it gets the range readings of all possible directly communicating nodes in the network along with the initial region information of all nodes.*

Proof. The minimum region of residence of a node u may be refined only when at least one of its neighbors, say v, is able to improve its region of residence. This, in turn, depends on whether any neighbor of v excluding u can improve its region of residence, and so on. The process can continue until all the nodes in the network are assessed for improvements in their respective regions of residences. Hence, if u has all the necessary information from all of these updates, u can eventually compute its stable region of residence. □

Theorem 15. *The computation of the stable regions of residence of all nodes in a network is functionally equivalent to an all-to-all broadcast of the range information of all node pairs in the network (the set RR) and the set Ref_n.*

Proof. To reconstruct the ad hoc network graph centrally, two pieces of information are required: i) the measured ranges of all node pairs and, ii) whether an individual node is a reference node or not. From Theorem 14, we see that if a node possessed the range values of all node pairs in the network (the set RR) and the set Ref_n, it could locally construct the network graph and then compute the stable regions of residence of all nodes. Since the possession of the set RR and the set Ref_n by a node in the network effectively implies a broadcast of these two sets, if every node were to locally compute the stable regions of residence, the problem maps to that of an all-to-all broadcast of the set RR_i and status (whether its a reference node) of each node i in the system. Each node, on receiving this information from a neighbor, would attach its own RR_i set and its status and broadcast the message again. □

4.5.2 Region Estimation Algorithm

Every node in a network maintains a local variable $status$ which is set to 1 if the node is a reference node and zero otherwise. Initially, the minimum regions of residence of all non-reference nodes are assumed to be infinity. Each node i performs ranging with its neighbors to obtain a set of measured ranges RR_i and i then computes the viewed region of residence for every node $j \in N(i)$. Node i then exchanges three pieces of information with each neighbor $j \in N(i)$: i) value of the status variable, $status_i$; ii) viewed region of residence of j, R_{ji} and iii) area of the current minimum region of residence of i, A_i.

Let $RR_i = \{r_{ij} : j \in N(i)\}$, $\mathcal{T}$ represent a set of viewed regions of residence R_{ij}. Once node i has its viewed region of residence R_{ij} from its every neighbor j, it computes its current minimum region of residence using the following algorithm:

Function compute_region: Boolean

var A_{old}, A_i : **Real**;

begin

  for each $R_{ij} \in \mathcal{T}$ such that $A_j \neq \infty$ **do**

    /* Compute the minimum region of residence of node i from the viewed regions */

    $\mathcal{R}_i \leftarrow \mathcal{R}_i \cap R_{ij}$;

  endfor;

  $A_i \leftarrow$ Area of $\mathcal{R}_i$;

  if $A_i < A_{old}$ **then return** `true`;   /* $\mathcal{R}_i$ improved */

```
    else return false;    /* No improvement in 𝓡_i */
end.
```

Once node i computes its minimum region of residence, it tries to improve the minimum region of residence of each of its neighbors as follows:

Procedure improve_region

begin

  for each $j \epsilon N(i)$ such that $status_j \neq 1$ **do**

    /* Construct R_{ji}, the viewed region of j */

    $R_{ji} = \mathcal{R}_i \oplus r_{ij}$;

    Transmit R_{ji} to node j;

  endfor;

end.

The following location identification algorithm is executed by each node i until the node attains its stable region of residence $\mathcal{S}_i$.

Algorithm location_region_identify

var $region\_change\_flag$: **Boolean**;

begin

  while (`true`)

    Get neighbor set $N(i)$;

    Generate RR_i : measure range with every neighbor $j \epsilon N(i)$;

    $region\_change\_flag$ = `false`;

    repeat

      Get $\mathcal{T}$: viewed regions of residence R_{ij} from every neighbor j;

      if $status_i = 0$ **then** $region\_change\_flag$ = compute_region(i, $\mathcal{T}$);

      improve_region(i, $N(i)$, RR_i);

until $region\_change\_flag$ = `false`; /* iterate until $\mathcal{R}_i = \mathcal{S}_i$ */

endwhile;

end.

4.5.3 Complexity Analysis

The operations *compute_region* and *improve_region* both need $O(\Delta)$ time in the worst case, where Δ is the maximum node degree in the network. Also, the operation *Get* $\mathcal{T}$ needs $O(\Delta)$ time. If we assign a distinct time slot to each node depending on its unique *id* number [15] to avoid collision, each iteration of the *repeat* loop would need $O(n)$ time slots. From Theorem 15, the algorithm terminates when an all-to-all broadcast of the sets RR and Ref_n is achieved. Assuming that the effect of the viewed range information of all node pairs can be transmitted in $O(1)$ time slots, the whole communication process will be completed in $O(D)$ rounds, where D is the diameter of the network (each round consists of $O(n)$ time slots for message communication from all nodes in a layer to the nodes in the next layer [15]). Hence, the algorithm *location_region_identify* will need $O(nD)$ time.

4.5.4 Simulation Results

The goal of the simulation studies performed by the authors of [6] was to evaluate the performance of the proposed algorithm *location_region_identify* in determining the stable regions of residence of all nodes in an ad hoc network. They experimented with various graph topologies by randomly generating static ad hoc graphs. A small percentage of the nodes in the graph were randomly chosen to be reference nodes that were assumed to have point locations of zero area. All other nodes were assumed to have initial regions of residence of infinite area. All communications were assumed to be symmetrical. A transmission range of 30 units of distance was assumed for all nodes in the system. All range estimates were assumed to be erroneous due to NLOS propagation errors. All errors were assumed to be positive, following Equation (4.1). An environment was considered with a mix of LOS and obstructed LOS (OLOS) signals. OLOS signals represent the scenario where no LOS communication is present between two nodes and hence all received communication between them is through NLOS propagation. The LOS range errors are drawn from a Gaussian distribution of mean μ and standard deviation σ, while the OLOS range errors are drawn from an exponential distribution with parameter λ [6]. The parameters μ, σ and λ characterize the channel characteristics and hence the magnitude of ranging errors. The performance of the above algorithm was evaluated with varying channel characteristics, percentage of LOS/OLOS signals and percentage of reference nodes in the system. For each set of parameters, the experiments were repeated for 100 randomly generated graphs and the areas of the stable regions of residence of all nodes in the system were measured. Figures 4.11 to 4.14 demonstrate the results of the simulation for random graphs of 100 nodes. For example, with $\lambda = 5.3$, $\mu = 0.0$, $\sigma = 0.3$, 80% LOS signals and 25% of the nodes designated as RNs, a median area of 1.98 sq. units and a mean area of 7.008 sq. units were obtained. Figures 4.11(a) and 4.11(b) illustrate the variation of the median and mean areas of the stable regions of residence (in square units) respectively, against λ. The median area was chosen to measure as it is less affected by outlier cases. The mean area, on the other hand, may be significantly affected by situations such as a non-reference node i having only one neighbor (say j), in which case $R_i = R_{ij}$ and the area of R_{ij} can be very small or large, depending on the ranged distance between the two nodes. Figures 4.12(a) and 4.12(b) show the variations of median and mean areas of stable regions of residence against σ. Figures 4.13(a) and 4.13(b) demonstrate that the median and mean areas of the stable regions of residence decrease exponentially with increases in percentages of LOS signals.

Similar behavior is observed by varying the number of reference nodes, as seen from Figures 4.14(a) and 4.14(b).

The regions of residence of every node in the system were approximated at every step of the algorithm with circumscribing convex polygons for the ease of implementation. Thus, the true areas of the stable regions of residence were smaller than those presented in this chapter.

From the graphs it appears that a small increase in the number of reference nodes would significantly improve the computed regions of residences for any given scenario. Also, using different technologies such as ultra-wideband (UWB), which would generally increase the percentage of observed LOS signals can also lead to significant improvement in the regions computed by the algorithm as seen from Figures 4.13(a) and 4.13(b).

Table 4.1 shows how the median areas of the stable regions of residence are affected by the node density measured in number of nodes per square unit. The median areas depicted are for $\lambda = 5.3$, $\mu = 0.0$, $\sigma = 0.03$, 80% LOS signals and 25% of the nodes designated as RNs.

TABLE 4.1: Variation of median area of stable regions of residence with node density

Node density (nodes/sq. units)	% of RNs	Median area of stable region (sq. units)
0.01	25	1.980
0.0075	25	2.656
0.005	25	5.249
0.003	25	10.275

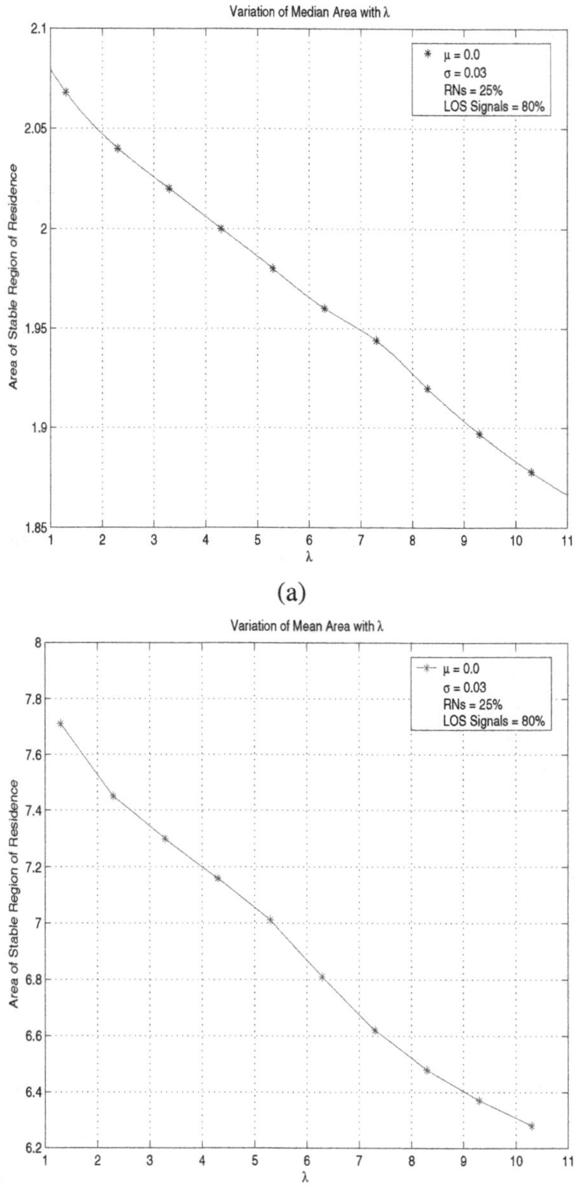

FIGURE 4.11: Variation of size of stable region of residence with λ

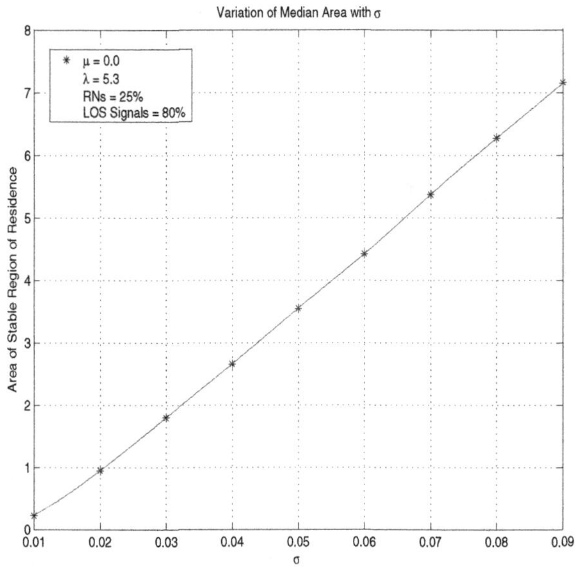

(a)

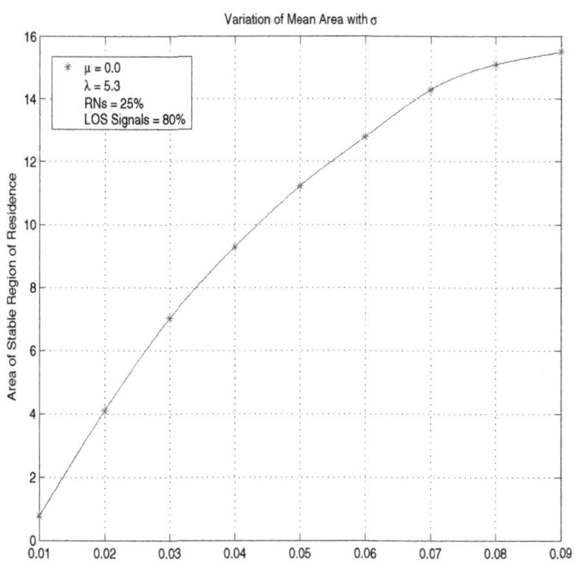

(b)

FIGURE 4.12: Variation of size of stable region of residence with σ

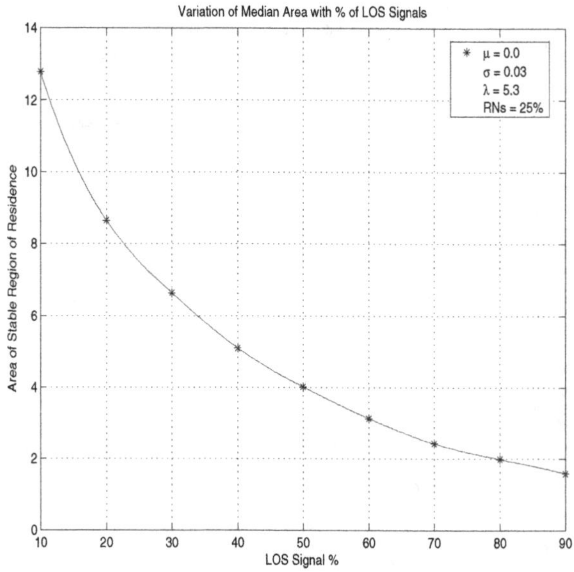

(a)

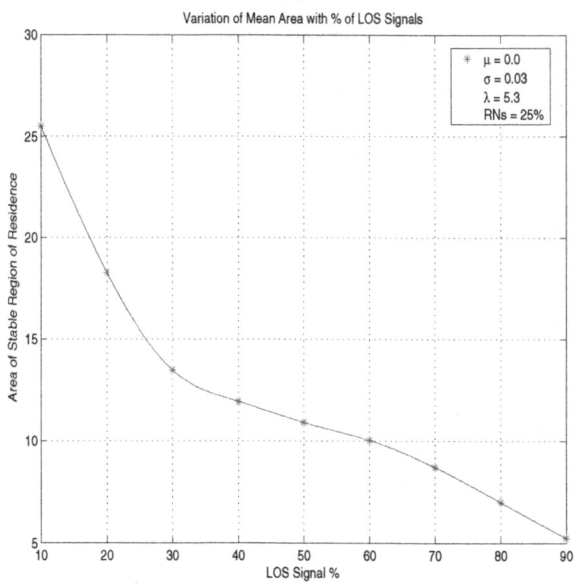

(b)

FIGURE 4.13: Variation of size of stable region of residence with percent of LOS signals

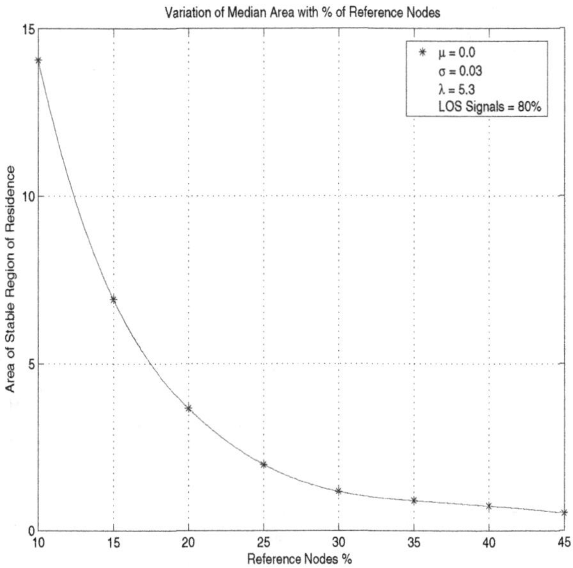

(a)

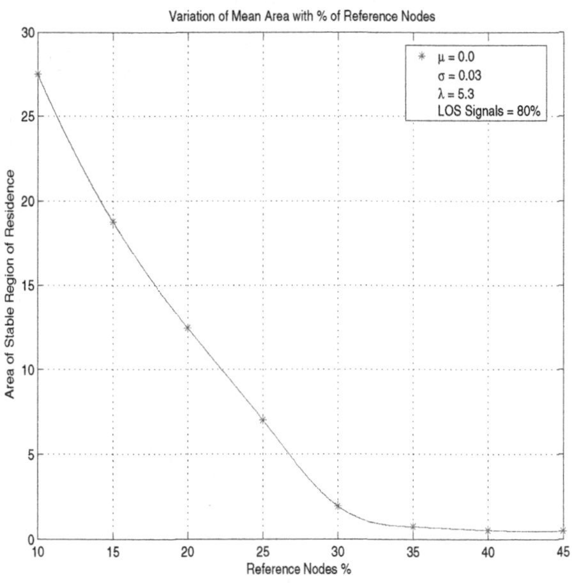

(b)

FIGURE 4.14: Variation of size of stable region of residence with percent of RNs

4.6 Exercises

1. Explain the method of triangulation used for finding the location of a node in the network on a two-dimensional plane. How many beacon nodes are needed in this technique? How can you extend this technique to find the location of a node in a three-dimensional space?

2. What are the possible sources of errors in finding the location of a node using triangulation?

3. Derive an expression for the error in the measured location of a node due to the errors in the beacon nodes.

4. Derive an expression for the error in the measured location of a node due to the errors in the measured range.

5. What do you recommend for the orientation of the beacon nodes to bound the error in finding the location of a node? Justify your answer with appropriate reasons. Also, derive the maximum possible error in finding the location with your recommended orientation.

6. Define the term *region residence* of a node in an ad hoc network.

7. What is meant by the *viewed region of residence* of a node v by another node u in the network?

8. Consider a rectangle of sides one unit and two units centered at the origin (0,0) and a circle of radius two units centered at the point (2,0). Find Minkowski's sum of this rectangle and the circle.

9. Describe a suitable beacon node selection algorithm which would minimize

the error in locating a node in an ad hoc network. Also, estimate the time complexity of your algorithm.

10. Describe an algorithm for range estimation of a mobile node and also compute the time complexity of your algorithm.

Bibliography

[1] M. P. Wylie-Green and S. S. (Peter) Wang, "Robust range estimation in the presence of the non-line-of-sight error," *Proc. 54th IEEE Vehi. Tech. Conf.*, vol. 1, pp. 101–105, 2001.

[2] J. J. Caffery and G. L. Stüber, "Subscriber location in CDMA cellular networks," *IEEE Trans. Vehi. Tech.*, vol 47, pp. 406–416, 1998.

[3] J. Hightower and G. Borriello, "Location systems for ubiquitous computing," *IEEE Computer*, vol. 34, No. 8, pp. 57–65, 2001.

[4] P. Bahl and V. Padmanabhan, "RADAR: an in-building RF-based user location and tracking system," *Proc. IEEE Infocom*, pp. 775–784, 2000.

[5] N. B. Priyantha, A. Chakraborty and H. Balakrishnan, "The cricket location-support system," *Proc. of MOBICOM 2000, ACM Press*, Boston, pp. 32–43, 2000.

[6] B. Alavi and K. Pahlavan, "Modeling of the distance error for indoor geolocation," *Proc. IEEE Wireless Commun. Networking Conf.*, vol. 1, pp. 668–772, 2003.

[7] G. Turin, "An introduction to matched filters," *IEEE Trans. on Information Theory*, vol. 6(3), pp. 311–329, 1960.

[8] R. E. Ziemer and W. H. Tranter, *Principles of Communications*, Wiley, New York, pp. 465–468, 1988.

[9] M. de Berg, M. van Kreveld, M. Overmars and O. Schwarzkopf, *Computational Geometry: Algorithms and Applications*. Springer, Heidelberg, 1997.

[10] K. Sinha and A. Datta Chowdhury, "A distributed location identification algorithm for ad hoc networks using computational geometric methods," *Proc. of 12th IEEE Intl. Conf. on High Performance Computing*, pp. 62–72, December 2005.

[11] K. Sinha and A. Datta Chowdhury, "A beacon selection algorithm for bounded error location estimation in ad hoc networks," *Proc. Intl. Conf. on Computing: Theory and Applications* IEEE Computer Society Press, pp. 87– 92, March 2007.

[12] S. Basagni, D. Bruschi and I. Chlamtac, "A mobility transparent deterministic broadcast mechanism for ad hoc networks," *IEEE Trans. Network.*, vol. 7, pp. 799–807, Dec. 1999.

Chapter 5

Message Communication in Ad Hoc Networks

5.1 Introduction

In this chapter, we discuss various communication protocols used in ad hoc networks (stationary or mobile) for different types of communication requirements, e.g., broadcast, gossiping and point-to-point communication among the nodes in the network. Broadcast is a process by which a source node transmits a message to be received by every node in the network. On the other hand, if every node has a different message to be sent to all other nodes in the network, the situation is called *gossiping* [1].

While discussing the protocols for all these communication requirements, we assume that the nodes in an ad hoc network all communicate over only one single channel. Under this assumption, it is likely that when one node is trying to send a message to its neighbors (nodes within hearing zone of the transmitting node), other nodes may also transmit through the common channel during the same time. This may lead to a *collision* of two or more messages at the receiving end. Because of mobility and limited battery power of the nodes, it is important to have fast, energy-efficient and deterministic (as opposed to randomized) broadcast and gossiping protocols for ad hoc networks by resolving collisions efficiently. Such protocols with efficient collision resolution mechanisms would provide low end-to-end latency and low overhead for retransmission of the collided messages. We first discuss the broadcast problem, followed by descriptions of gossiping and point-to-point routing protocols.

5.2 Broadcast in Ad Hoc Networks

Broadcast protocols in ad hoc networks may be centralized or distributed and may be based on deterministic or randomized algorithms. A broadcast protocol should have the following desirable properties:

- The protocol should be *mobility independent* or *mobility resilient* so that the broadcast is always completed correctly without knowledge of the identities of the neighbors of a node or the mobility of the nodes.
- It should be *deterministic* so that an upper bound on the broadcast completion time can be ascertained.
- It should be *distributed* rather than centralized so that the nodes can execute the protocol without *a priori* knowledge of the topology of the entire network. This property would successfully handle the dynamic network topology of an ad hoc network in a more efficient way than a centralized protocol.
- It should be *simple* so that the computational overhead at each node is minimized.

We first describe some deterministic distributed broadcast protocols for which we assume a system model as given below.

5.2.1 System Model

Following the ideas of [2, 3], let us model an ad hoc network consisting of n nodes by an undirected graph $G = (V, E)$, where $V = 0, 1, 2, \cdots, n-1$ is the set of vertices and E is the set of bidirectional edges. We assume that there exists an edge between two vertices if the corresponding nodes in the network are in the hearing zone of each other. Note that it is possible that a node u can hear some other node v, but v is unable to hear u for some reasons (e.g., transmission power of u may be too small, sensitivity of the receiver at node v may not be sufficiently high, etc.). In that case there will be only a directed edge from v to u, but no directed edge from u to v. We, however, exclude this possibility in this chapter unless stated otherwise.

We denote the set of neighbors of a node i by $N(i)$, the maximum degree of a node in the network by Δ, and the diameter of the network graph G by D. Thus, $D = max_{i,j \in V} d(i, j)$, where $d(i, j)$ is the distance between two nodes i and j of G.

There will be one distinguished node in the network called the source node s which carries the message to be broadcast to all other nodes in the network. S initiates the broadcast process. Any node $i, 0 \leq d(s, i) = \ell \leq D$ is said to belong to the layer ℓ of the network starting from the initiator source node s.

We assume that the time axis is divided into several *phases*, and each phase will consist of k rounds or time slots (numbered from 0 to $k - 1$), each round having the same time duration which is long enough to complete one hop of communication from a node to its neighbor. Phase 1 will start only after s initiates the broadcast process.

During any round, a node can be in the transmitting state (transmitting a message to its neighbors), or in the receiving state (ready to receive a message from any of its neighbors), or it may remain idle (neither transmitting nor receiving). However, a node cannot be in the transmitting state unless it receives some message correctly; and once it receives the message correctly, it will switch to the transmitting state or may remain idle during any round until the broadcast is completed. A node u can receive a message correctly during a round if and only if only one of the nodes belonging to $N(u)$ is transmitting the message during that round; if more than one node belonging to $N(u)$ is in the transmitting state during a round, u experiences a collision and no message can be received by u during that round.

We also assume in this model that all nodes are synchronized on a round-by-round basis. Every node will have a phase counter. When a node receives a message m, it waits for the beginning of the next phase when it sets its phase counter to 0. From that time onward, the phase counter is incremented by 1 at the end of every k rounds. Also, there will be a round counter at each node which will be incremented by 1 (modulo k) at the end of every round. We assume that the network always remains connected and the topology does not change during one broadcast process.

5.2.2 Deterministic Transmission Schedule

To have a distributed deterministic protocol for broadcasting a message originated at the source node s, we need to find a transmission schedule of every node i *a priori* irrespective of its current set of neighbors $N(i)$ at the moment, given the values of only n, Δ and the identifier (i) of the node. Note that the neighbor set $N(i)$ of a node i may change due to factors like mobility of the nodes, reduced battery power of the nodes and other network conditions. Despite these dynamic situations in the network, we need to find a transmission schedule for every node, with an ultimate upper bound on the number of phases p, each phase having an appropriate number of rounds k, so that the message is correctly broadcast among all nodes of the network. In other words, once we can give a protocol requiring at most p phases each of k rounds, the broadcast will be guaranteed to be completed in pk rounds of transmissions. Every node will execute some function, e.g., Get_Rounds to get the round numbers in a phase during which it should transmit, and will store these round numbers as its transmission schedule. After receiving a message m, a node should mechanically follow this computed transmission schedule to determine whether it should transmit the message or remain idle during any round in a given phase.

5.2.2.1 Collision Avoidance Approach

The simplest way of setting the transmission schedule of a node achieving the above goal can be as follows, where my_id is the identifier of the node [3]:

Function Get_Rounds (n, Δ) : integer

begin

return my_id

end

There will be a total of n rounds in every phase and only one node will be transmitting during any round. Hence, there will be no collisions and the broadcast will be completed in D phases. The upper bound on the number of rounds required for broadcasting by following this protocol will thus be nD. However, this simple protocol has the drawback that even when the maximum node degree Δ is very small (as is usually the case), only one node will be allowed to transmit in a round although it might have been possible to allow more than one node to transmit in a round without causing a collision. That is, for low values of Δ and large values of n, the number of rounds needed to complete the broadcast may be much smaller than nD.

5.2.2.2 Collision Resolution Approach

Another approach to compute the transmission schedule of a node which may lead to smaller number of rounds than nD to complete the broadcast, particularly with very large n and small Δ, may be to allow collisions to occur. When such a collision occurs at the receiving node, it should be detectable by that node so that the colliding messages are all discarded by the receiving node. Eventually, after the completion of all the rounds in a phase, it should be guaranteed that the message has been successfully broadcast from a node i to at least one of its neighbors, say the node $u \in N(i)$ that was yet to receive the message. This can be possible only when amongst all neighbors $N(u)$ of u, the node $i \in N(u)$ will be the only node in transmit mode during that round, so that u does not experience collision in that round with a message from any node other than $i \in N(u)$. Let us then pose the following question:

Is it possible to find some transmission schedule for a given node u so that during all the rounds in a phase, there will be at least one round in which only one node belonging to the set $N(u)$ will be in transmit state?.

The answer to the above question is "yes" by virtue of the results in [3, 2]. We describe below the protocol proposed in [2] and then compare its performance with that in [3]. To do this, we define a node as *covered* if it has already received the message, and *uncovered* if it is yet to receive the message. At any time instant, the sets of covered and uncovered neighbors of a node i will be denoted by $N_c(i)$ and $N_u(i)$, respectively. A set C of covered nodes will be called a **conflicting set** if there is at least one neighbor common to all the nodes in C that has not yet received the message. By following the protocol in [2], the broadcast proceeds in a layer-by-layer fashion, i.e., all nodes at layer $\ell, 1 \leq \ell \leq D-1$, become informed of the message m before any node at layer $\ell+1$. Also, the transmission slots for every node will be independent of the identities of the current neighbors of the node, i.e., independent of the topology of the network. These transmission slots will also be independent of the mobility of the nodes subject to the conditions that the topology does not change during the broadcast of a single message and the network always remains connected, as assumed in the system model mentioned above.

5.2.2.3 Successive Partitioning Scheme

The transmission slots of a given node i will be computed by using the successive partitioning scheme based on radix encoding of the node identifiers (*id*) in the set V. To simplify our discussions, let us assume a radix $r, r \geq 2$ and $n = r^m$ for some integer m. Each node $id \in V = \{0, 1, 2, \cdots, n-1\}$ is converted into an m-digit code in radix r number system as $id = d_{m-1}d_{m-2}\cdots d_0$, where $0 \leq d_i \leq r-1$ for all $i, 0 \leq i \leq m-1$. Consider the d_i values of each node for a specific i; the set V is partitioned into r disjoint blocks $B_i(j), 0 \leq j \leq r-1$, i.e., $d_i(id) = j \implies id \in B_i(j)$. Since there are m digits (m different values of i), we get m different partitions of V each into r different blocks, each partition induced by a digit of id. The total number of blocks generated by these partitions is thus equal to rm. A given id value is a member of exactly m of these rm blocks.

Definition 4. *The blocks $B_i(j), 0 \leq i \leq m-1, 0 \leq j \leq r-1$ will be called the* **blocks of level 1 partitioning** *and will be denoted by $P(1,k), k = 0, 1, \cdots, rm-1$, where $P(1,k) = B_i(j)$, if $k = ri + j$.*

Note that $P(1,k), 0 \leq k \leq rm-1$ of the above level 1 partitioning will have exactly n/r elements (nodes).

Example 20. *Let $n = 64$ and $r = 4$ (see Figure 5.1). Hence, $m = 3$, i.e., each of the 64 node id values will be encoded in a three-digit code in radix 4 system. The 0th digit (least significant digit) induces the partition of V in four blocks $B_0(0), B_0(1), B_0(2)$ and $B_0(3)$, where $B_0(0) =$ (0, 4, 8, 12, 16, 20, 24, 28, 32, 36, 40,44, 48, 52, 56, 60), $B_0(1) =$ (1, 5, 9, 13, 17, 21, 25, 29,33, 37, 41, 45, 49, 53, 57, 61), $B_0(2)$= (2, 6, 10, 14, 18, 22, 26, 30,34, 38, 42, 46, 50, 54, 58, 62) and $B_0(3) =$ (3, 7, 11, 15, 19, 23, 27, 31,35, 39, 43, 47, 51, 55, 59, 63). Similarly, blocks $B_1(0), B_1(1), B_1(2)$ and $B_1(3)$ (induced by the next significant digit position are given by $B_1(0) =$ (0, 1, 2, 3, 16, 17, 18, 19,32, 33, 34, 35, 48, 49, 50, 51),$B_1(1) =$ (4, 5, 6, 7, 20, 21, 22, 23,36, 37, 38, 39, 52, 53, 54, 55),$B_1(2) =$ (8, 9, 10, 11, 24, 25, 26, 27,40, 41, 42, 43, 56, 57, 58, 59) and $B_1(3)$ = (12, 13, 14, 15, 28, 29, 30, 31,44, 45, 46, 47, 60, 61, 62, 63). Likewise, $B_2(0) =$ $(0, 1, 2, \cdots, 15)$, $B_2(1) = (16, 17, 18, \cdots, 31)$, $B_2(2) = (32, 33, 34, \cdots, 47)$ and $B_2(3) = (48, 49, 50, \cdots, 63)$. Thus, the level 1 partitioning is computed as $P(1,0)$*

$= B_0(0)$, $P(1,1) = B_0(1)$, $P(1,2) = B_0(2)$, $P(1,3) = B_0(3)$, $P(1,4) = B_1(0)$, $P(1,5) = B_1(1)$, $P(1,6) = B_1(2)$, $P(1,7) = B_1(3)$, $P(1,8) = B_2(0)$, $P(1,9) = B_2(1)$, $P(1,10) = B_2(2)$, $P(1,11) = B_2(3)$.

Lemma 16. *For an arbitrary subset R of the set V, there exists at least one $P(1,k), 0 \leq k \leq rm-1$, which contains at most ($\lfloor |R|/2 \rfloor$ elements of R.*

Proof: Consider the m-digit radix-r codes of the elements in R. Since all these codes are distinct, there exists at least one digit position where there will be at least two different values v_1 and $v_2, 0 \leq v_1, v_2 \leq r-1$ in all these codes in R. This means that the elements of R will be placed in at least two different blocks of the partitions at level 1. One of these blocks must contain at most ($\lfloor |R|/2 \rfloor$ elements.

Lemma 17. *Each element in any arbitrary block $P(1,k), 0 \leq k \leq rm-1$ of level 1 partitioning can be uniquely re-coded by using only $m-1$ digits in the radix-r number system.*

Proof: Let $j = \lfloor k/r \rfloor$. It follows from the definition that the block $P(1,k)$ is induced by j^{th} digit, d_j, $0 \leq j \leq m-1$ of the radix-r codes for the elements in V. Consider an element $id \in P(1,k)$. Let the m-digit code for id be $d_{m-1}d_{m-2}\cdots d_{j+1}d_jd_{j-1}\cdots d_1d_0$. Since $\forall\ id \in P(1,k)$, the value of d_j is the same, it does not play any role in distinguishing the elements in $P(1,k)$. Thus, if we have a new $(m-1)$-digit code for each id in $P(1,k)$ given by $d_{m-1}d_{m-2}\cdots d_{j+1}d_{j-1}\cdots d_1d_0$ by removing the jth digit in each code, these $m-1$ digit codes are sufficient to uniquely identify the elements in $P(1,k)$. It follows from this process of re-coding that each element id in $P(1,k)$ is mapped to a unique element $id' \in V_1 = \{0,1,2,\cdots,n/r-1\}$ by the following one-to-one and onto mapping: $id' = (\lfloor \alpha/r^{j+1} \rfloor) \times r^j + \alpha\ MOD\ r^j$, where $\alpha = id - d_j \times r^j$.

Example 21. *Let $n = 64$ and $r = 4$ as in the previous example and consider the partition $P(1,7)$ given by $P(1,7) = (12,13,14,15,28,29,30,31,44,45,46,47,60,61,$

62, 63). *We have* $j = \lfloor 7/4 \rfloor = 1$. *Considering the* m*-digit radix-*r *codes of each of these ids in* $P(1,7)$, *we see that* $d_1 = 3$ *for all these ids. Hence, the set of* α *values for these elements is* $\{0,1,2,3,16,17,18,19,32,33,34,35,48,49,50,51\}$. *The set of* id' *values for these elements is* $\{0,1,2,3,4,5,6,7,8,9,10,11,12,13,14,15\}$, *which can be uniquely identified by 2-digit radix-4 codes as follows: 12: (0 0), 13: (0 1), 14: (0, 2), 15: (0 3), 28: (1 0), 29: (1 1), 30: (1 2), 31: (1 3), 44: (2 0), 45: (2 1), 46: (2 2), 47: (2 3), 60: (3 0), 61: (3 1), 62: (3 2), 63: (3 3).*

Consider now any one of the blocks $P(1,k), 0 \leq k \leq rm-1$ where each element (integer) has a unique radix-r code using $(m-1)$ digits $d'_{m-2}d'_{m-3}\cdots d'_1 d'_0$ obtained through the above process. We apply the partitioning process on each of these blocks using the digits similar to the level-1 partitioning we applied on the entire set V. We call this as **level 2 partitioning**.

Definition 5. *If we partition the elements of* $P(1,i), 0 \leq i \leq rm-1$ *into* r *blocks in* $(m-1)$ *different ways, each induced by a digit of the new* $(m-1)$*-digit code, the new blocks are called the blocks of level 2 partitioning. If a level 2 partitioning block is obtained from* $P(1,i)$ *induced by its* j^{th} *digit code* $d'_j, 0 \leq j \leq m-2$ *such that* $d'_j = \beta, 0 \leq \beta \leq r-1$, *we denote this block as* $P(2,k)$ *where* $k = r \times (m-1) \times i + r \times j + \beta$.

Example 22. *Consider the previous example again. We have* $r(m-1) = 8$. *Hence, the blocks from level 2 partitioning are given as* $P(2,56) = (12,28,44,60), P(2,57) = (13,29,45,61), P(2,58) = (14,30,46,62), P(2,59) = (15,31,47,63), P(2,60) = (12,13,14,15), P(2,61) = (28,29,30,31), P(2,62) = (44,45,46,47), P(2,63) = (60,61,62,63)$.

Remark 15. *Each block* $P(1,k)$ *of level 1 partitioning generates* $r(m-1)$ *blocks of level 2 partitioning. The total number of blocks in level 2 partitioning is given by* $r^2m(m-1)$.

Remark 16. *Each level 2 block,* $P(2,k), 0 \leq k \leq r^2m(m-1)-1$ *is of size* n/r^2.

Remark 17. *An arbitrary node* $id \in V$ *will belong to exactly* $m(m-1)$ *blocks of level 2 partition.*

Lemma 18. *Given any subset* R *of* V *such that* $|R| \leq 2n/r$, *there exists a block* $P(2,k), 0 \leq k \leq r^2m(m-1)-1$ *which contains at most* $\lfloor |R|/4 \rfloor$ *elements of* R.

.

Proof. By Lemma 16, there exists a partition $P(1,k), 0 \leq k \leq rm-1$ which contains at most $\lfloor |R|/2 \rfloor \leq n/r$ elements of R. This level 1 block $P(1,k)$ is now subjected to level 2 partitioning and by similar arguments as in the proof of Lemma 16, we get a level 2 block of size $\lfloor |R|/4 \rfloor$ at most. □

The partitioning process can now be generalized. We generate blocks of level $(i+1)$ partition from the blocks of level i partition, $i \geq 1$, as follows. For each element $id \in P(i,k)$, we compute $j = \lfloor k/r \rfloor$ and $\alpha = id - d_j \times r^j$, where $d_{m-i-1} \cdots d_0$ is the $(m-i)$-digit radix-r code of the element id. We then map id to $id' = (\lfloor \alpha/r^{j+1} \rfloor) \times r^j + \alpha\ MOD\ r^j$, so that all the elements in $P(i,k)$ are now mapped to a set of integers $\{0, 1, \cdots, r^{m-i}-1\}$ via a one-to-one and onto mapping. These id' values, corresponding to the elements of $P(i,k)$, are used to re-code the elements of $P(i,k)$ into $(m-i-1$-digit radix-r numbers to generate the blocks of level $(i+1)$ partition. This successive partitioning scheme is illustrated by a *partition tree* as in Figure 5.1 for $r = 4$ and $n = 64$ where V is the root (assumed to be at level $i = 0$) with blocks $P(1,k)$ as its children, and at every successive level i, $P(i,k')$ is a child of $P(i-1,k), 1 < i \leq m$, iff $P(i,k')$ is obtained from the block $P(i-1,k)$ by the above process. At level i, we get the blocks $P(i,k), 0 \leq k \leq r^i \times m(m-1)(m-2)\cdots(m-i+1)-1$ corresponding to level i partitioning. As before, blocks $P(i+1,k')$ obtained from $P(i,k), i \geq 1$, are numbered with k' from $r(m-i)k$ to $r(m-i)(k+1)-1$.

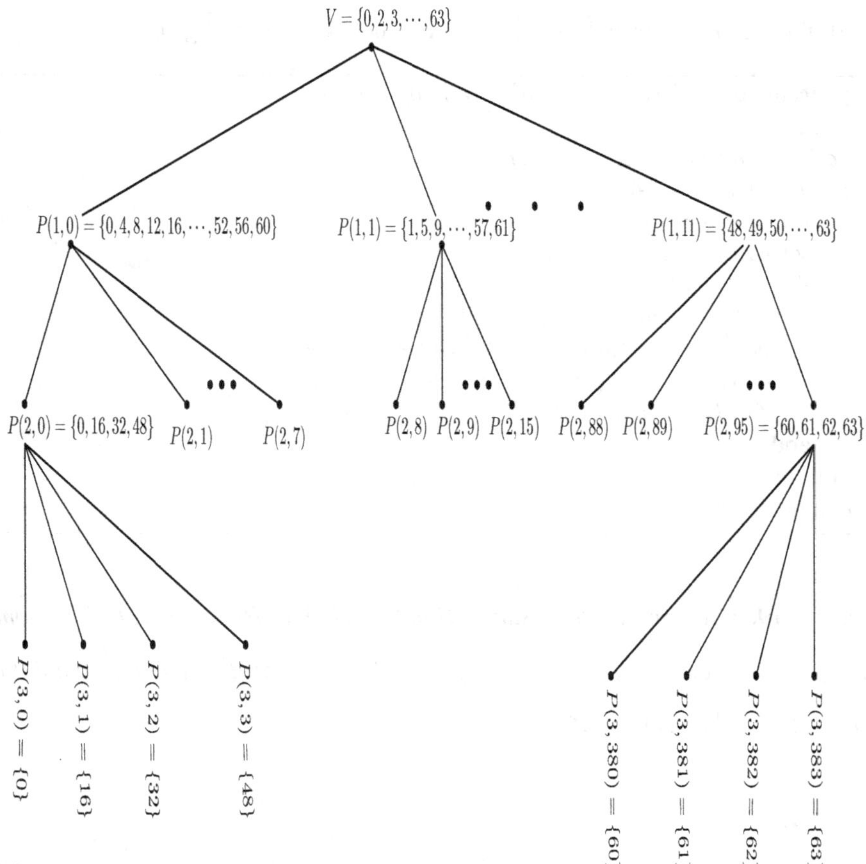

FIGURE 5.1: Partition tree for $n = 64$ and $r = 4$

The partitioning process is formalized by the $find\_partition$ procedure as given below, where we use a simple primitive procedure $radix\_code(value, q, r, D)$ in which $value, q$ and r are integer inputs, and D is a q-dimensional output array of integers. The $radix\_code(value, q, r, D)$ procedure converts the integer $value$, $(0 \leq value \leq r^q - 1$ to a q-digit radix-r number and stores the digits in locations $D(0)$ through $D(q)$ (from least significant to most significant digit).

```
procedure find_partition(value, i, h, offset, Flag);
begin;
radix_code(value, m − i + 1, r, D);
for j = 0 to m − i do
    temp ← r × (j − 1) + D(j);
    set Flag(i, offset + temp) ← 1;
    if i ≤ h − 1 then
        value ← value − D(j) × r^j;
        value ← (valueDIVr^(j+1)) × r^j + valueMODr^j;
        offset ← r × (m − i) × temp;
        find_partition(value, i + 1, h, offset, Flag);
    end if
end for
end procedure;
```

Lemma 19. *Given an arbitrary subset R of V such that* $|R| \leq n \times (2/r)^i$ *for some* $i, i \geq 0$, *there exists a block* $P(i+1, k)$ *at level* $(i+1)$ *partitioning, that contains at most* $\lfloor |R|/2^{i+1} \rfloor$ *elements of R.*

Proof. Since $|R| \leq n/(r/2)^i, \forall i \geq 0$, we can apply level 1 partitioning on elements of R to get at least one block, say R_s, of size $(n/2)/(r/2)^i$ at most. If $i \geq 1$, then $(n/2)/(r/2)^i \leq n/r$, (for $r \geq 2$) and hence we can apply level 2 partitioning on block R_s. We use induction on $j, 1 \leq j \leq i+1$ to show that it is possible to successively apply level 1 through level $(i+1)$ partitioning on the smallest block at each step. The induction hypothesis is true for $j = 1$ and $j = 2$. Assume that it holds up to level j partitioning. Hence, after level j partitioning, $(j < i+1)$, we get at least one block, say R'_s, which contains at most $\lfloor |R|/2^j \rfloor$ elements of R. Now $|R'_s| \leq 2^{i-j}n/r^i \leq n/r^j$, since $2^{i-j} \leq r^{i-j}$. It is now possible to apply level $(j+1)$ partitioning on R'_s. Thus, the induction hypothesis is verified to be true for all values of j up to $i+1$. □

Theorem 16. *Given any arbitrary subset R of V*, $2^h \leq |R| \leq 2^{h+1} - 1$, $h \geq 0$ *and* $n \geq r^h(2^{h+1} - 1)/2^h$, *there exists at least one block of level* j *partitioning*, $j \leq h$, *which contains only one element of R.*

Proof. It follows from the previous discussions that every time we apply the partitioning scheme corresponding to a specific level on a given set of elements, there will be at least one block with no more than half the number of elements in the given set. If $2^h \leq |R| \leq 2^{h+1} - 1$, it follows that h levels of partitioning are sufficient. From Lemma 19, we see that for h successive levels of partitioning to be applied to the smallest block derived from the previous partitioning step, $|R| \leq 2^h n/r^h$, i.e., $2^{h+1} - 1 \leq 2^h n/r^h$. Hence the theorem. □

Example 23. *Let* $R = \{0, 1, 2, 4, 5, 6, 16, 17, 18, 20, 22, 32, 33, 34, 36\}$ *with* $n = 64$ *and* $r = 4$. *Since there are* 15 *elements in R,* $h = 3$ *for this example. Applying level* 1 *partitioning, the partition blocks derived from* R *are* $(0, 4, 16, 20, 32, 36)$, $(1, 5, 17, 33)$, $(2, 6, 18, 22, 34)$; $(0, 1, 2, 16, 17, 18, 32, 33, 34)$, $(4, 5, 6, 20, 22, 36)$; $(0, 1, 2, 4, 5, 6)$,

$(16, 17, 18, 20, 22)$, $(32, 33, 34, 36)$. *The smallest cardinality of all these blocks is* 4. *Consider a smallest block, say* $(32, 33, 34, 36)$. *The elements in this block will first be mapped as* $32 \rightarrow 0$, $33 \rightarrow 1$, $34 \rightarrow 2$, *and* $36 \rightarrow 4$. *Then these will be recoded in radix-*4 *as* $32 : (00)$, $33 : (01)$, $34 : (02)$, $36 : (10)$. *Level* 2 *partitioning then generates the following blocks from* $(32, 33, 34, 36)$: $(32, 36), (33), (34); (32, 33, 34), (36)$. *Thus, another level of partitioning is not even necessary to generate a singleton block from R.*

5.3 Transmission Schedule

We use the successive partitioning scheme developed in the previous section to design the transmission schedule of an arbitrary node in an ad hoc network. Consider a node with identification number my_id; the node knows only its own identification number n (total number of nodes in the network) and Δ, the maximum node degree of the network. Given Δ, we compute h such that $2h \leq \Delta < 2h + 1$. The node then computes the blocks up to the level h partitioning, $P(h, k)$, $0 \leq k \leq r^h \frac{m!}{(m-h)!} - 1$. The details of the pseudo-code for the protocol to be used at each node to compute its own transmission slot or round are given in the algorithm below. The $find\_my\_slots$ and $compute\_slot$ procedures needed for this purpose are also given below.

/* **code to be executed by each node having** my_id **as identifier** */

```
h = ⌊log2 Δ⌋;
max_slot_number ← r^h × m × (m − 1) × ··· × (m − h + 1) − 1;
find_my_slots(my_id, n, Δ);
for phase_number = 1 to D − 1 do
    for i = 0 to max_slot_number do
        if T(i) = 1 then
            transmit the message;
        end if
    end for
end for
```

```
procedure find_my_slots(my_id, n, Δ);
begin;
h = ⌊log2 Δ⌋;
offset ← 0;
Flag ← 0;
T ← 0;
find_partition(my_id, 1, h, offset, Flag);
compute_slot(Flag, h, T);
end procedure;
```

```
procedure compute_slot(Flag, h, T);
begin;
for k = 0 to [r^h × m × (m − 1) × ··· × (m − h + 1) − 1] do
    if Flag(h, k) = 1 then
        T(k) ← 1;
    end if
end forend procedure
```

Note that $Flag[\,]$ is a two dimensional array and $T[\,]$ is a one-dimensional array; $Flag(i, j)$ is set to 1 iff my_id value of a node belongs to the block $P(i, j)$, and $T(k)$ is set to 1 iff $Flag(h, k)$ is set to 1. $Flag$ and T are both initialized to all zero values. Thus, in phase i, the node with identification my_id will transmit during a time slot or round k iff $Flag(i, my\_id) = 1$. The variable offset is used for properly numbering the partition blocks at any level.

Theorem 17. *Consider a node x which has not received the message at the beginning of a phase and at least one node y has the message among the set R of neighbors of node x. During this phase, there exists at least one time slot $T(k)$, $0 \le k \le r^h \frac{m!}{(m-h)!} - 1$, when exactly one of the informed neighbors of node x will transmit the message and node x will receive it correctly (without collision).*

Proof. Let R be the set of neighbors of node x that are informed (i.e., already have the message to transmit) at the beginning of the phase. Clearly, $|R| \le \Delta$. Consider the blocks $P(h, k), 0 \le k \le r^h \frac{m!}{(mh)!} - 1$, generated by level h partitioning of the vertex set V; by Theorem 16, there exists at least one k such that block $P(h, k)$ contains only one element of the set R. During time slot $T(k)$, exactly one node from the set R will transmit and node x will correctly receive the message. □

Consider the entire broadcast process. The source node s transmits the message. All its neighbors, set to *receive_mode*, receive the message from s. Suppose the phases start after this. That means all nodes of layer 1 received the message at the beginning of phase 1. These nodes will then be set to $transmit\_mode$ so that they now transmit the message in their respective time slots of phase 1 as specified above. Since any node in layer 2 will have at most Δ neighbors in layer 1, it follows that after the completion of phase 1 transmissions, all nodes in layer 2 will receive the message. This process continues for the successive layers so that at the beginning of phase i, $i \geq 2$, all nodes in layer i have received the message successfully and are ready to transmit the message during phase i at their respective time slots. By Theorem 17, it follows that all nodes in layer $i + 1$ will receive the message at the end of phase i. Thus, at the end of phase $\mathcal{D}1$, all nodes of the layer $\mathcal{D}$ of the network will receive the message and the broadcast is complete. Hence, we have the following result.

Theorem 18. *The broadcast process completes in* $(\mathcal{D} - 1)$ *phases consisting of a total of* $(\mathcal{D} - 1) \times \tau$ *rounds (where* $\tau = r^h \frac{m!}{(m-h)!}$*), or in* $O(\mathcal{D} r^h \log_r{}^h n)$ *time.*

Remark 18. *In light of the* $\Omega(\mathcal{D} \log n)$ *lower bound [5] for a deterministic distributed broadcast protocol in mobile multi-hop networks, the bound given in Theorem 18 is tight when* $h = 1$*, i.e., for networks with* $\Delta = 3$.

Remark 19. *It is to be noted that for radix* $r = 2$ *the frame length needed by the above protocol (Theorem 18) is exactly the same as that needed by the protocol Division of [3].*

Theorem 19. *The above broadcast protocol is resilient to node mobility as long as the network is not disconnected and at least one node from a conflicting set always remains in the hearing range of any neighboring uncovered node.*

Proof. By assumption, when the nodes are mobile, the network never gets disconnected, and at least one node from a conflicting set always remains in the hearing range of any neighboring uncovered node. Now, consider the situation after phase i, $i \geq 1$. At this stage, because of the layer-by-layer broadcast, the nodes in the conflicting sets constitute the last layer of nodes that have received the message. The neighboring uncovered nodes of each conflicting set will be members of the next layer of nodes. Hence the claim. □

The above protocol has the advantage of tuning the value of r to suit a particular value of Δ to reduce the number of rounds to complete the broadcast process. Table 5.2 shows the required number of rounds (τ) by this scheme for different values of n, r and Δ (the values of τ shown for $r = 2$ are also equal to the number of rounds needed by the *division* protocol [3]. The minimum value of τ for a given value of n and Δ (over possible choices of r) is shown by bold entries in the table. The minimum values that are smaller than the corresponding values of n are boxed by rectangles. We make the following observations:

- As noted in [3], the *division* protocol remains better than the *simple* protocol as long as $h < \log_2 n/(\log_2 \log_2 n + 1)$ (since for larger h values the frame length τ exceeds n); note that this bound on h is approximate and not exact. The above protocol remains better than the *simple* protocol as long as $h < \log_r n/(\log_r \log_r n + 1)$ (the bound is approximate, not exact) by suitably choosing r. For higher values of n, the range of h and hence that of Δ over which the protocol remains better than *simple* is larger in this protocol. For example, for $n = 16384$, and $\Delta = 15$ (corresponding to $h = 4$), the value of τ for the above protocol is $13440 (< n)$ (by choosing $r = 4$) compared to 17472 for the *division* protocol (corresponding to $r = 2$).

- For lower values of Δ (corresponding to when $h < \log_r n/(\log_r \log_r n + 1)$ after proper choice of r), the value of τ in the above protocol is always less than or equal to that in the *division* protocol and in most cases substantially less. For example, for $n = 256$ and $\Delta = 7$, the minimum τ is 192 (for $r = 4$) which is smaller than the corresponding value of 224 for the *division* protocol. This improvement is more prominent for larger values of n and Δ. For example, for

			values of τ for							values of τ for			
n	r	$\lceil \log_r n \rceil$	Δ=3	Δ=7	Δ=15	Δ=31	n	r	$\lceil \log_r n \rceil$	Δ=3	Δ=7	Δ=15	Δ=31
256	2	8	16	224	2688	26880	16384	2	14	28	728	17472	384384
256	3	6	18	270	3240	29160	16384	3	9	27	648	13608	244944
256	4	4	16	192	1536	6144	16384	4	7	28	672	13440	215040
256	5	4	20	300	3000	15000	16384	5	7	35	1050	26250	525000
256	6	4	24	432	5184	31104	16384	6	6	36	1080	25920	466560
256	7	3	21	294	2058	–	16384	7	5	35	980	20580	288120
256	8	3	24	384	3072	–	16384	8	5	40	1280	30720	491520
512	2	9	18	288	4032	48384	65536	2	16	32	960	26880	698880
512	3	6	18	270	3240	29160	65536	3	11	33	990	26730	641520
512	4	5	20	320	3840	30720	65536	4	8	32	896	21504	430080
512	5	4	20	300	3000	15000	65536	5	7	35	1050	26250	525000
512	6	4	24	432	5184	31104	65536	6	7	42	1512	45260	10888640
512	7	4	28	588	8232	57624	65536	7	6	42	1470	41160	864360
512	8	3	24	384	3072	–	65536	8	6	48	1920	61440	1474560
1024	2	10	20	360	5760	80640	4194304	2	22	44	1848	73920	2808960
1024	3	7	21	378	5670	68040	4194304	3	14	42	1638	58968	1945944
1024	4	5	20	320	3840	30720	4194304	4	11	44	1760	63360	2027520
1024	5	5	25	500	7500	75000	4194304	5	10	50	2250	90000	3150000
1024	6	4	24	432	5184	31104	4194304	6	9	54	2592	108864	3919104
1024	7	4	28	588	8232	57624	4194304	7	8	56	2744	115248	4033680
1024	8	4	32	768	12288	98304	4194304	8	8	64	3584	172032	6881280

FIGURE 5.2: Number of rounds τ in frame for different n, r, Δ

$n = 4194304$ and $\Delta = 31$, the minimum value of τ with the above protocol is 1945944 ($r = 3$), while the number of rounds needed with the *division* protocol is 2808960.

5.4 Slot Assignment Based on Location Information

We now introduce another strategy for assigning time slots to the nodes in an ad hoc network based on the location information of the individual nodes. The technique in [6] significantly improves slot utilization by an elegant technique of *re-using* the time slots by sufficiently distant nodes, avoiding collisions during transmission. The basic idea is that if the number of nodes in an ad hoc network is very large and spread over a large geographical region and we follow the technique of determining the transmission schedule as discussed in the preceding section, the number of transmission cycles needed for a successful broadcast of a message may be very large. Instead, we can first partition the deployment zone into regular hexagonal cells similar to the cellular networks with the assumption that *a given node in some cell i may hear from the nodes which lie in the same cell or any other cell j which is at Distance 1 from cell i (i.e., adjacent to cell i), but not beyond that.* Thus the nodes in some cell i cannot hear anything from any node in some cell at Distance 2 from cell i. Note that with this assumption, a node p lying in some cell i may be in the *hearing zones* of two nodes q and r lying in cells j and k, respectively, where j and k are both adjacent (intercell distance is equal to one) to the cell i, but they are Distance 2 apart from each other. Under such a situation, it is likely that nodes q and r are both trying to send messages to node p in the same transmission cycle, causing collision at node p. To avoid this situation while allowing a time slot used by a node in some cell to be re-used by some other node in some other cell, we put the restriction that *a time slot used by the nodes in some cell i may be re-used* only *by the nodes in other cells which are at least Distance 3 away from cell i.* Thus, basically, we define an *interference zone* of the nodes in a cell i as the region which spreads over only cells within Distance 2 from cell i. Note that although the hearing zone of a given cell i in our assumption extends up to cells at Distance 1 from cell i, the interference zone of cell i extends up to cells at Distance 2 from cell i. Hence, for scheduling the transmission slots, at a time we need to consider *only* the nodes lying in a cell and all other cells within Distance 2 from it, but not the entire set of nodes in the whole ad hoc network. This may result in using a much smaller number of communication slots for effecting broadcast, as discussed below.

Using the location information of the nodes, the number of active nodes and the actual *demand* of each *cell* at that instant of time are computed. We use this cell demand information to assign time slots to each mobile node by a clever *re-use* of the time slots by cells lying farther than Distance 2 as explained above, by exploiting the hexagonal symmetry of the imposed cellular structure and avoiding interference among the nodes. The technique described below ensures an *optimal collision-free* assignment for every node of the network in $O(m)$ time, m being the number of nodes in the network. This problem of finding an *optimal* time slot assignment schedule for the ad hoc network is called the *slot assignment problem* (SAP). The slot assignment algorithm presented here supersedes the existing algorithms [3, 7] with respect to optimality, and requires $O(\log\ \log\ m + mD + n)$ time to determine an optimal, collision-free slot assignment schedule for the entire network where n is the number of cells in the overlaid cellular graph and D is the diameter of the ad hoc network.

Mobility of the nodes is also handled by invoking the assignment algorithm whenever a node moves from one cell to an adjacent cell. Appropriate protocols for identifying such a situation through the use of special *control slots* and broadcasting the *id* of the leader of every cell to all nodes within that cell during these control slots, have been presented.

To describe this technique, we first present the assumed system model, followed by some preliminaries.

5.4.1 System Model

We assume the pre-existence of a partitioning of the MANET deployment area into a number of disjoint hexagonal cells, i.e., a hexagonal cellular network is overlayed on the MANET deployment area. To implement this, the nodes in the network are assumed to possess location information; they are either GPS enabled or are able to determine their locations relative to the deployment zone through some location discovery technique. A mapping is then used to convert the location information (geographical coordinates) to hexagonal grid cells [8, 9]. The nodes need to be synchronized in time. GPS can provide highly accurate and synchronized global time and accurate location information.

5.4.2 Preliminaries

We first consider the static model of the slot assignment problem, where the number of slots required for each cell is known *a priori*. The available time space is partitioned into equal length time slots and the slots are numbered as $0, 1, 2, \ldots$ from the lower end. The interference between two assigned time slots is represented in the form of *co-slot constraints* that ensure the same slot cannot be assigned to certain pairs of cells simultaneously.

We consider a *2-cell buffering* slot assignment problem (similar to two-band buffering discussed in Chapter 3) for a hexagonal cellular network overlaid on an ad hoc network, in which a slot can be reassigned to a cell more than Distance 2

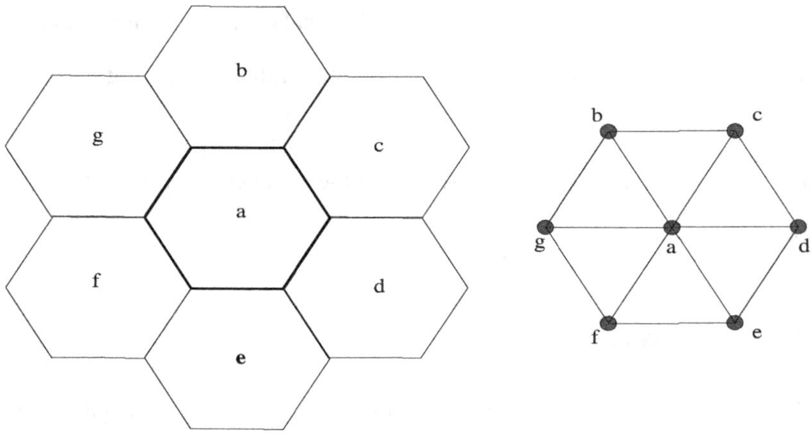

FIGURE 5.3: Conversion of hexagonal grid to hexagonal cellular graph

away. Following the notations used in Chapter 3, let s_0, s_1 and s_2 be the minimum slot separations between assigned slots in the same cell, at Distances 1 and 2 apart, respectively. In the present case of slot assignment in a *time-division multiplexing (TDM)* environment, $s_0 = s_1 = s_2 = 1$. A *cellular graph* is a graph $G = (V, E)$, where each cell of the hexagonal grid is represented by a node and an edge exists between two nodes if the corresponding cells are adjacent to each other, i.e., they share a common cell boundary. Cells i and j are Distance k apart if the minimum number of hops to reach node i from j in the graph G corresponding to the overlaying hexagonal cellular network is k. All edges are assumed to be symmetrical.

Figure 5.3 shows a cell a and its six adjacent cells. The diagram on the right models this scenario as a *hexagonal cellular graph* of seven nodes. The notation $N_i(u)$ denotes the set of all cells that are at a distance $\leq i$ from cell u.

Definition 6. *Suppose $G = (V, E)$ is a cellular graph. A subgraph $G' = (V', E')$ of G is said to be a Distance k clique if every pair of nodes in G' is connected in G by a path of length at most k and V' is maximal.*

Definition 7. *A Distance 2 clique of seven nodes in a hexagonal cellular network is defined as a complete Distance 2 clique. The node that is at a Distance 1 from all*

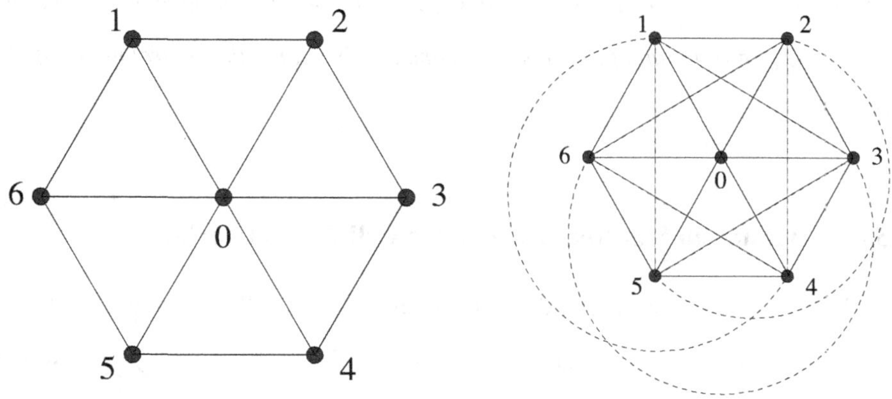

(a) A complete Distance 2 clique

(b) A cellular Distance 2 clique

FIGURE 5.4: Complete Distance 2 clique and corresponding cellular Distance 2 clique

other nodes in the complete Distance 2 clique is termed its central node or central cell and the remaining nodes are peripheral nodes or peripheral cells.

In a two-cell buffering environment, the co-slot interference may extend up to cells at Distance 2 apart. In view of this, we define a *cellular Distance 2 clique* as follows.

Definition 8. *A cellular Distance 2 clique* $G_2 = (V_2, E_2)$ *is a graph generated from a complete Distance 2 clique* G_1 *by adding edges to* G_1 *between every pair of nodes that are at Distance 2 in* G_1*.*

Figures 5.4(a) and 5.4(b) illustrate a complete Distance 2 clique and the corresponding cellular Distance 2 clique. Cell 0 is the central node of the graph. The dashed edges in the cellular Distance 2 clique are the edges joining the Distance 2 neighbors.

Definition 9. *If* G_1 *is a cellular Distance 2 clique with node* u *as the central node, then a cellular Distance 2 clique* G_2 *is said to be adjacent to* G_1 *iff, i)* u *is a periph-*

eral node of G_2, ii) the central node of G_2 is also a peripheral node of G_1, and iii) G_1 and G_2 have a total of four nodes in common, including the central nodes of G_1 and G_2.

5.4.3 Minimum Slot Requirement for Cellular Networks

Let $\mathcal{D}_7^{(2)}(G)$ be the sum of demands of all cells of a cellular Distance 2 clique, $G = (V, E)$, where the cardinality of V, $|V| \leq 7$. Then, $\mathcal{D}_7^{(2)}(G) = \sum_{i \in G} w_i$, where w_i is the demand from the cell i.

Definition 10. *A seven-node cellular Distance 2 clique G or its subgraph is called a critical block CB_7 which is composed of a maximum of seven cells, such that the sum of the demands of the cells in CB_7 is maximal over all possible cellular Distance 2 cliques in the network.*

We denote the demand of a critical block by $\mathcal{D}_7^{(2)^*}$. Thus $\mathcal{D}_7^{(2)^*} = \max_{\forall G} \mathcal{D}_7^{(2)}(G)$. Note that there may be multiple cellular Distance 2 cliques. We first consider the simpler case of homogeneous cell demand where all cells have the same demand.

5.4.4 Homogeneous Cell Demand

Let w represent the homogeneous demand for all cells in the network. For $w = 1$, the critical block demand $\mathcal{D}_7^{(2)^*}$ would be seven time slots. Referring to Figure 5.5(a), we see that due to structural symmetry, any Distance 2 clique can be chosen as the critical block. Without loss of generality, let the cellular Distance 2 clique $abcdefg$ be designated as the seven-node critical block, with node g as the central node. Considering now the cellular Distance 2 clique $gbpqrdc$, centered at c, we note that

1. Node p is at Distance 3 from nodes e and f of CB_7. Hence p can be assigned the same time slot as either of these two nodes.

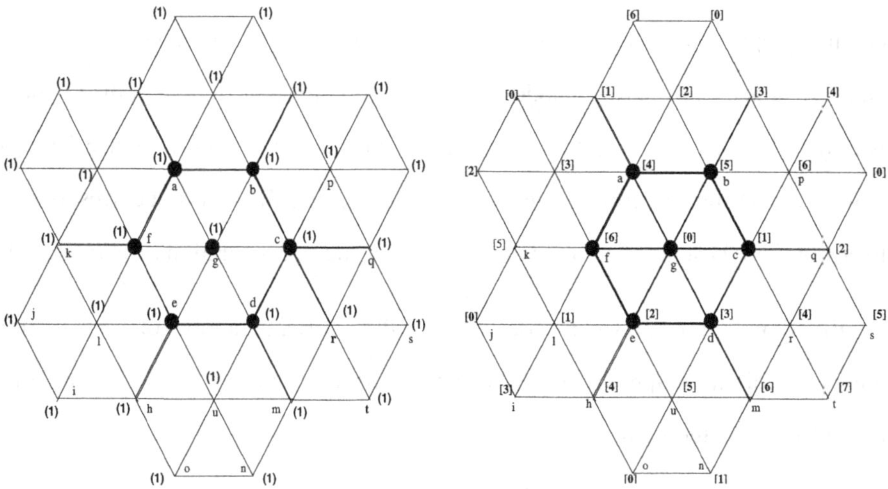

(a) Homogeneous demand of unit slot (b) An optimal assignment scheme

FIGURE 5.5: Slot assignment for a cellular graph with homogeneous unit demand

2. Node q is at Distance 3 from nodes a, f and e. Thus, q's demand can be satisfied from any of these nodes.

3. Nodes a and f are at Distance 3 from r.

A particular assignment may assign node p the same time slot as that of node e, while nodes q and r may be assigned the time slots of nodes f and a, respectively.

Thus, we find that the demand of the cellular Distance 2 clique $gbpqrdc$ can be satisfied completely by the time slots assigned to the critical block. Figure 5.5(b) depicts a possible assignment scheme for the cellular graph of Figure 5.5(a). We now state the following results.

Lemma 20. *For any given unsatisfied node u adjacent to one or more satisfied cellular Distance 2 cliques, it is always possible to find a satisfied node v at a Distance 3 from u such that the slot assigned to v is unused within Distance 2 of u.*

Proof. By definition, a cellular Distance 2 clique is a seven-node subgraph of the cellular graph such that no node in the clique is more than Distance 2 from one another. Hence, considering a node u that is adjacent to some node v of the clique, it is easy to see that there must be at least two nodes in the cellular Distance 2 clique that are at a Distance 3 from u. Hence, u can be assigned the time slot corresponding to any of these Distance 3 nodes without violating the slot separation constraints. □

Based on Lemma 20 we now demonstrate the minimum number of slots requirement of a cellular graph with homogeneous demand of one slot per cell in the form of the following theorem.

Theorem 20. *The minimum slot requirement of a cellular graph with two-cell buffering having $s_0 = s_1 = s_2$ and a homogeneous demand of one time slot per cell is equal to the demand of the seven-node critical block of the graph.*

Proof. Let G^* be the critical block of the network with node g as the central node. Let G be a cellular Distance 2 clique adjacent to G^*, whose nodes have to be assigned time slots. As G is adjacent to G^*, the four common nodes of G and G^* have already assigned time slots. From Lemma 20, for the remaining nodes in G we observe that the symmetry of the cellular graph implies that for any node i in G, it is always possible to find at least one satisfied node j at Distance 3 from i such that the time slot assigned to j is unused within Distance 2 of node i. Hence, G can be satisfied using only the time slots from G^*.

The above procedure can be repeated for all cellular Distance 2 cliques adjacent to G^* to obtain a collision-free schedule using only seven time slots. By exploiting the symmetry of the cellular graph, this pattern can be repeated over the network by considering unsatisfied cellular Distance 2 cliques adjacent to each satisfied cellular Distance 2 clique. □

Corollary 2. *The optimal slot requirement for a cellular graph having a demand of unit slot per cell is seven time slots.*

We now present an algorithm that generates a regular slot assignment pattern for a hexagonal cellular graph with the optimal number of slots.

5.4.4.1 Algorithm homogeneous_unit_slot_assignment

Step 1: Assign slot number 0 to the center cell of the critical block.

Step 2: Assign slot number $i \bmod 7, i \geq 1$ to the ith cell to the right of the center cell along a particular direction, say, along the horizontal line as shown in Figure 5.6. That is, starting from the center cell, we assign the increasing order slot numbers $0, 1, 2, \cdots, 6$ repeatedly to the cells to the right of the center cell.

Step 3: Assign slot number $-i \bmod 7, i \geq 1$ to the ith cell to the left of the center cell. That is, we assign the decreasing order slot values $6, 5, \cdots, 0$, repeatedly to the cells to the left of the center cell.

Step 4: For horizontal rows below the center cell, shift the 0 slot value 3 cells to the *left* and then repeat Steps 2 and 3 to obtain a slot assignment for each such row.

Step 5: For the horizontal rows above the center cell, shift the 0 slot value 3 cells to the *right* and then repeat Steps 2 and 3 for each such row.

Figure 5.6 shows the assignment scheme obtained using the above *homogeneous_unit_slot_assignment* algorithm. The numbers within square braces indicate the slot assigned to a cell. For example, cell g is assigned slot 0, a is assigned slot 4, cell p is assigned slot 6 and so on. The cellular Distance 2 clique $abcdefg$ was chosen as the critical block with g as the central node.

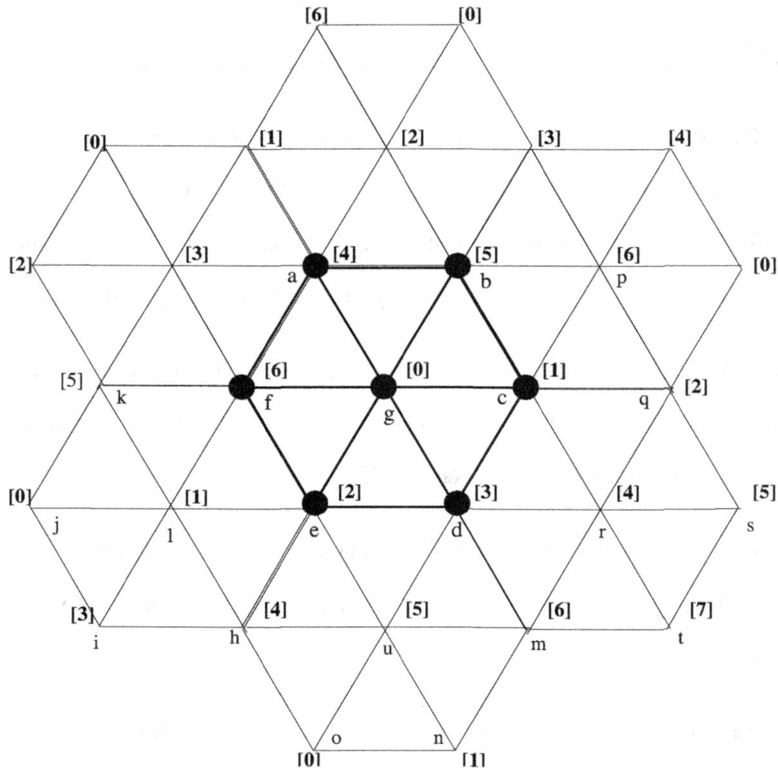

FIGURE 5.6: Optimal slot assignment schedule for homogeneous demand of unit slot

We now extend the results of homogeneous demand with $w = 1$ to the general case of $w > 1$ by simply assigning blocks of w consecutive slots to each node instead of a single slot, leading to the following result.

Lemma 21. *The optimal number of slots required for a cellular graph with homogeneous demand of w slots per cell is $7w$ time slots.*

For all positive and negative integer values (including 0) of m and n, we define the operation $(m, n) \bmod k$ as returning the slot numbers starting from $m \bmod k$ to $n \bmod k$, (including both m and n). The algorithm to handle w slots per cell demand is presented below; it uses only the optimal number of required slots.

5.4.4.2 Algorithm homogeneous_slot_assignment

Step 1: Assign slot numbers $(0, w-1)$ to the central cell of the critical block.

Step 2: Assign slot numbers $(iw, (i+1)w-1) \bmod 7w$, $i \geq 1$ to the ith cell to the right of the central cell along a particular direction, say along the horizontal line as shown in Figure 5.5(b). That is, we assign the increasing order slot numbers $(0, w-1)$, $(w, 2w-1)$, ..., $(6w, 7w-1)$ repeatedly to the cells to the right of the central cell along the horizontal direction.

Step 3: Assign slot numbers $(-iw, -(i-1)w-1)) \bmod 7w$, $i \geq 1$ to the ith cell to the left of the central cell. That is, we assign the decreasing order slot values $(7w-1, 6w)$, $(6w-1, 5w)$, ..., $(w-1, 0)$ repeatedly to the cells to the left of the central cell.

Step 4: For rows below the central cell, shift the $(0, w-1)$ slot value 3 cells to the *left* and then repeat Steps 2 and 3 to obtain a slot assignment for each such row.

Step 5: For rows above the central cell, shift the $(0, w-1)$ slot value 3 cells to the *right* and then repeat Steps 2 and 3 for each such row.

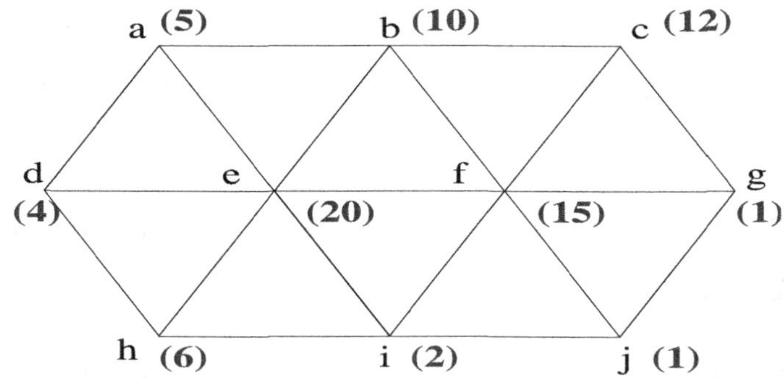

FIGURE 5.7: Heterogeneous demand. Seven-node CB fails to give minimum number of slots

5.4.5 Heterogeneous Cell Demand

We now consider the general case of SAP, where cells have different demands, i.e., $\exists\, w_i, w_j, i \neq j$, such that $w_i \neq w_j$. The seven-node critical block is insufficient to determine the optimal number of slots of the cellular graph, as demonstrated below.

Example 24. *Consider the cellular graph in Figure 5.7. The numbers in parentheses beside each cell denote the demand of the cell. The cellular Distance 2 clique $abfihde$ has a demand of 62 slots. The subgraphs $bcgjief$ and $abcfe$ have demands of 61 and 62 time slots, respectively. Thus, we see two candidate critical blocks in the network: subgraph $abfihde$ or subgraph $abcfe$. We arbitrarily choose $abfihde$ as a seven-node critical block. For the Distance 2 clique $bcgjief$ adjacent to the critical block, cells g and j can have their demands satisfied from the slots assigned to the cells a and d. However, the demand of cell c ($w_c = 12$) is greater than the slots assigned to its two Distance 3 neighbors, d and h of the seven-node critical block. The demand sum of cells d and h, $w_d + w_h = (4 + 6) < w_c = 12$. Hence, it is necessary to assign slots in addition to those assigned to the critical block to satisfy the demand of cell c. Thus, for heterogeneous demands, the seven-node critical block will not always give the optimal number of slots of the cellular network. Figure 5.8*

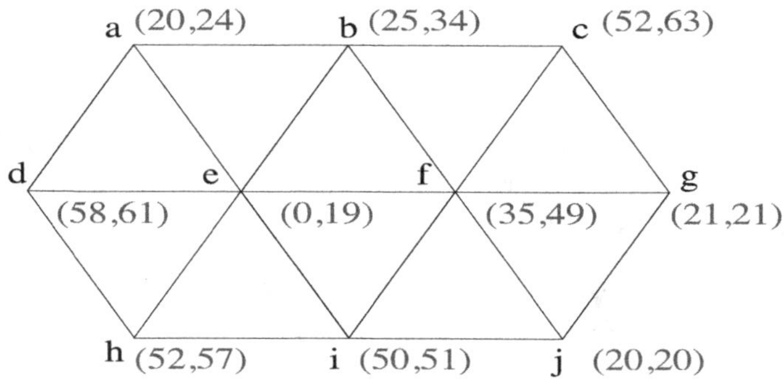

FIGURE 5.8: Assignment scheme requiring 64 slots

shows a possible slot assignment scheme for the graph in Figure 5.7. The two-tuple beside each cell denotes the slots assigned to that cell - (m, n) *indicates the slots in the range* m *to* n*, both inclusive.*

The seven-node critical block fails to give the optimal number of slots as it is possible for one of the nodes adjacent to a node of the critical block (but not a part of it), to have a demand that exceeds the sum of the demands of its Distance 3 neighbors in the critical block. From the cellular graph we see that for every peripheral node of the critical block, there are three neighbors at Distance 3 from some other peripheral node of the critical block. Consider, for example, node f in Figure 5.7 with neighbors c, g and j. The slots assigned to node d can be re-used to satisfy the demands of all three nodes while the slots assigned to node a can be re-used to satisfy the demands of only j and g, and the slots assigned to node h can only be re-used to satisfy the demands of only c and g. Hence, each of these three neighbors is a potential source of excess demand over that of the $\mathcal{D}_7^{(2)^*}$, either individually or in combination with the others. This suggests that it is necessary to include these three nodes in computing the optimal number of slots. Using an 8- or 9-node critical block would also fail to obtain a lower bound on the number of slots for the same reasons as for a seven-node critical block. So, we consider a ten-node block consisting of a seven-node Distance

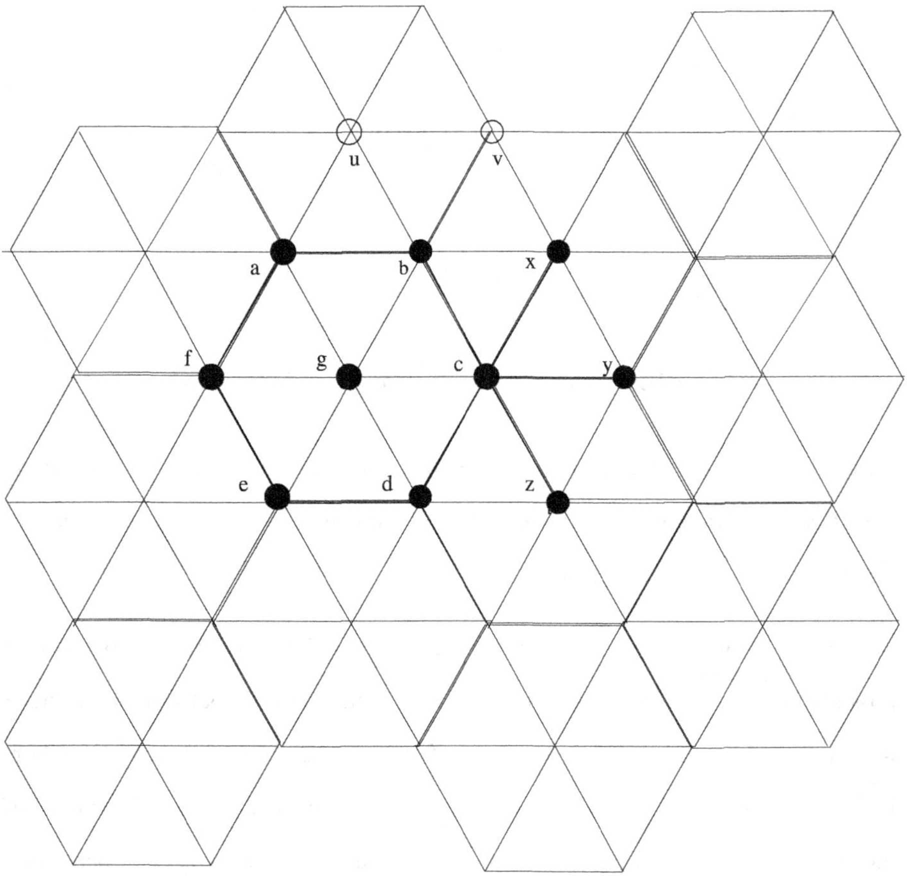

FIGURE 5.9: Ten-node critical block

2 clique and three other nodes outside this Distance 2 clique which are neighbors of a peripheral node of this Distance 2 clique. We thus get the following result.

Lemma 22. *For a cellular network with a heterogeneous demand vector, to find the optimal bandwidth requirement of the network, it is necessary to consider a ten-node critical block, as using a critical block with fewer than ten nodes would not be sufficient to compute the minimum slot requirement of the network.*

To compute the demand of the ten-node critical block for which the number of slots will be maximum among all such ten-node blocks, let $C = (V, E)$ be a cellular Distance 2 clique. Let $free_u$ denote the number of slots of node $u \in \mathcal{C}$ that can be used by a node which is at a Distance 3 from u and not a part of $\mathcal{C}$, and $used_u(j)$ be the number of slots assigned to $u \in \mathcal{C}$ that are re-used by node $j \not\in \mathcal{C}$ at a Distance 3 from u. Noting that $N_3(u)$ is the set of all Distance 3 neighbors of node u, we define *residual demand* res_j of node $j \in N_1(i)$, $i \in \mathcal{C}, j \not\in \mathcal{C}$ as $res_j = \max(0, w_j - \sum_{u \,\in\, N_3(j) \cap V} used_u(j))$. For $i \in \mathcal{C}$, the sum of residual demands of $N_1(i)$ which are not in $\mathcal{C}$ will be designated the *residual sum* of neighbors of i and defined as $Res_i = \sum_{j \,\in N_1(i), j \not\in \mathcal{C}} res_j$.

We demonstrate the procedure for computing the ten-node critical block with the help of the following example.

Example 25. *Consider the cellular graph shown in Figure 5.9. Let $abcdefg$ be a candidate critical block. Without any loss of generality, we consider the three neighbors x, y and z of node c. Initially, $free_a = w_a$, $free_f = w_f$ and $free_e = w_e$. The computation of Res_c would be as follows,*

Step 1: *Assign slots to node x using maximum number of slots from node e, and the rest, if any, from the node f.*

$used_e(x) = min(w_x, free_e);\ free_e = free_e - used_e(x).$

$used_f(x) = min(w_x - used_e(x), free_f);\ free_f = free_f - used_f(x).$

$res_x = \max(0, w_x - (used_e(x) + used_f(x))).$

Step 2: *Assign slots to node z using maximum number of slots from node a, and the rest, if any, from the node f.*

$used_a(z) = min(w_z, free_a);\ free_a = free_a - used_a(z).$

$used_f(z) = min(w_z - used_a(z), free_f);\ free_f = free_f - used_f(z).$

$res_z = \max(0, w_z - (used_a(z) + used_f(z))).$

Step 3: *Assign slots to y using available number of slots from nodes e, a and f.*

$$res_y = \max(0, w_y - (free_a + free_e + free_f)).$$

Step 4: *Sum the residual demands of x, y and z, i.e., $Res_c = res_x + res_y + res_z$.*

Let $Res_{max}(\mathcal{C}) = \max_{i \in \mathcal{C}} [Res_i]$. Referring to Figure 5.9, let $\mathcal{D}_{10}^{(2)}(\mathcal{G})$ represent the demand of the ten-node subgraph, $\mathcal{G} \equiv abcdefgxyz$, where $\mathcal{D}_{10}^{(2)}(\mathcal{G}) = \mathcal{D}_{7}^{(2)}(\mathcal{C}) + Res_{max}(\mathcal{C})$. The demand of the ten-node critical block, $\mathcal{D}_{10}^{(2)^*}$ is then defined as the demand of a ten-node subgraph that has the maximal $\mathcal{D}_{10}^{(2)}(\mathcal{G})$ in the network, i.e., $\mathcal{D}_{10}^{(2)^*} = \max_{\forall\ mathcalG}[\mathcal{D}_{10}^{(2)}(\mathcal{G})]$.

Let $\mathcal{R}_\mathcal{C}$ represent the set of nodes outside $\mathcal{C}$ but adjacent to some peripheral node of $\mathcal{C}$, corresponding to $Res_{max}(\mathcal{C})$. We call $\mathcal{R}_\mathcal{C}$ the *maximum residual set* of $\mathcal{C}$.

Theorem 21. *The demand sum $\mathcal{D}_{10}^{(2)^*}$ is the optimal bandwidth requirement of a hexagonal cellular network having a heterogeneous demand vector.*

Proof. We established from Lemma 22 that it is necessary to consider at least a ten-node critical block to compute the minimum slot requirement of a cellular network. We now prove that the demand of a ten-node critical block gives the minimal bandwidth requirement of a hexagonal cellular network with a heterogeneous demand vector.

The nodes adjacent to the critical block have the least number of contributing nodes in the critical block to satisfy their demands. The further away a node is from the critical block, the higher is the number of contributing nodes in the critical block to satisfy its demand. Hence, proving that the demand of any Distance 2 clique adjacent to the critical block can be satisfied using only the slots assigned to the critical block, implies that the demand of all other Distance 2 cliques in the cellular graph can be satisfied using only the slots assigned to the ten-node critical block.

Suppose the nodes 1 to 10 in Figure 5.10 form the ten-node critical block in a

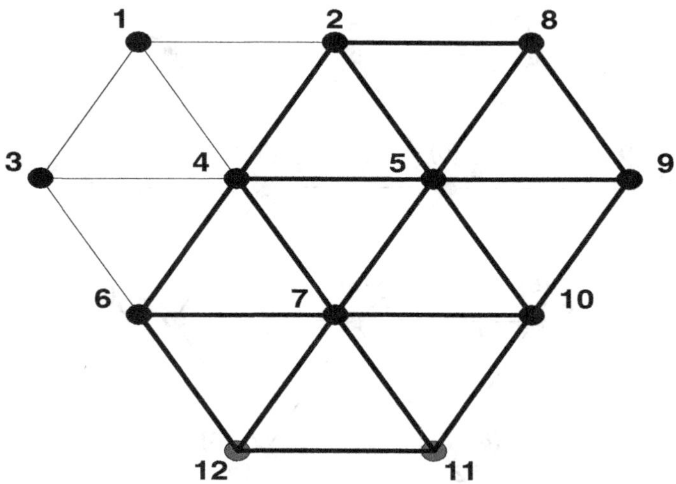

FIGURE 5.10: Actual critical block for Case 1

cellular graph. The nodes 8, 9 and 10 form the maximum residual set corresponding to the critical block. Without any loss of generality, let nodes 11 and 12 be adjacent to the critical block whose demands need to be satisfied.

Let S_i denote the set of slots allocated to node i.

Let N_i denote the set of extra slots allocated to node i after re-using the slots from the critical block.

The objective is to prove that the cardinality $|N_{11} \cup N_{12}|$ is always 0. Two scenarios can arise when attempting to assign slots to the nodes 11 and 12 while re-using the slots assigned to the critical block:

Case 1: All slots of cells 1, 3 and 6 are utilized to satisfy demands of cells 8, 9 and 10.

Case 2: Some slots of cells 1, 3 and 6 remain unused after satisfying demands of cells 8, 9 and 10.

We now prove that $|N_{11} \cup N_{12}| = 0$ under all assignment conditions. For this we consider the above cases separately.

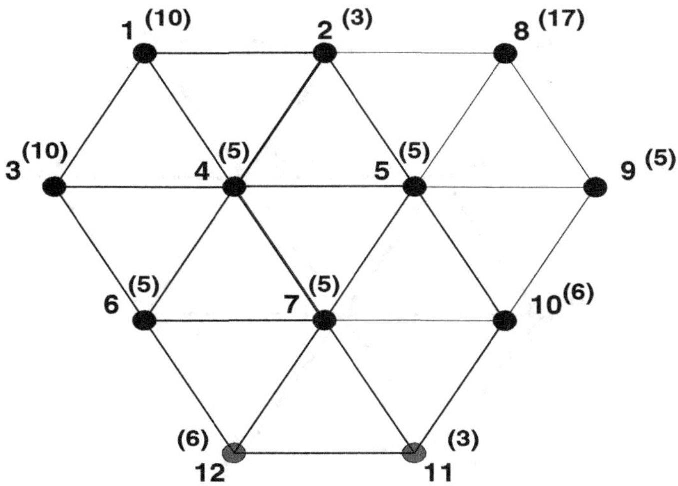

FIGURE 5.11: Example graph corresponding to Case 1

Proof of Case 1: Assuming all slots of nodes 1, 3 and 6 are re-used to satisfy the demands of nodes 8, 9 and 10, we have:

$$\begin{aligned} S_8 &\subset S_3 \cup S_6 \cup N_8 \\ S_9 &\subset S_1 \cup S_3 \cup S_6 \cup N_9 \\ S_{10} &\subset S_1 \cup S_3 \cup N_{10} \end{aligned}$$

$$\left.\begin{aligned} S_{11} &\subset S_1 \cup S_2 \cup S_3 \cup S_8 \cup N_{11} \\ S_{12} &\subset S_1 \cup S_2 \cup S_8 \cup S_9 \cup N_{12}. \end{aligned}\right\} \quad (5.1)$$

The worst case scenario, where the minimum number of slots from the critical block can be re-used for nodes 11 and 12 arises when:

1. S_1 is fully used in S_{10}.
2. S_3 is fully used in satisfying S_9 and S_{10}.

3. $S_2 = \emptyset$.

Under such a scenario, the only contributor to S_{11} from node 8 can be the slot N_8. Similarly for S_{12}, only N_8 and N_9 can be reused from nodes 8 and 9. Furthermore, the slots of S_1 and S_3 cannot be re-used in either S_{11} or S_{12} due to the two-cell buffering constraint. The modified expressions for S_{11} and S_{12} thus become:

$$\left.\begin{aligned} S_{11} &\subset S_2 \cup N_8 \cup N_{11} \\ S_{12} &\subset S_2 \cup N_8 \cup N_9 \cup N_{12}. \end{aligned}\right\} \tag{5.2}$$

$S_2 = \emptyset$, and $N_{11} \cup N_{12} \neq \emptyset$ imply that the demand sum of the ten-node subgraph formed by the nodes of 2, 4, 5, 6, 7, 8, 9, 10, 11 and 12 is greater than that of the ten-node subgraph formed by the nodes 1 to 7 and nodes 10, 11 and 12. According to the above definition of a ten-node critical block, the ten-node subgraph formed by the nodes 2, 4, 5, 6, 7, 8, 9, 10, 11 and 12 should have been the ten-node critical block, which is a contradiction to our original assumption. The ten-node subgraph formed by the thick edges in Figure 5.10 depicts this and Figure 5.11 shows an example graph for Case 1.

We now prove that even for Case 2 it is possible to satisfy the demands of nodes 11 and 12 by re-using only the slots allocated to the critical block.

Proof of Case 2: Under the condition that not all slots of nodes 1, 3 and 6 are utilized to satisfy the demands of nodes 8,9 and 10 we have:

$$\begin{aligned} S_8 &\subset S_3 \cup S_6 \\ S_9 &\subset S_1 \cup S_3 \cup S_6 \\ S_{10} &\subset S_1 \cup S_3 \end{aligned}$$

$$\left.\begin{aligned} S_{11} &\subset S_1 \cup S_2 \cup S_3 \cup S_8 \cup N_{11} \\ S_{12} &\subset S_1 \cup S_2 \cup S_8 \cup S_9 \cup N_{12}. \end{aligned}\right\} \tag{5.3}$$

From Figure 5.12 we see that S_8 cannot be used for node 12. Also only the contribution from S_3 to S_8 can be utilized for S_{11}. Similarly, only the slots of node 1 assigned to S_9 can be re-used for S_{12}. The worst case scenario with the minimum number of re-usable slots from the critical block for nodes 11 and 12 will arise when $S_2 = \emptyset$. The expressions for S_{11} and S_{12} would then look like

$$\left.\begin{aligned} S_{11} &\subset S_1 \cup S_2 \cup S_3 \cup N_{11} \\ S_{12} &\subset S_1 \cup S_2 \cup N_{12}. \end{aligned}\right\} \tag{5.4}$$

$S_2 = \emptyset$ and $N_{11} \cup N_{12} \neq \emptyset$ implies that the residual demand of nodes 11 and 12 is greater than the residual demand sum of nodes 8 and 9. According to the definition of a ten-node critical block, nodes 1 to 7 and nodes 10, 11 and 12 should have been the ten-node critical block of the cellular graph, which is a contradiction to our original assumption. Figure 5.12 represents this scenario where the subgraph formed by the thickened edges should have been the actual ten-node critical block and Figure 5.13 shows an example graph for Case 2.

Hence we see that under all conditions, $|N_{11} \cup N_{12}| = 0$, implying that any Distance 2 clique adjacent to the critical block can be allocated slots using only the slots assigned to the critical block. By reasoning along similar lines, we can show that for any other Distance 2 clique in the cellular graph, it is sufficient to use only the slots assigned to the ten-node critical block, thus proving that the demand sum of the ten-node critical block is optimal for a hexagonal cellular network having a heterogeneous demand vector. □

Note that the cellular Distance 2 clique of a ten-node critical block may not be a seven-node critical block, as may be seen from Figure 5.14. We see that the seven-

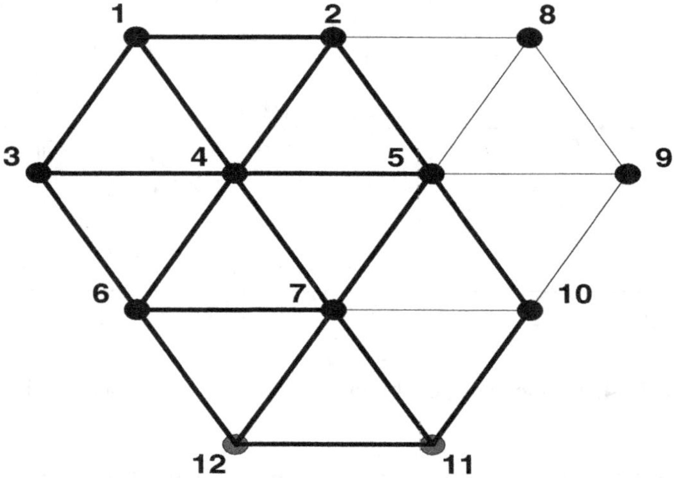

FIGURE 5.12: Actual critical block for Case 2

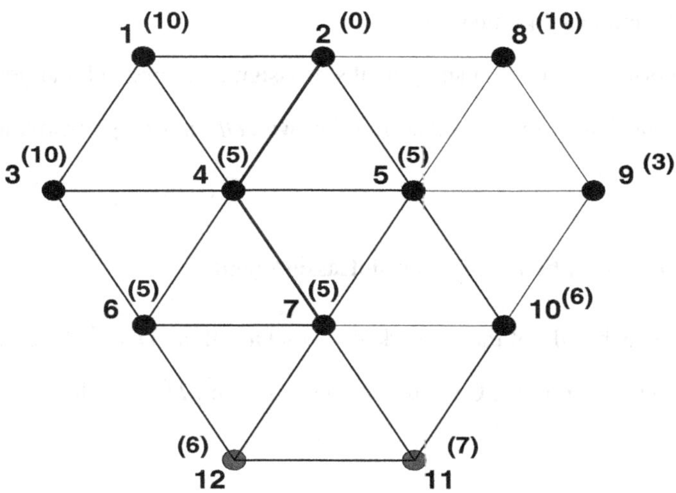

FIGURE 5.13: Example graph corresponding to Case 2

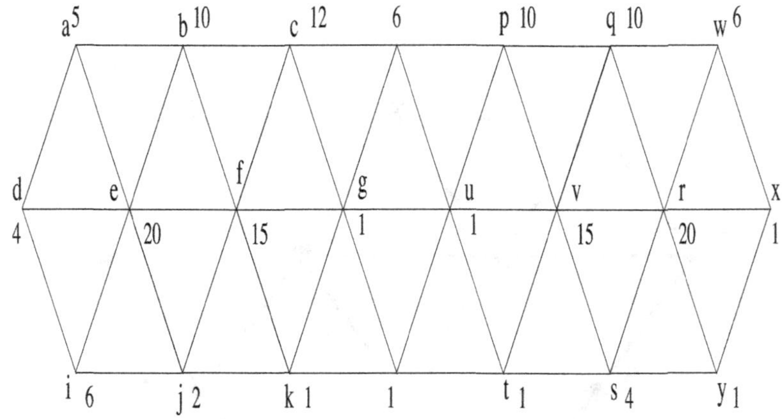

FIGURE 5.14: Ten-node critical block not formed by seven-node critical block

node critical block demand is 62 slots, while the ten-node critical block demand is 65 slots. Subgraphs $abfjide$ and $abcef$ both have demands of 62 slots (corresponding to a seven-node critical block), while the subgraphs $pqrstuvwxy$ and $pqwvrstu$ both have demands of a ten-node critical block. If, $abfjide$ ($abcef$) is chosen as the seven-node critical block, then the demand of the ten-node subgraph $abfjidecgk$ ($abcefdij$) would be 64 time slots.

The algorithm for finding an optimal slot assignment for a cellular network with heterogeneous demand while satisfying the *two-cell buffering* constraint is as follows:

5.4.5.1 Algorithm heterogeneous_slot_assignment

Step 1: For each cell i of the network, construct a cellular Distance 2 clique, $\mathcal{C}$ with i as the central node. Compute the demand sum, $\mathcal{D}_7^{(2)}$, of the cells belonging to $\mathcal{C}$.

Step 2: For each peripheral node $j \in \mathcal{C}$, compute the *residual sum set*, Res_j.

Step 3: The *maximum residual sum set*, $\mathcal{R}_\mathcal{C}$ corresponds to the set of neighbors of a peripheral node $k \in \mathcal{C}$ such that $Res_k = Res_{max}(\mathcal{C}) = \max_{j \in \mathcal{C}} [Res_j]$. Let $\mathcal{G}$ denote the ten-node subgraph corresponding to central node i of $\mathcal{C}$. Then, $\mathcal{G} = \mathcal{C} \cup \mathcal{R}_{\mathcal{C}}$.

Step 4: Compute the demand of $\mathcal{G}$, $\mathcal{D}_{10}^{(2)}(\mathcal{G}) = \mathcal{D}_{7}^{(2)} + Res_{max}(\mathcal{C})$.

Step 5: Repeat step 1 to 4 to obtain the demand $\mathcal{D}_{10}^{(2)}(\mathcal{G})$ of all ten-node subgraphs in the network. The maximum of these demands is the ten-node critical block demand. $\mathcal{D}_{10}^{(2)^*} = \max_{\forall \mathcal{G}}[\mathcal{D}_{10}^{(2)}(\mathcal{G})]$.

Step 6: Arbitrarily choose one of the ten-node candidate critical blocks as the ten-node critical block of the cellular network.

Step 7: Satisfy the demand of the nodes of CB_{10} under the *two-cell buffering* constraint.

Step 8: Satisfy the demands of all Distance 2 cliques in layer 1 adjacent to CB_{10}. Begin with the one formed by the nodes of maximum residual set of CB_{10}.

Step 9: Continue the process of assigning slots to Distance 2 cliques in layer 2, layer 3 and so on in a spiral, layer-by-layer fashion as described in Theorem 21.

5.4.6 Centralized Optimal Slot Assignment Algorithm

We present in this section a centralized slot allocation algorithm for assigning slots as per demand of each cell in the cellular network, while utilizing the minimum number of slots required for generating a collision-free transmission schedule that satisfies the *two-cell buffering* constraint, $s_0 = s_1 = s_2 = 1$.

Each MT is assigned a unique identifier (id) from the set $\{1, 2, 3, \ldots, m\}$, where m is the total number of mobile terminals. Initially, each mobile terminal (MT) knows its positional coordinates.

For each cell, a mobile terminal is also elected as the *cell leader* to handle mobility of mobile terminals, joining of new mobile terminals and MTs leaving the cell or network. For simplicity, we arbitrarily assume the lowest id MT in a cell to be the cell leader. An advantage of such a leader election protocol is that no additional communication is required among the MTs to elect the cell leader. The cell leader must:

- Compute the slot demand of the cell.
- Control admission of new mobile terminals into the network.
- Handle handoff situations and MTs leaving the network.
- Communicate to the *network leader* the demand of the cell and request a re-computation of the allocated slots whenever necessary.

To handle mobility of the mobile terminals, each cell keeps a few slots for transmitting control messages and some unused slots for handling new MTs joining the network and handoff scenarios. In general, a cell i computes its demand w_i as the sum of the number of mobile terminals in the cell, the number of slots allocated for control messages and an additional few unused slots. We assume the number of unused slots to be some fraction f of the number of mobile terminals currently in the cell. If m_i is the number of MTs currently in cell i and c slots are used for control purposes, then the demand, w_i of cell i is, $w_i = m_i + c + \max(1, \lceil fm_i \rceil), \quad 0 \leq f \leq 1$.

5.4.6.1 Algorithm COSA

The steps of the algorithm are:

Step 1: Elect an MT as the *network leader* through some leader election protocol [10, 11] and call it L.

Step 2: L broadcasts the mapping to convert the geographical region into a hexagonal grid structure to all the nodes of the network. It transmits a message M consisting of:

- The id of the MT transmitting the message
- A mapping to convert the geographical region into a hexagonal grid structure
- A list of the ids of all MTs heard by L

Each node, on receiving this message, appends it with its own location coordinates to be known to all other nodes. An MT i transmits its message in the ith slot to avoid collisions during this step.

Step 3: For each cell i, a *cell leader* L_i is elected from the MTs residing in cell i, based on some metric such as remaining battery power, load, location, etc. [11].

Step 4: The demand of each cell i, w_i is communicated by each cell leader L_i to the network leader L. L produces an optimal, collision-free transmission schedule by executing the *homogeneous_slot_assignment* or the *heterogeneous_slot_assignment* algorithm.

Step 5: L broadcasts the slot assignment schedule of the network to each cell leader. The slot assignment schedule details the slots assigned to each cell i, that demanded w_i slots. Once a cell leader L_i of cell i receives the information about the slots assigned to it from L, it generates a transmission schedule for the MTs in cell i and transmits a periodic local broadcast of this schedule within cell i.

For Step 2 of the *COSA algorithm*, assuming it takes $O(1)$ time to transmit M, the broadcast algorithm will complete in $O(mD)$ time slots where D is the diameter of the ad hoc network. At the end of the broadcast, each MT j would have the following information:

- The *id* i of the hexagonal cell in which it is located
- The set of its neighbors, $N(j)$
- The ids of the MTs in cell i and hence the number of MTs currently in cell i

In Step 5, if d_i is the diameter of the ad hoc subgraph in cell i, after $O(d_i w_i)$ time slots, each MT j of the cell i will know the time slot allocated by L_i in which it can transmit.

Definition 11. *A round of transmission for a cell i consists of w_i slots, where w_i is the number of slots requested by cell i.*

Each cell leader periodically transmits in one or more control slots a *cell status message* M_s containing the following information:

- The id of the cell leader
- The list of the control slots
- The list of used and unused slots in the corresponding cell and which slot is assigned to which MT so that when a new node joins or enters the cell, it knows the available slots in which it can transmit without causing collisions

Remark 20. *We assume that the lowest numbered c slots assigned to a cell are marked as the control slots. Also, the highest value control time slot is always assumed to be unused so that in the event that there are no unused time slots currently in the cell, a new mobile terminal in the cell can use this time slot to communicate with the cell leader.*

HELLO	Sender ID	Slot value used

FIGURE 5.15: Structure of HELLO message

5.4.6.2 Complexity Analysis

The leader election process in Step 1 of *algorithm COSA* takes $O(\log\ \log\ m)$ time [10, 11, 12]. Steps 2 and 5 take $O(mD)$ time each for round-robin broadcast, assuming $\forall\ i,\ d_i\ \ll\ D$ and $w_i\ =\ O(m)$ for Step 5 of *algorithm COSA*. Step 3 of *algorithm COSA* takes O(1) time. Computation of an optimal slot assignment schedule by algorithm *homogeneous_slot_assignment* or algorithm *heterogeneous_slot_assignment* takes $O(n)$ time where n is the number of cells in the cellular network. Thus, Step 4 takes $O(mD) + O(n)$ time. Hence, the complexity of the *algorithm COSA* is $O(\log\ \log\ m + mD + n)$ time.

5.4.6.3 New Mobile Terminal Joining the Network

In each cell, the cell leader periodically transmits a *cell status message* M_s as stated above. This message is locally broadcast within the cell. When a new MT j joins the network in some cell i, it first waits for the M_s message. Subsequently it executes the following steps to join the network in cell i:

Step 1: If there exists at least one unused time slot in cell i, mobile terminal j randomly selects an unused time slot k of cell i and sends a HELLO message to the cell leader L_i. In the event that there are no free slots in cell i that can be used by j, j transmits a HELLO message in the highest value control slot, which according to the protocol is always unused. The HELLO message contains the following information and its structure is shown in Figure 5.15.

1. Message Type is set to HELLO

2. Id of the new MT joining the cell
3. The slot value used by the new MT to transmit the HELLO message

Step 2: The HELLO message, if heard by its neighbors, is retransmitted by each neighbor of j in cell i.

Step 3: If L_i hears a HELLO message from a new MT j in an unused time slot k of cell i, execute Steps 3.1 and 3.2, otherwise go to Step 4.

Step 3.1: L_i marks slot k to be allocated to j.

Step 3.2: L_i initiates a local cell broadcast of updated cell status message M_s reflecting the slot allocated to j. The cell status message is also an acknowledgment (ACK) from L_i to mobile terminal j that it has been assigned slot k.

Step 4: If the HELLO message from j was transmitted in a control slot,

Step 4.1: L_i initiates a *garbage collection routine* to reclaim slots assigned to MTs that have left cell i. The garbage collection process works as follows:

Step 4.1.1: L_i transmits a local cell broadcast requesting each MT in cell i to send an ACK to L_i.

Step 4.1.2: The ACKs from neighbors are piggy-backed on the ACK transmission of each MT.

Step 4.2: If after garbage collection L_i fails to find sufficient unused slots to accommodate all the new nodes in cell i,

Step 4.2.1: L_i computes the new demand w_i of cell i, taking the new nodes in i into account.

Step 4.2.2: L_i sends a request to the *network leader* L to reassign slots using the new w_i.

Step 5: If j fails to hear an ACK message from L_i within some timeout interval T, it assumes that it was not allocated any time slot by L_i. This can occur under the following conditions:

1. Case 1: L_i was unable to allocate a slot to j.
2. Case 2: Some new nodes tried to transmit in the same time slot as j and hence the transmission was garbled due to collision.

On timeout, j uses an exponential backoff algorithm and tries to transmit again. It continues this until it receives an ACK from L_i.

5.4.6.4 Mobile Terminal Leaving Cell or Network

If a mobile terminal u other than the cell leader or the network leader leaves the current cell, no immediate update action is taken. The slot assigned to u may be reclaimed by the cell leader if the cell leader fails to find sufficient unused slots in the cell to satisfy a new incoming mobile terminal in the cell. If a cell leader leaves the network or the cell, according to this protocol no transmission will be heard by the nodes of the corresponding cells in the control slots. When such a situation occurs, a new *cell leader* is elected from the remaining MTs.

If the network leader has left the network, the existing cell leaders elect a new network leader.

5.4.6.5 Handoff Situations

A *handoff* is required when a mobile terminal u moves from one cell i to another cell j. The handoff is treated in the same way as a new MT joining cell j is treated with an additional message sent from L_j to L_i indicating the new cell in which u

MOV	Sender ID	Old Cell Leader ID	Slot value used

FIGURE 5.16: Structure of MOVE message

can be found. For mobile terminal u, when it moves from a cell i into a cell j, it waits to hear the M_s message from the cell leader L_j of cell j. u then sends a control message MOVE to L_j instead of a HELLO message. The structure of a MOVE message is shown in Figure 5.16 where the field MOV contains a code to indicate that the mobile terminal is moving from one cell to another. The procedure by which a MOVE message is sent and the actions taken by L_j are the same as when a new MT joins a cell. In addition, a control message is sent by L_j to L_i, informing L_i of the new cell in which u currently resides.

5.5 Deterministic Broadcast and Gossiping

Broadcasting protocols can be classified into *deterministic* and *probabilistic* algorithms. In the probabilistic approach, each node, upon receiving a broadcast packet, retransmits it with a probability p. However, the probabilistic algorithms cannot guarantee full coverage of the network. In contrast, the deterministic approach guarantees full coverage by the broadcast algorithm.

In the *broadcast problem* one node distinguished as the source node has a message to send to all other nodes; in the *gossiping problem*, each node has a different message that has to be distributed to all other nodes. Gossiping finds application in routing, spreading updates in networked databases and multicasting [1].

Because of the mobility and limited battery power of the nodes, it is important to have fast, energy-efficient and deterministic broadcast and gossiping protocols for wireless ad hoc networks by resolving collisions efficiently. Such protocols with efficient collision resolution mechanisms would provide low end-to-end latency and low overhead for retransmission of the collided messages. Most existing broadcast and gossiping algorithms do not address these issues together.

In this section, we present deterministic broadcast and gossiping algorithms as proposed [13] for mobile ad hoc networks that have low broadcast latencies. The algorithms presented here assume collision detection capabilities at the mobile terminals.

The deterministic broadcast algorithm is based on breadth-first traversal of the nodes of the network. The algorithm allows *simultaneous transmission* by multiple nodes. Although this may result in collisions at the receiving nodes, the algorithm can detect and take corrective measures to eventually avoid collisions. The idea of breadth-first broadcast was first proposed in [7, 15] where the two different types of rounds of transmission, *control round* and *message round*, made the protocol somewhat complex. However, in [13], only one type of round of transmission has been used. The algorithm presented in citesinha08 works for scenarios where network topologies change frequently due to high mobility of the nodes and takes $O(n \log\ n)$ time in the worst case. Both the average case and worst case performances of this breadth-first algorithm are better than the *polylog broadcast algorithm* proposed in [3]. Simulation results show that the time complexity of this breadth-first algorithm on an average is upper bounded by $8D \log n$. The reported simulation results were obtained for networks having 100 nodes to 1024 nodes. It may be noted that the asymptotic lower bound for deterministic broadcast in networks where spontaneous transmission by the nodes is not allowed is on average $\Omega(D \log n)$ [5]. Also, based on simulation results, the time to complete broadcast is almost insensitive to the mobility speed, although it is dependent on the specific mobility model assumed for the nodes of the network.

Extending the above idea of breadth-first broadcast technique, a mobility-resilient, deterministic gossiping algorithm has also been presented [13]. This algorithm completes in $O(Dn \log n + D^2 \log n)$ time which is better than the $O(n^{3/2} \log^2 n)$ time needed by the algorithm in [16]. For large values of Δ, the time complexity of this algorithm is comparable to the $O(D\Delta^2 \log^3 n)$ time algorithm in [17].

To describe these algorithms, we assume the following system model.

5.5.1 System Model

We consider an ad hoc network consisting of n nodes that may be stationary or mobile. Initially the nodes do not possess information about the network topology. All communication links are assumed to be bidirectional and all communication occurs over a single channel. The nodes are assumed to be synchronized in time. In a specific time slot, a station can transmit and/or listen to the channel. A node v is a *neighbor* of u if they are within each other's hearing zones. The network is assumed to be connected at all times. In the broadcast problem, one node is distinguished as the *source node* (also known as the *root*), which has a message M to be broadcast. It is assumed that the mobile terminals possess *collision detection capabilities*. In the rest of our discussions in this chapter, we will use the terms node *id* and address interchangeably to identify a particular node.

5.5.2 Breadth-first Broadcast

We now present a mobility-resilient broadcast algorithm that allows simultaneous transmission by nodes by detecting and resolving collisions. We use the concept of *layers* to explain how the broadcast message propagates in the network and how collisions are resolved. The source node is the only node in layer 0; all nodes within the hearing zone of the source node form layer 1, and in general all nodes within the hearing zone of the nodes in layer i but not in layers $(i - 1)$ and i, form the layer $(i + 1)$, for $i \geq 1$.

5.5.2.1 Preliminaries and Basic Idea

A transmission of M by the source node will be received by all nodes in layer 1. The nodes in layer 1 then transmit the message to the nodes in layer 2. However, at this point, a node in layer 2 may receive a collided message due to simultaneous transmission from more than one node in layer 1. This situation can also arise from simultaneous transmissions by nodes in different layers in the same time slot. The proposed protocol proceeds in the way that i) a receiving node experiencing such a collision sends some appropriate acknowledgment signals to the transmitting nodes, and ii) one or more of the transmitting neighboring nodes are selectively dropped (dropped nodes do not transmit the message) at each of the succeeding stages (or *rounds*), so that eventually collision is resolved at each of the receiving nodes. Essentially, we drop out all nodes except the one with the maximum *id* value among the transmitting nodes causing collision at a receiving node. This is done by successively scanning the bits of *id* values of each such transmitting node from the most significant bit (msb) side. As an example, consider the scenario in Figure 5.17 in which a node u which has experienced a collision due to messages simultaneously transmitted from its three neighboring nodes having the id values 11111001, 11100011 and 10110110.

In the next round, the nodes whose msb values are will be dropped. Since there is no such node, all of these nodes will again transmit in this second round. In the next round, the nodes whose $(msb - 1)$-th bits are 0 will be dropped and only those with bit value equal to 1 will be allowed to transmit. Thus, the node 10110110 will

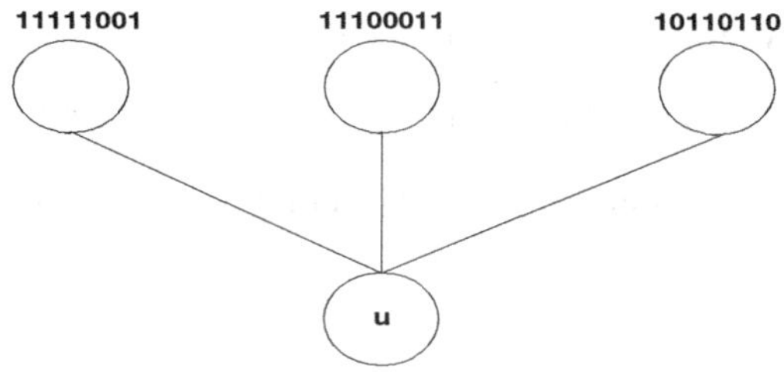

FIGURE 5.17: Collision resolution example 1

be dropped. But the node u will still experience a collision due to messages coming from the nodes 11111001 and 11100011. At this point, the $(msb - 2)$-th bit of *id* will be tested and those with this bit value equal to 1 will be allowed to transmit only in the next round. Since both the transmitting nodes have this bit value equal to 1, the node u will again experience a collision. In the next round, however, the node 11100011 will be dropped since its $(msb - 3)$-th bit is equal to 0, and only the node 11111001 will be allowed to transmit so that u receives a clean message.

However, the above mechanism of successively dropping out the nodes will give rise to a problem when all the transmitting nodes have the corresponding scanned bit value equal to 0 and thus all of them will be dropped out. We explain in a later subsection how this problem can also be tackled to resolve the collision.

5.5.2.2 Node States and Acknowledgment Signals

At any instant of time, a node may be in any one of the four states described below.

receive: node is ready to receive data transmitted by any other node

collision: node has received a collided message

active: node has received the message correctly, but is not transmitting it

transmit: node will transmit the message

A *round* of message transmission consists of several control slots, a message slot for transmitting the actual message M, and three slots for receiving acknowledgments from the neighboring nodes as explained below. The control slots are used to guarantee collision-free reception of the message by all receiving nodes and their usage is defined later. The acknowledgment slots are:

Slot S for receiving a single-bit acknowledgment from the nodes which received the message correctly without any collision

Slot C for receiving a single-bit acknowledgment from the receiving nodes, if the corresponding receiving nodes detected a $collision$ during message slot

Slot N for receiving a single-bit acknowledgment to indicate that it did not receive any message (clean or collided) in this round although it received a collided message in the previous rounds

If for a node u, the channel status is *null* during any of these three slots, we assume that none of the neighbors of u have responded in that slot. The presence of a clean or collided signal due to one or multiple nodes responding simultaneously in any of these S, C and N slots is interpreted as the appropriate acknowledgment signal for that slot by the node u.

5.5.2.3 Collision Resolution

Every transmitting node makes a decision depending only on the type of acknowledgments, clean or collided, received by it in the three slots S, C and N, respectively, associated with the previous message transmission slot. If a transmitting node receives an acknowledgment (which may come from a number of nodes in *receive* state) in slot S, it implies that the corresponding nodes in *receive* state received the message correctly without any collision. These receiving nodes become *active* in the successive rounds. If an acknowledgment is received by the transmitting node in

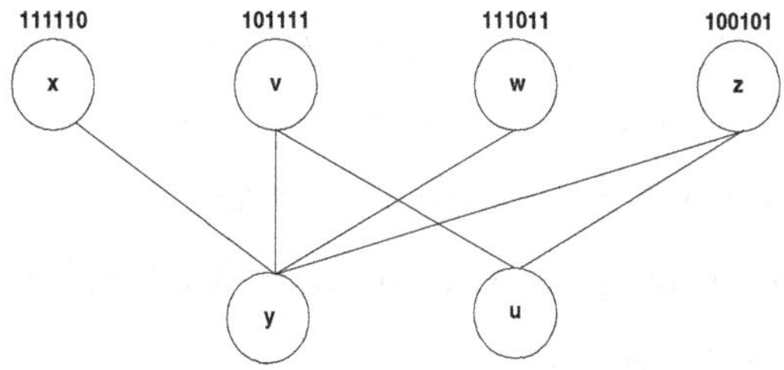

FIGURE 5.18: Collision resolution example 2

the C slot, it implies one or more of the receiving nodes experienced a collision during the previous message slot. Collision is resolved by preventing (dropping) some of the nodes from transmitting in the subsequent rounds. This is done by scanning the bit pattern (starting from the msb) of the id value of each of the transmitting nodes receiving a C ACK signal and allowing only those nodes to transmit whose corresponding bit value is equal to 1.

However, as we indicated earlier, a receiving node which experienced a collision may not be receiving the message in the immediate next round when all the neighboring transmitting nodes of a receiving node are dropped. At this point, the corresponding receiving node sends an acknowledgment in the N slot. The transmitting nodes receiving N ACKs, try to transmit again in the succeeding rounds. However, there may be some complex situation at this stage if a transmitting node receives an N signal and also a C signal from other nodes. As an example, consider the situation shown in Figure 5.18, where the node y experienced a collision due to messages from nodes v, w, x and z, and node u experienced a collision as a result of simultaneous transmissions from nodes v and z. Let the *id* values of the nodes v, w, x and z be 101111, 111011, 111110 and 100101, respectively.

In the first round, nodes v, w, x and z all receive C signals. Since all these nodes have msb values equal to 1 in the next (second) round as well, all of v, w, x and z will transmit the message and each will receive a C signal. Nodes v and z will be dropped in the third round as both have $(msb - 1)$-th bits equal to 0. Hence, in this

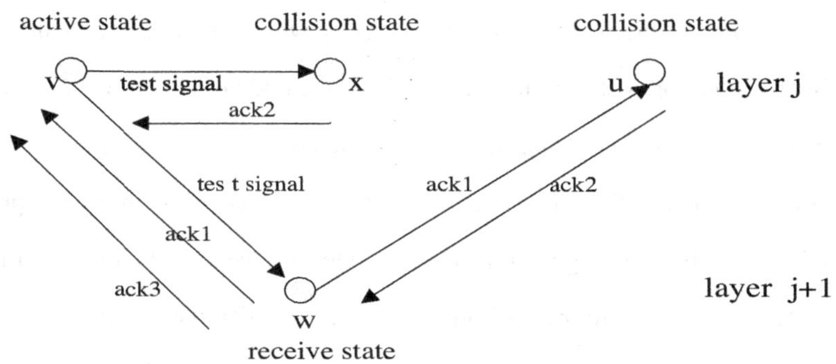

FIGURE 5.19: Role of control signals

round, both v and z will receive N ACKs from u and C ACKs from node y. Neither v nor z can be allowed to resume transmission in the fourth round, as they would have caused a collision at y, jeopardizing the collision resolution at y. Collision at y will be resolved in the fifth round as the $(msb - 3)$-th bit of w is 0 and hence it will be dropped in the fifth round; leaving only node x to transmit. At this point, nodes v and z will get no C ACKs in this (fifth) round, but only an N ACK from node u. Now, v and z both can be allowed to resume transmission in the next (sixth) round and this will cause a collision at u. In the next (seventh) round, v and z will test their $(msb - 2)$-th bits (as they were dropped based on the value of their $(msb - 2)$-th bits), and at this time only v will be allowed to transmit since its $(msb - 2)$-th bit is equal to 1; a clean message is eventually received by node u.

5.5.2.4 Moving from *active* State to *transmit* State

It is possible that after a particular round, some nodes in a given layer, say j, will receive a message correctly and hence will become *active*, while others in the same layer experienced collisions and are in *collision* state. Two possible scenarios may arise:

1. As an *active* node in layer j starts transmitting a message, the message can also

reach some of the nodes of the same layer j which are still in *collision* state. Such a situation is shown in Figure 5.19, where node v is in *active* state in layer j, and node x is in *collision* state in the same layer j and lies within the hearing zone of node v. While the collision resolution at x is going on (by the process of selectively dropping its transmitting nodes in layer $j-1$), the transition of node v from *active* to *transmit* state during this collision resolution period will cause further collision at node x due to the message transmitted from v which may not be resolved. This is illustrated by the following example.

Example 26. *Assume that the node x in Figure 5.19 has two transmitting neighbors with id values equal to 11110001 and 11011011, differing first in the third bit from the msb side, while the node v has an id value equal to 10110001. Thus, in the third round after experiencing the collision for the first time, node x could get a clean message from the node 11110001 as the other node 11011011 would have been dropped based on the testing of the $(msb-2)$-th bit. However, if in this same round, node v moved from* active *to* transmit *state, x would instead experience a collision due to messages coming from both 11110001 and v (= 10110001). Note that the id value of v is identical to that of this node from the $(msb-2)$-th bit onward (toward the least significant bit side). Also, v would try to scan the bits of its own id from its msb, while node 11110001 would scan its bits from the $(msb-3)$-th bit, which will completely jeopardize the collision resolution at node x.*

To avoid such a situation and to simplify the collision resolution algorithm, an active node in layer j would first determine through some additional control signals (described below), whether it is a neighbor of any node in *collision* state. If so, then this active node does not immediately move to the *transmit* state.

test ack1 ack2 ack3	message	S C N

FIGURE 5.20: Structure of round for algorithm *bf_broadcast*

2. Referring to Figure 5.19, consider another situation in which node u in collision state in layer j is not in the hearing zone of node v, but a node w in layer $j+1$ in *receive state* is adjacent to both the nodes v and u in layer j. Suppose v starts transmitting a message to w even when u is in the collision state. w may receive a collided message (due to transmission from some other active nodes like v in layer j), and some of these nodes in layer j will be dropped from transmitting, depending on the bit patterns of their ids. When this dropping process continues, node u may receive a clean message from some node in layer $j-1$ and change its state from *collision* to *active*. Subsequently, node u (because of its independent decision to transmit) may cause collision at w. To prevent this situation, we keep the node v in *active* state until no node like w that has some parent in *collision* state exists.

To handle the two scenarios, we utilize four additional control signals in a round to test whether an *active* node v in layer j can move to *transmit* state after satisfying the following two conditions:

1. No collision is caused to any other node in the same receiving layer j.
2. If the message from v is received by a node w (in *receive* state) at layer $j+1$, then w cannot be in the hearing zone of some node, say u (in *collision* state) at layer j.

Thus, we augment a round with four additional control slots as follows:

test: 1-bit test signal sent by an *active* node only

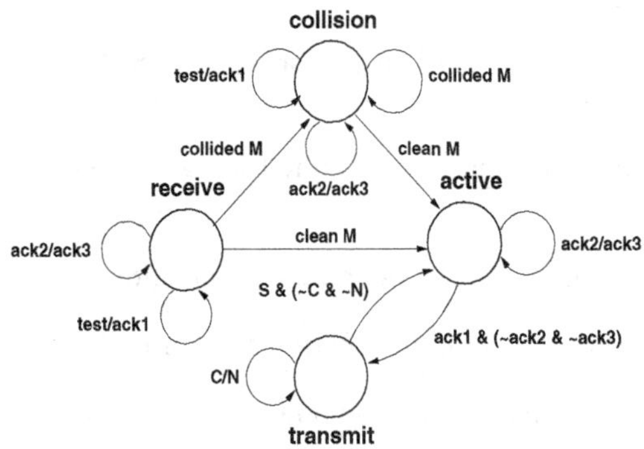

FIGURE 5.21: State transition diagram

ack1: 1-bit acknowledgment signal sent by a node in *receive* state which received the test signal above

ack2: 1-bit acknowledgment signal sent by a node in *collision* state which received either the *test* signal or *ack1* signal

ack3: 1-bit signal sent by a node in *receive* state which got the *test* and *ack2* signals

The overall structure of a round with the control signals for avoiding collisions is shown in Figure 5.20. The four control signals and the three acknowledgment signals S, C and N are 1-bit long. The state transition diagram in Figure 5.21 depicts the transitions between the four possible states based on the various signals.

From the above discussions, we satisfy following properties by the broadcast protocol.

Property 1. *The above mechanism of changing an* active *state of a node to the* transmit *state guarantees that while a node u in* collision *state is resolving the collision arising from a set $\mathcal{T}$ of its transmitting neighbors, no new transmitting node $\notin \mathcal{T}$ can join the nodes in $\mathcal{T}$ to send the message to u.*

Property 2. *With the property 1 above satisfied, the collision resolution mechanism described in the above section guarantees that a collision experienced by a node u*

arising from a set $\mathcal{T}$ of its transmitting neighbors can always be resolved so that eventually a clean message is received by u.

To minimize the overhead, we would like to use the four control slots to test whether an *active* node can move to the *transmit state* only once every $\lceil \log n \rceil$ rounds. However, this may cause a situation in which the nodes in layer 1 get the message in round 1 and then wait for the next $\lceil \log n \rceil - 1$ rounds until they become eligible for transmitting the message to the next layers. To avoid this unnecessary waiting, we propose using these control slots in two successive rounds in every $\lceil \log n \rceil$ rounds. Additionally, for mobility resilience, the source node will continue transmitting the message every $\lceil \log n \rceil$ rounds.

5.5.2.5 Algorithm bf_broadcast

We use a local variable $round\_count$ initialized to 0 at each node to keep track of the round number of transmissions. Subsequently, each node would increment its respective $round\_count$ variable at the start of successive rounds. The following steps are executed by all nodes in parallel for broadcasting a message M.

Step 1: (Initialization) The source node sets its state to *transmit*. All other nodes set their states to *receive*. The following steps are executed in each of the successive rounds.

Step 2: (Actions taken by the source node) If $round\_count$ MOD log $n = 0$, then the source node transmits M.

Step 3: (Test for eligibility of an *active* node to move to *transmit* state) If the state is *active* and $round\_count$ MOD log $n \leq 1$, execute Steps 3.1, 3.2 and 3.3.

Step 3.1: Send a *test* signal (one of the four control signals) and wait for *ack2* or *ack3* signals.

Step 3.2: If neither *ack2* nor *ack3* signals were received in respective slots, node state is changed to *transmit*, set $transmission\_count$ = 0 and set $scanned\_bit\_position = \log\ n - 1$. Increment $round\_count$ by 1 and go to Step 4 for the next round.

Step 3.3: If either *ack2* or *ack3* signal is received, increment $round\_count$ by 1 and go back to Step 3 for the next round.

Step 4: (*Transmit* state actions) If node state is *transmit*, execute steps 4.1 to 4.5.

Step 4.1: If $transmission\_count$ = 0, transmit the message, increment the value of $transmission\_count$ by 1, and wait for acknowledgment signals in the S, C and N slots. Increment $round\_count$ by 1 and go back to Step 4 for the next round.

Step 4.2: If an acknowledgment was received in the C slot of the preceding round:

Step 4.2.1: If the id bit value at the current $scanned\_bit\_position$ is 1 and the node transmitted the message in the immediate previous round, retransmit M, increment $transmission\_count$ by 1, decrement the value of $scanned\_bit\_position$ by 1, and look for S, C and N signals again. Increment $round\_count$ by 1 and go back to Step 4 for the next round.

Step 4.3: If an acknowledgment was received in the N slot but not in C, /* some node in *receive* state has *only* the dropped nodes (including this node) from which it received the message; hence we need to restart transmission from this node */

Step 4.3.1: if the node's id bit at the $scanned\_bit\_position$ is 0, transmit the message, increment $transmission\_count$ by 1, decrement the value of *scanned_bit_position* by 1, and look for S, C and N signals. Increment *round_count* by 1 and go back to Step 4 for the next round.

Step 4.4: If an acknowledgment was received only in the slot S but not in C and N, the state of the node is changed back to *active*, i.e., the transmission was successful. Increment *round_count* by 1 and go back to Step 3 for the next round.

Step 5: (*receive* state actions) If node state is *receive*, execute Steps 5.1 to 5.4.

Step 5.1: If a signal is received in *test* slot, send *ack1* in this round. If signal received in *ack2* slot, send *ack3* signal in this round. Increment *round_count* by 1 and go back to Step 5 for the next round.

Step 5.2: If clean M is received, send an acknowledgment in S slot and change state to *active*. Increment *round_count* by 1 and go back to Step 3 for the next round.

Step 5.3: If collision occurs in *message* slot, send acknowledgment in C slot and change state to *collision*. Increment *round_count* by 1 and go to Step 6 for the next round.

Step 6: (*Collision* state actions) If node state is *collision*, execute Steps 6.1 to 6.5.

Step 6.1: If a signal is received in *test* slot or *ack1* slot, send the *ack2* signal in this round. Increment *round_count* by 1 and go back to Step 6 for the next round.

Step 6.2: If a collision is encountered in the *message* slot, send an acknowledgment in C slot. Increment *round_count* by 1 and go back to Step 6 for the next round.

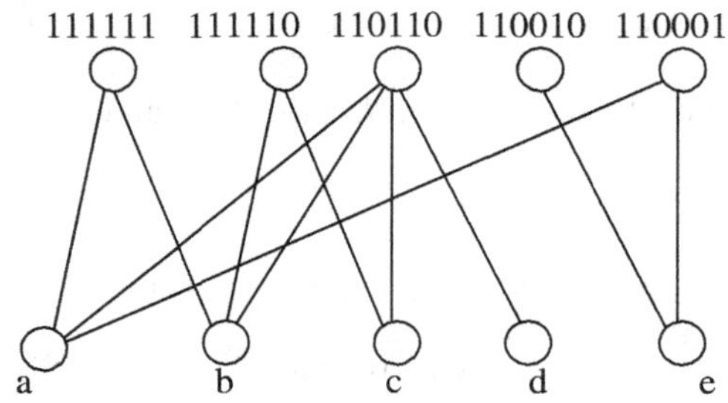

FIGURE 5.22: An example of collision resolution by the broadcast algorithm

Step 6.3: If no signal is received in the message slot, but a collided message was received in some previous round, send an acknowledgment signal in the N slot. Increment $round\_count$ by 1 and go back to Step 6 for the next round.

Step 6.4: If a clean signal is received in the *message slot*, send an acknowledgment in S slot and change state to *active*. Increment $round\_count$ by 1 and go back to Step 3 for the next round.

To illustrate the basic ideas of the above algorithm, we consider the following two examples.

Example 27. *Consider the network in Figure 5.22, where nodes 111111, 111110, 110110, 110010 and 110001 are the transmitting nodes, and nodes* a, b, c, d *and* e *are the receiving nodes. Following Step 4.1, all of the nodes 111111, 111110, 110110, 110010 and 110001 send the message simultaneously. Node* d *receives the clean message from the node 110110, but nodes* a, b, c *and* e *detect a collision. Following the Step 4.2 of the algorithm, nodes 110110 and 110001 are dropped as sender nodes after the testing of the third bit from the most significant bit side and* c *receives the clean message from node 111110 in the next round. Subsequently, node 111110 is dropped after the sixth bit so that nodes* a *and* b *get clean messages from 111111 in the next round. Node 110010 receives an acknowledgment signal in slot* N *from*

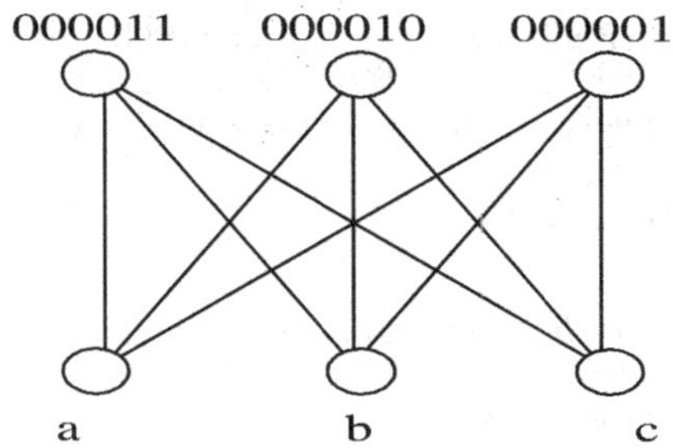

FIGURE 5.23: Another example for collision resolution by the broadcast algorithm

node e corresponding to the third bit position so that, by virtue of Step 4.3, node 110010 transmits in the next round. Thus, node e would receive the message from node 110010.

Example 28. *Consider the network in Figure 5.23, where nodes 000011, 000010 and 000001 are the transmitting nodes, and nodes a, b and c are the receiving nodes. Due to Step 4.1, all these nodes transmit the message in the first round (i.e., in the round immediately after their state was changed to transmit). Due to collision, none of the nodes will transmit in the second round (as the second most significant bit of all these nodes is 0). In the third round, due to Step 4.3, all of them will start retransmitting the message. Eventually node 000001 is dropped after the fifth bit (from the msb side) testing, and node 000010 is dropped out after the sixth bit testing. Finally, nodes a, b and c receive the clean message from node 000011.*

Based on the above examples, Steps 4.2 and 4.3 play a crucial role in selectively dropping a node from transmission (if there were a collision) and re-vitalizing it for transmitting the message (when this node cannot cause collision to any receiving node, but it has at least one receiving node as its neighbor which is not reachable from any of the transmitting nodes in the previous round). Thus, due to dropping of nodes, a node in the *collision* state can eventually get the message within $2 \log n$ rounds. Also, an active node will eventually change its state to *transmit* due to Step

3.2, when there will be no *ack2* or *ack3* signals caused by the nodes in *collision* state (as a node in *collision* state will eventually become active by the above argument). Hence, we get the following result.

Theorem 22. *The algorithm bf_broadcast correctly executes the one-to-all broadcast among all the n nodes of the network.*

5.5.2.6 Mobility Resilience

The layer-by-layer transmission of the message as in [3] and periodic retransmission of the message by the source node after every log n rounds ensures that even if a node moves from a layer j to a layer i before it has received the message M and after all nodes in layer i have finished transmitting M, where $i < j$ (i.e., it is now at a smaller distance from the source node s), it will still be able to receive the broadcast message correctly at some point. Thus, the algorithm *bf_broadcast* works correctly even when the network topology changes frequently.

5.5.2.7 Complexity

Lemma 23. *At least one node is guaranteed to have received the message correctly, after every* $2\log\ n$ *rounds of transmission.*

Proof. Because of the typical nature of dropping successive transmitting nodes causing a collision, it is guaranteed that for every $2\log\ n$ rounds, at least one node of the receiving layer j (assuming that transmitting layer is $j-1$) gets the message correctly. The factor of 2 is included to take care of the fact that if a bit tested for 1 causes a *no signal* to the receiving nodes, this bit is flipped to 0. Since there are a maximum of log n such flippings, the claim is justified. □

Theorem 23. *The algorithm bf_broadcast takes $O(n\log\ n)$ rounds to complete the broadcast in the worst case.*

Proof. Since an active node will be tested for eligibility for transmission every log n rounds, the maximum waiting time to start transmission to the next layer after all nodes in a layer correctly received the message, is log n. Effectively each layer j completes its broadcast in $(2p_j + 1) \cdot \log\ n$ rounds in the worst case (assuming no overlap in the broadcast from different layers), (where p_j is the number of nodes in layer j), after which the next layer starts transmitting the message. Hence, the worst case number of rounds for broadcast is $\sum (2p_j + 1) \log\ n = 2(n-1) \log\ n + D \log\ n$, where D is the diameter of the graph. □

Considering eight slots per round, the total number of required slots by algorithm *bf_broadcast* is $8(2n + D - 2) \log\ n$ in the worst case.

5.5.2.8 Simulation Results

To evaluate the performance of the above algorithm *bf_broadcast*, Sinha et al. [13] simulated the algorithm on both static and mobile scenarios. They used ns-2 [18] simulator for the purpose of random graph generation in their simulation studies. They modeled their round structure similar to that of the GSM TDMA frame, assuming eight consecutive slots each of duration 0.577 ms for a total of 4.615 ms frame duration. All message transmissions were assumed to be completed in a single time slot.

For the static scenarios, the broadcast algorithm was simulated on 100 different random graphs for each graph size. For every such randomly generated graph, each node was considered as the source node that initiated the broadcast and the relevant performance data was collected. The averages of all such runs are presented as the performance results of the *bf_broadcast* algorithm in Figures 5.24 to 5.26.

For the results depicted in Figures 5.24 and 5.25, the average and maximum node degrees (δ_{avg} and Δ) were fixed at 7 and 15, respectively. Figure 5.24 demonstrates the average case time and message complexity of the *bf_broadcast* algorithm under

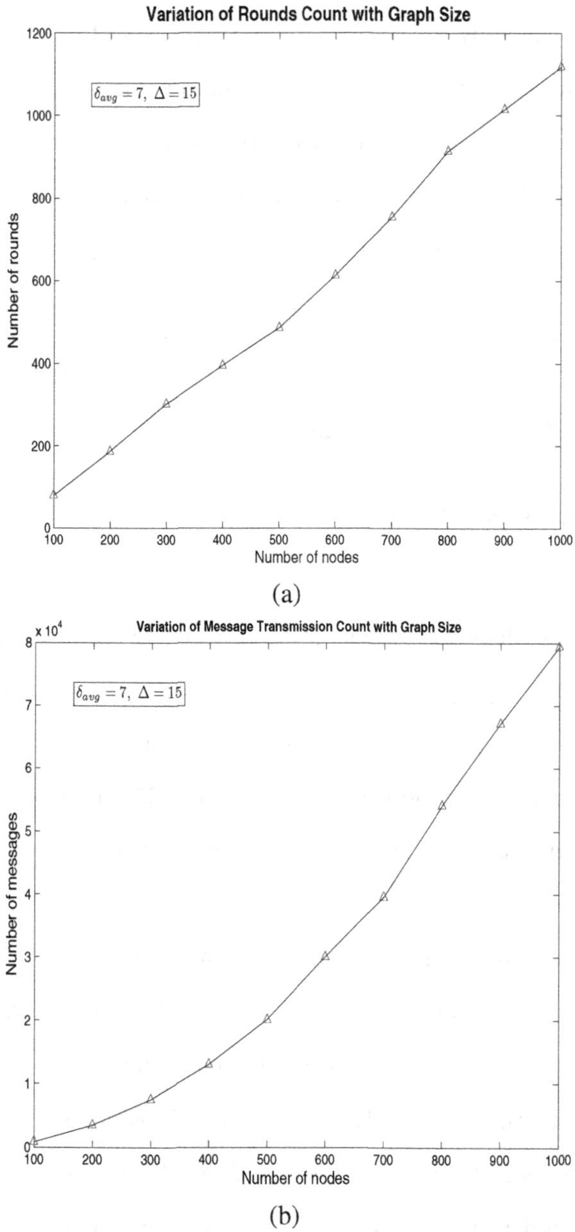

FIGURE 5.24: Execution time of algorithm *bf_broadcast*

these constraints. Message complexity (Figure 5.5.2.8) refers to the number of transmissions of the broadcast message M, while time complexity (Figure 5.5.2.8) is the number of rounds required to complete the broadcast. It can be recalled that each round consists of eight time slots. Figure 5.5.2.8 shows the number of retransmissions of the broadcast message M that occurred on average. Retransmissions can occur due to two reasons: i) collision detected by the receiving nodes and, ii) a node in *transmit* state receives an N message. Simulation results showed that the number of actual message collisions suffered was slightly less than the total number of retransmissions of M. Figure 5.5.2.8 depicts the total number of transmission attempts on average including the transmission of the control signals and the broadcast message.

Figure 5.26 shows the variation of average case time and message complexity with graph diameter for graphs having 256, 512 and 1024 nodes. Figure 5.27 shows the variation of the number of collisions and retransmissions on average against graph diameter.

The graphs in Figures 5.28 to 5.32 show the performance of the above algorithm for mobile scenarios. Sinha et al. [13] selected two mobility models: the widely used *random way point* model and the *city section* model, also known as the *Manhattan mobility* model (as introduced in Chapter 2) for the purpose of studying the performance of the *bf_broadcast* algorithm. The city section mobility model provides realistic movements for a section of a city as the mobile nodes must follow predefined paths and behavior guidelines such as traffic laws. The mobility trace files were generated using the bonnmotion tool [19] and then ns-2 was used to obtain the results. We considered graphs of 100, 200 and 300 nodes for comparison of results. As in the case of the static scenario, each node was in turn considered as the source node and the averages of 100 different simulation trials for each graph size presented as the performance results. The simulation grid sizes chosen were 200×200, 500×500 and

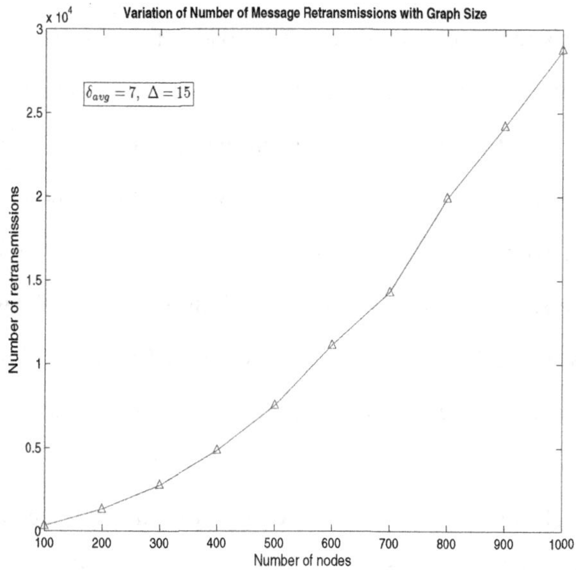

(a)

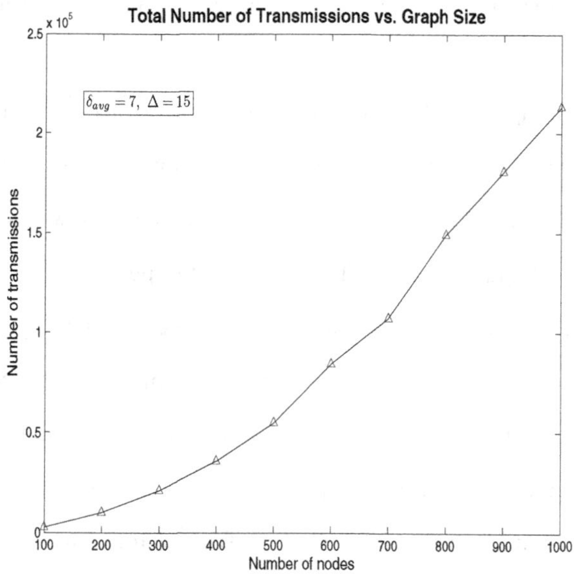

(b)

FIGURE 5.25: Execution time of algorithm *bf_broadcast*

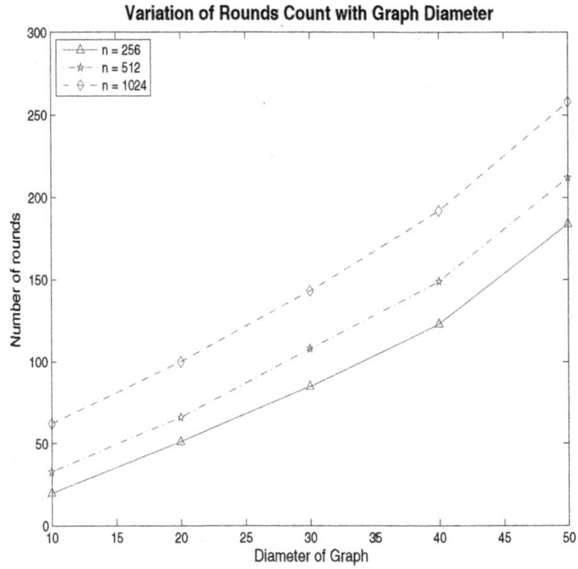

(a)

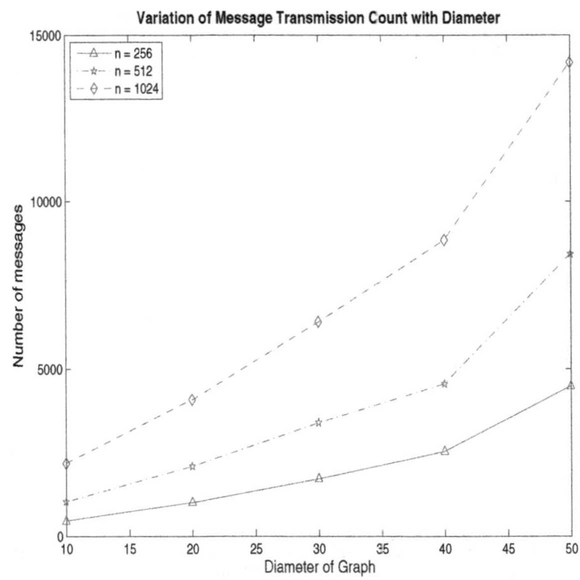

(b)

FIGURE 5.26: Execution time of algorithm *bf_broadcast*

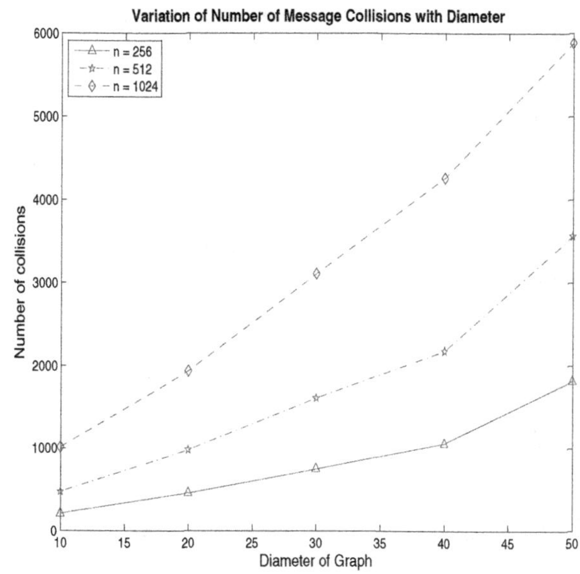

(a)

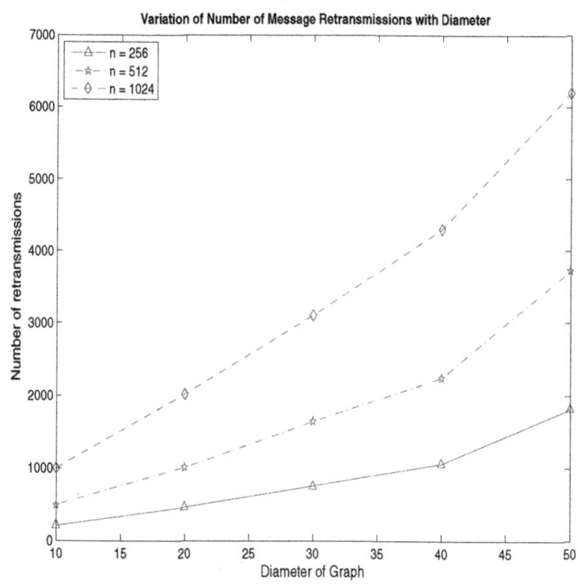

(b)

FIGURE 5.27: Execution time of algorithm *bf_broadcast*

TABLE 5.1: Slots requirement comparison between *bf_broadcast* and *polylog algorithm*

n	D	Δ	**Algorithm Polylog**	**Algorithm bf_broadcast**	
				Worst Case	**Average Case**
128	22	15	36969	15456	773
256	35	20	940800	34880	1240
512	40	36	19353600	76464	1513
1024	58	36	56125440	168320	2518

700×700 square units for graph sizes of 100, 200 and 300 nodes, respectively. Pauses between two displacements were used for both the mobility models. The pause time was chosen randomly from the interval $[0, 60]$ seconds. Each simulation trial was run for 1200 seconds to ensure that the mobile graphs remained fully connected for sufficient time to ensure that every node received the broadcast message at some point of time.

Figures 5.28 and 5.29 show the average case time and message complexity of the algorithm for the Manhattan grid and the Random Waypoint mobility models, respectively. The most significant observation from these simulation trials is that the performance of this algorithm is nearly *independent* of the speed of the mobile terminals but *dependent* on the mobility model. The increase in the average time and message complexity is very small for an increase in maximum possible speed of the nodes from 10 m/s to 30 m/s, as can be seen from Figures 5.28 and 5.29. Moreover, from Figures 5.30, 5.31 and 5.32 we see that on an average, more time slots and messages are required to complete broadcast in the Manhattan grid mobility model than in the Random Waypoint model, for the same maximum speed, grid dimensions and graph size.

Table 5.1 provides a comparison of the performances of the breadth-first broadcast algorithm with the *polylog broadcast algorithm* [3]. From the table we observe that even the worst case performance of the algorithm *bf_broadcast* described above is better than that of the one proposed in [3]. Also, for the simulation trials, the av-

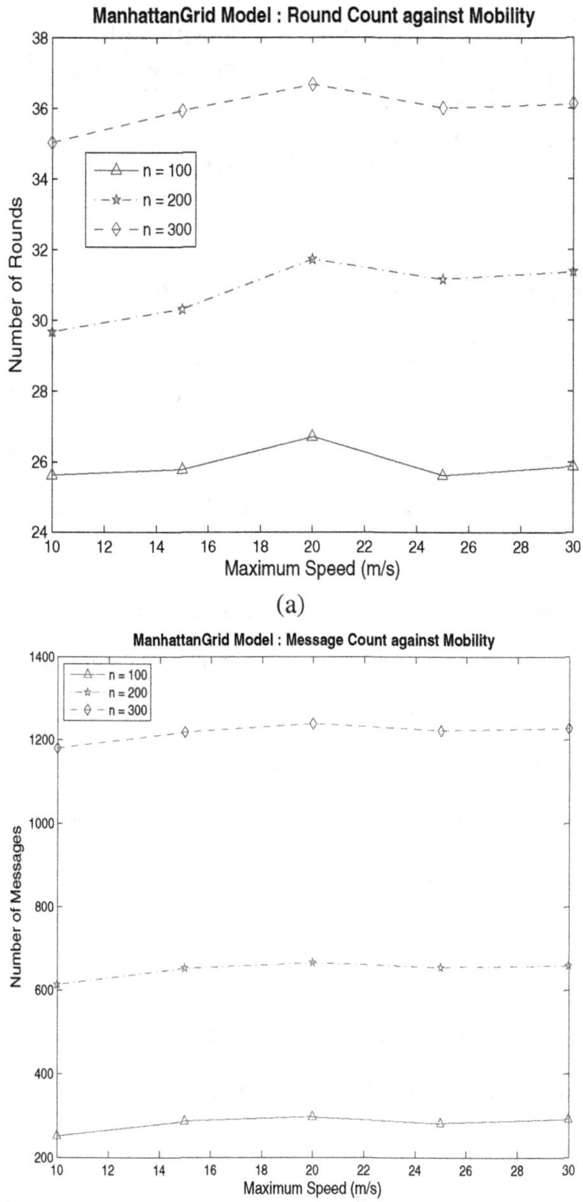

(a)

(b)

FIGURE 5.28: Execution time of algorithm *bf_broadcast*

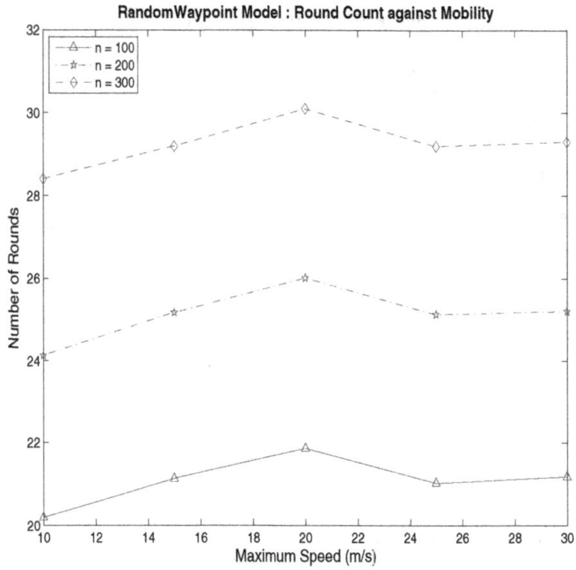

(a)

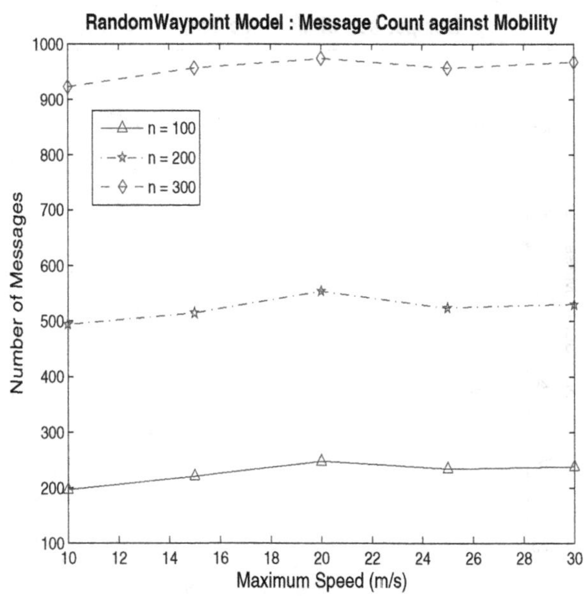

(b)

FIGURE 5.29: Execution time of algorithm *bf_broadcast*

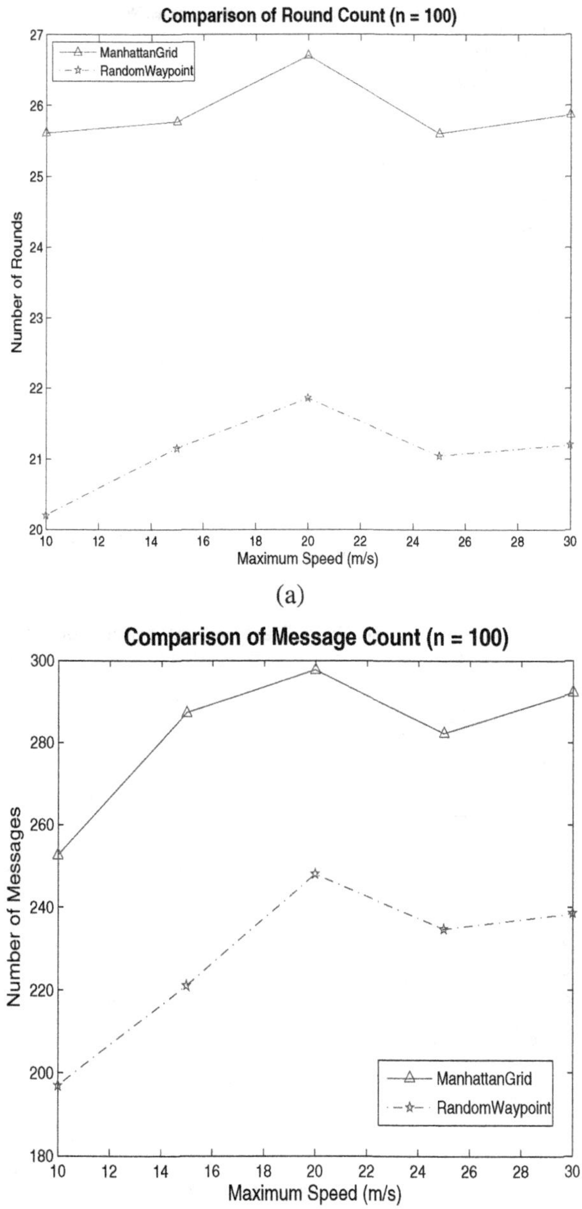

(b)

FIGURE 5.30: Execution time of algorithm *bf_broadcast*

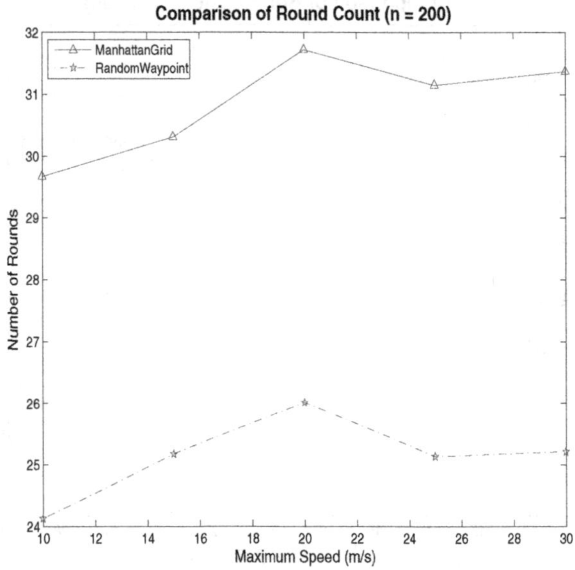

(a)

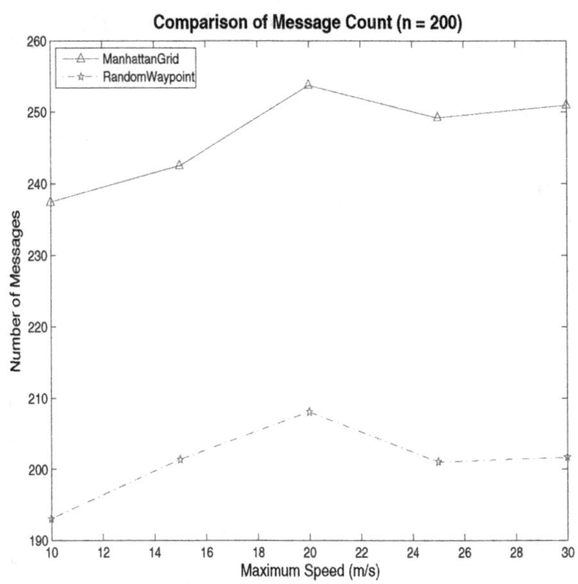

(b)

FIGURE 5.31: Execution time of algorithm *bf_broadcast*

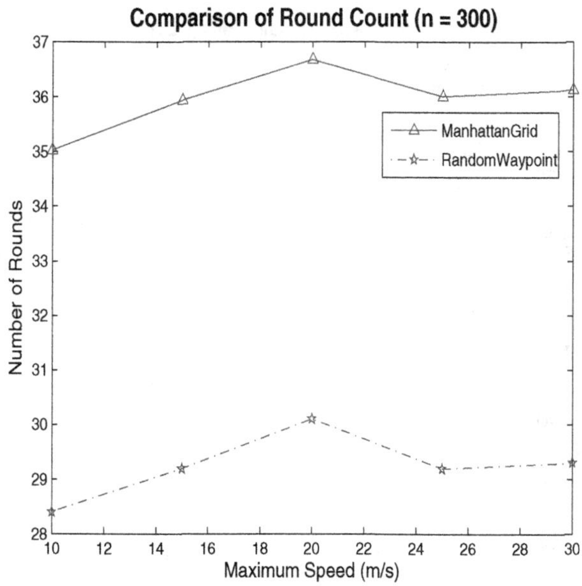

(a)

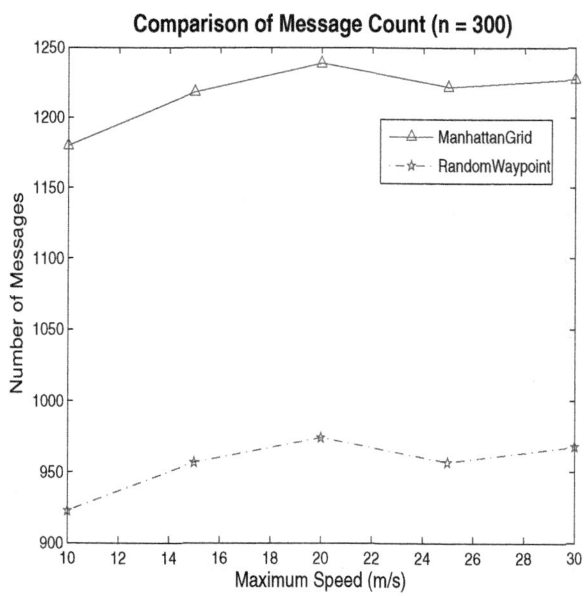

(b)

FIGURE 5.32: Execution time of algorithm *bf_broadcast*

erage case performance of this algorithm is upper bounded by $8D \log\ n$. It may be noted that the asymptotic lower bound for deterministic broadcast in networks where spontaneous transmission by the nodes is not possible is given by $\Omega(D \log\ n)$ [5].

5.5.3 Gossiping

Under the collision detection model, we extend the above idea of breadth-first broadcast to devise an algorithm for gossiping among the nodes of ad hoc mobile network so that every node u has some message m_u to be broadcast to all other nodes of the network.

5.5.3.1 Basic Idea

In the algorithm *bf_broadcast*, some of the nodes in *transmit* state at, say, layer $j-1$, are selectively dropped so that eventually a receiving node in the layer j would have exactly one transmitting node as its neighbor. A node in layer j receiving the message without any collision sends an *ack* signal in the S slot.

Let m_u denote the message that node u wishes to broadcast. Each node maintains a message array of length n. The message from node i is stored in the ith location of this message array. Let $\mathcal{M}_u$ denote the message set of a node u, which consists of all clean messages that u has received so far. The source node s that initiates the gossip algorithm initially has only one message (m_s) in its message set. Each neighbor of s subsequently appends its own message m_u and all other messages it has received from its neighbors to form its message set $\mathcal{M}_u$. When u moves to the *transmit* state, it transmits $\mathcal{M}_u$. The algorithm terminates when every node in the network has received the message of every other node in the network.

Specifically, consider the situation in which the nodes $u_1, u_2, \cdots, u_k$ all receive the message set $\mathcal{M}_u$ without any collision from only one single transmitting parent node u, as shown in Figure 5.33.

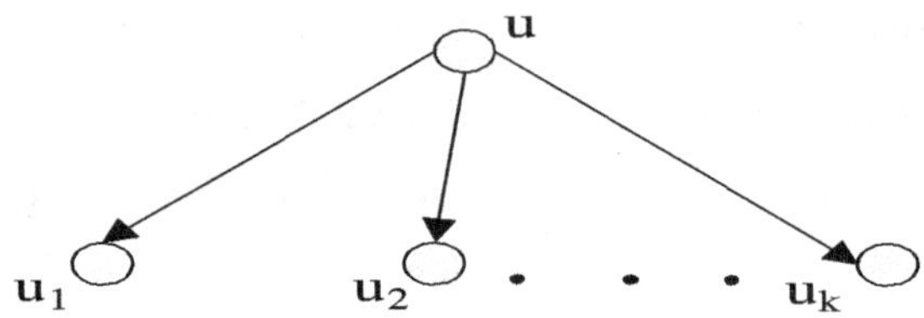

FIGURE 5.33: Collision-free message transmission from node *u* to its neighbors

Subsequently u receives an *ack* signal in the S slot. Now, the node u, instead of immediately changing its state to *active* as in the *bf_broadcast* algorithm, would first execute a procedure *gather_messages* to collect the message sets to be broadcast by each of the nodes $u_1, u_2, \cdots, u_k$, and then will change its state to *active*. Also, each of the nodes $u_1, u_2, \ldots, u_k$ receiving a clean message will change its state to a new state *back transmit*, in which it would respond to all control signals generated by u for message gathering, instead of immediately going to *active* state as in the *bf_broadcast* algorithm. By executing this modified algorithm, which we call *bf_gossip*, all the messages that each of $u_1, u_2, \ldots, u_k$ in layer j have heard, move one layer up, i.e., to the node u in layer $j - 1$. Hence, if D is the diameter of the network, D number of such calls to algorithm *bf_gossip*, would cause the messages of all nodes to reach the root node s. One more call to the algorithm *bf_broadcast* by s would complete the required message transfers for gossiping.

The procedure *gather_messages* called by a node u will successively select the neighboring node with the maximum *id*, that is yet to transmit its message list $\mathcal{M}_{u_i}$ to u, from the nodes $u_1, u_2, \cdots, u_k$, by using the function *find_next_max_id* as described later. Node u collects all such messages. Note that when u will send a control message for this purpose, only its neighbors in *back transmit* state would respond. After a node u_i sends its message set $\mathcal{M}_{u_i}$ to u, it changes its state to *active*. Finally, on return from the procedure *gather_messages*, node u changes its state to *active*.

To implement the above idea, we augment the round structure described in the

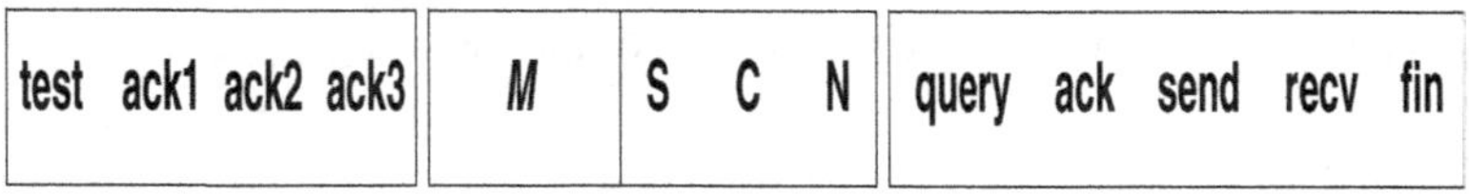

FIGURE 5.34: Structure of round for algorithm *bf_gossip*

bf_broadcast algorithm (Figure 5.20) with five additional slots: i) *query*, ii) *ack*, iii) *send*, iv) *recv* and, v) *fin*. In the *query* slot, a node in *transmit* state that has received an S signal only will try to find its neighbor in *back transmit* state with next maximum *id* by executing the *find_next_max_id* procedure. Each node in *back transmit* state that hears a signal during the *query* slot either responds by sending its own *id* or remains silent during the *ack* slot, depending on the bit pattern of its address. The *find_next_max_id* procedure can last for one or more rounds. A neighbor j selected thus by a node will change its state to *active* after transmitting its message set $\mathcal{M}_j$ and not respond to further *find_next_max_id* calls. A node u will send a command in the *send* slot to its neighbor j, selected by executing the *find_next_max_id* procedure, to send its message set during the *recv* slot. During the *recv* slot, j will send its message set back to u. Node j indicates the end of transmission of its message set by sending a signal in the *fin* slot. When node u hears a clean or collided signal in the *fin* slot, it resumes execution of the *find_next_max_id* in the subsequent round, until all neighbors in *back transmit* state have been touched. The purpose of the *fin* slot is to enable the use of multiple rounds of *query* and *recv* slots to execute the *find_next_max_id* procedure and to receive the message set of chosen neighbor j. Figure 5.34 shows the structure of a round used for implementing the gossip algorithm.

We introduce a fifth possible state of a node, called *back transmit*, in addition to the *receive*, *collision*, *active* and *transmit* states as mentioned in the previous section.

For the function *find_next_max_id*, we use the following notations: Suppose the id of a node is j. Let $m = \lceil \log\ n \rceil - 1$. j will then be represented by $\lceil \log\ n \rceil$ bits as $j = j_m j_{m-1} \cdots j_0$. We denote the portion of the bit string $j_m j_{m-1} \cdots j_k$ by the symbol $[j]_{m,k}$.

5.5.3.2 Algorithm bf_gossip

We describe in this section the modifications required to *bf_broadcast* to convert it to the algorithm *bf_gossip*. First, a node i gathers the clean messages received by all its neighbors by executing the procedure *gather_messages*.

Next, for a node in state *transmit*, Step 4.4 of algorithm *bf_broadcast* is modified as follows:

Step 4.4: If an ack was received only in slot S but not in C and N, execute procedure *gather_messages* and change the state back to *active*. Increment $round\_count$ by 1 and return to Step 3 for the next round.

Also, if the state of a node is *receive*, Step 5.2 for the case when a clean signal is received in the *message* slot will now be as follows:

Step 5.2: If a clean signal is received in *message slot*, send an acknowledgment in S slot, and change the state of the node to *back transmit*. Increment $round\_count$ by 1 and go back to Step 3 for the next round.

Further, if the state of a node is *collision*, Step 6.4 will be changed as follows:

Step 6.4: If a clean signal is received in *message slot*, send an acknowledgment in S slot, and change the state of the node to *back transmit*. Increment $round\_count$ by 1 and go back to Step 3 for the next round.

Finally, a procedure *back_transmit_actions* is needed which will perform the actions corresponding to the *back transmit* state.

5.5.3.3 Complexity

It follows from the above discussions that the procedure *gather_messages* will need at the most $2\lceil \log n \rceil \Delta$ rounds, where Δ is the maximum degree of a node. However, when we sum over all such nodes u during one complete pass of the algorithm *bf_gossip*, the actual time needed to execute all the calls to procedure *gather_messages* will be limited to $2\lceil \log n \rceil (n-1)$ rounds.

Hence, the overall worst case time complexity of the algorithm for gossiping is equal to $[13(2n+D-2)\log n + 26\lceil \log n \rceil (n-1)](D+1)$ time slots = $O(Dn \log n + D^2 \log n)$, which is better than the $O(n^{3/2} \log^2 n)$ time in [16]. For large values of Δ, the time complexity of the above algorithm is comparable to the $O(D\Delta^2 \log^3 n)$ time algorithm in [17].

5.6 Point-to-Point Routing in Ad Hoc Networks

The point-to-point routing protocols in ad hoc networks can be broadly classified as:

1. Proactive routing protocols
2. Reactive routing protocols
3. Hybrid (both proactive and reactive) routing protocols

In proactive protocols, the routing information from a source to a destination is maintained in advance. Thus, every node computes and stores a route to every other node in the network *(route creation activity)*. These routes may be updated *(route maintenance activity)* periodically or may be triggered by some event (e.g., a link error when a previously existing link is removed due to mobility of nodes or reduction in transmission power of nodes, other changes in the network affecting the topology, addition of new nodes, changes in traffic load, etc.). Most commonly used proactive routing protocols include *Destination-Sequenced Distance Vector Routing (DSDV)* [21] and *Optimized Link State Routing (OLSR)* [20].

In reactive routing protocols, routes are not computed in advance. Rather, routes are discovered only when needed. If a source node wants to send a message to some destination node, it first examines its local routing table to see whether a route to the desired destination already exists. If no such route exists, it initiates a *route discovery* process to find a routing path to the destination node. Examples of such reactive routing protocols include *Ad Hoc On-Demand Distance Vector Routing (AODV)* [24], *Dynamic Source Routing (DSR)* [23], and *Temporally-Ordered Routing Algorithm (TORA)* [25].

Reactive routing protocols perform better than the proactive routing protocols due to reduced overhead in route discovery and route maintenance for low to moderate traffic loads in the network. However, if the traffic load is heavy, the signaling overhead in the route discovery of reactive protocols tends to become equal or even surpass that in proactive routing protocols. On the other hand, if the nodes in a MANET have a high mobility or network conditions change very frequently due to addition or removal of nodes or changing transmission conditions, the route maintenance overhead in proactive protocols may rise significantly (being triggered by specific events) and in these situations, reactive protocols may be better choices. Reactive protocols suffer from an additional drawback of a *latency* period of acquiring the route from the instant when the source node needs a route to a given destination. In proactive protocols, there is no such latency as the routes to all destinations are computed in advance at each source node. Due to these advantages and disadvantages of both proactive and reactive protocols, hybrid protocols have been proposed by some authors to benefit a mix of features such as *Zone Routing Protocol (ZRP)* [22] and *Distance Routing Effect Algorithm for Mobility (DREAM)* [4].

Below, we briefly describe the basic features of the following routing protocols:

- DSDV
- AODV
- DSR
- TORA
- ZRP

5.7 Destination-Sequenced Distance Vector Routing (DSDV) Protocol

In this protocol, every node maintains the most recent routing information to every other node in the network in the form of a routing table. The routing table at a node contains the following fields: destination address, next-hop node, hop count (number of hops needed to reach the destination), destination sequence number, and install time (to indicate the time when this entry was last updated in the table). The destination sequence number helps to store the information regarding the most recent route to the destination. Thus, if information regarding two routes to a destination D reach a given node S, S chooses the route having the larger destination sequence number. Further, if two sets of routing information to the same destination D reach S with the same destination sequence number, S chooses the one with the smaller hop count. The install time may be used to delete stale entries from the routing table.

This protocol allows both periodic and event-driven updates in the routing table of a node in a proactive manner. For periodic changes, every node broadcasts routing table updates to its neighbors along with the current sequence number. On each advertisement (broadcast to neighbors), the destination sequence number is always increased by two (only even numbers are used). If a node is not reachable (the node may have left the network), the corresponding destination sequence number is increased by unity (an odd sequence number) and the corresponding hop count entry is set to ∞. Routing table update information can be broadcast by replacing the full routing table or issuing incremental updates in the routing table covering only the entries which changed. The latter scheme reduces the overhead in terms of communication time and bandwidth consumed for route maintenance.

There may be a problem called *route fluctuation* associated with the $DSDV$ protocol. Consider two nodes A and D with an entry in the routing table of node A corresponding to the destination node D as $[D, P, 14, D - 50]$ (disregarding the *install time* entry for the example). P is the next-hop node from A to reach D with 14 hops and the destination sequence number for D is 50. Suppose now D makes a broadcast with sequence number 52. Sometime later, node A receives this update for reaching the destination D from one of its neighbors, say Q, with hop count 12 and destination sequence $D - 52$, so that the corresponding update for the entry at node A will be $[D, Q, 12, D - 52]$. A may decide to broadcast this change to its neighbors. After some time, A receives an update message from another neighbor, say R, with hop count 11 and destination sequence $D - 52$. A has to make changes in the corresponding routing table entry as $[D, R, 11, D - 52]$, as the hop count is smaller for this route. Thus, A needs to rebroadcast the route to D with the same destination sequence number. This can happen every time D makes a broadcast and A receives messages from different neighbors at different times, leading to *fluctuation* in $DSDV$. This is undesirable as it may cause a lot of overhead by consuming network bandwidth. Such fluctuations can, however, be damped if A records the new route to D whenever it receives this information from any of its neighbors, but A does not broadcast this information immediately to other nodes in the network. Rather, after receiving the first update message from one of its neighbors, A waits for possible arrival of the new route information from its other neighbors as well, all having the same destination sequence number. After this settling time, A broadcasts final route information to all of its neighbors.

The advantages of the DSDV protocol are that it guarantees loop-free routes and using the destination sequence number, it always chooses the most recent route with the minimum hops. The disadvantages are that it does not support multipath routing-incurs large overhead due to maintaining a routing table at each node and updating the routing information, particularly in low traffic conditions.

5.8 Dynamic Source Routing (DSR)

Dynamic source routing (DSR) is an example of a reactive routing protocol. This is a source routing protocol , i.e., each packet to be routed from a source node carries in its header the complete ordered list of nodes through which the packet must pass. This protocol has two distinct phases: *route discovery* and *route maintenance*. When a source node S wants to send a packet to a destination node D and S has no information regarding the route to D, node S has to find a suitable route and this is done in the *route discovery* phase. In the *Route Maintenance* phase, a new route is found if there appears a need for updating an already existing route between a given source and destination pair due to a special situation such as a link failure.

Every node u maintains a route cache to store the information regarding all the discovered routes to all other nodes, and also the routes from other nodes to u. Thus,

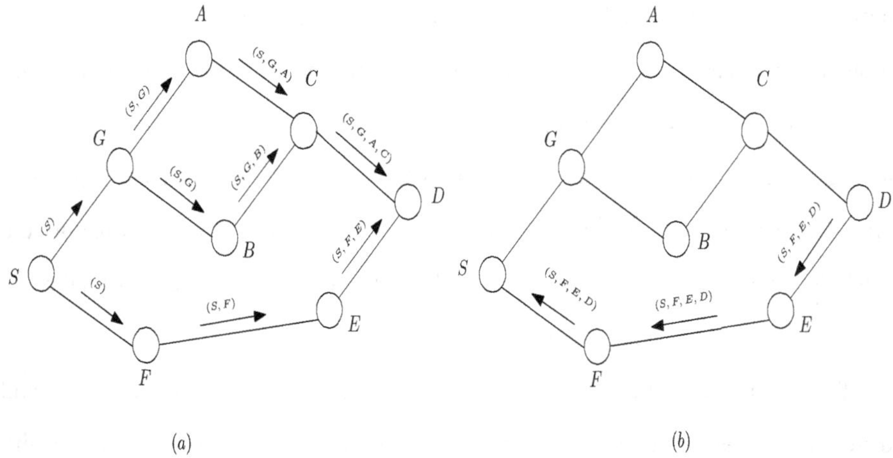

FIGURE 5.35: Dynamic source routing

when S needs to send a packet to a destination node D, and S does not find an entry in its route cache for a valid (unexpired) route to D, S sends a *route request* message $RREQ$ to all of its neighbors. The $RREQ$ contains the source id S, destination id D and a unique identification number for the request from S. When node u receives this $RREQ$ message, it looks into its own route cache to see whether it has an entry for a route from u to node D. If no such entry exists and node u is not the destination node D, u forwards this $RREQ$ to all of its neighbors, appending the currently known route from the source S to u. If an entry for a route from u to D is found in the route cache of u or u is the destination node D, the accumulated route information from node u to D is appended to the $RREQ$ message to generate a route from S to D and the route is sent from u to S (unicast) in the form of a route reply message $RREP$, by traversing the discovered route in the backward direction.

Refer to Figure 5.35(a). The $RREQ$ message (S) containing the id of the source node S reaches nodes G and F. Node G sends the $RREQ$ message containing the information (S, G) regarding the traversed nodes S and G to its neighbors A and B. Similarly, node F sends the $RREQ$ message to node E containing the information

(S, F) about the traversed nodes S and F. Node C receives two $RREQ$ messages, one containing (S, G, A) from node A and later a message containing (S, G, B) from node B. C ignores the latter message and sends only (S, G, A, C) to the node D. D also receives another $RREQ$ message from S with the same identification number containing the information (S, F, E) of the traversed nodes. D sends a $RREP$ message to S containing the route information (S, F, E, D), traversing this route backward through the nodes E and F in order, as shown in Figure 5.35(b).

After receiving an $RREQ$ message from a source node S, if any node u (including D) receives the $RREQ$ message from S with the same identification number once again through some other route, it can store this new path information from S to u in its cache and can also send another $RREP$ message back to S if u has route information from u to D (note that in the special case, u can also be D). This allows multiple routes to be known between a source and destination pair which enables multipath routing of packets from a source to a destination and also facilitates easy route maintenance. For example, if an existing route to the destination breaks, the node upstream of the link break, say node w, sends a *route error* message $RERR$ to the source node S. If S has an alternative route to D stored in its route cache, this new route can be chosen by S without the need for initiating a fresh $RREQ$ message by S. If an intermediate node (in the reverse route from w to S) receiving this $RERR$ message also has alternative route information from w to D without involving the broken link, that node can easily perform *route salvaging* by appending this alternative route from w to D to the route from S to w. However, even in this case, the $RERR$ message is to be sent back to S to let S know about this error and the corresponding salvaging action taken by w.

This protocol uses a reactive approach which eliminates the need to periodically flood the network with table update messages that are required in a proactive approach like $DSDV$. In contrast, in this protocol, a reactive (on-demand) approach is followed since a route is established only when it is required and hence the need to find routes to all other nodes in the network required by a proactive approach is eliminated. The intermediate nodes also utilize the route cache information efficiently to reduce the control overhead during route discovery and route maintenance. The disadvantage is that the route maintenance mechanism may not allow local repair of a broken link if there is no information for an alternative route to D at node w (on the upstream side of the broken link). Stale route cache information could also result in inconsistencies during route reconstruction. The latency or connection setup delay is higher than in proactive protocols, particularly in large networks. Even though the protocol performs well in static and low mobility environments, the performance degrades rapidly with increasing mobility.

5.9 Ad Hoc On-Demand Distance Vector Routing (AODV)

The *Ad hoc on-demand distance vector (AODV)* routing protocol belongs to the class of reactive routing protocols that blends some of the features from both $DSDV$ and DSR protocols. The protocol is based on an on-demand route discovery. However, every node also maintains a routing table for storing the information regarding the most recent routing path to a destination node if such a route has already been discovered. Each such route has an associated lifetime. When a source node has data packets to be sent to a destination node, it first checks its routing table to see whether there already exists a valid or unexpired (i.e., within the lifetime) route from this source node to the required destination. If a route already exists, the source node uses it. Otherwise, the route discovery process is initiated.

The route discovery process starts with broadcasting a route request message $RREQ$ to all the nodes in the network. The message has several fields containing the addresses of the source node and the destination node, a request identification number which uniquely identifies the $RREQ$ message from this source node, hop count (initialized to zero) and the last known sequence number for the destination. The request identification number is used in order to avoid processing duplicate $RREQ$ messages at any intermediate node or destination node when the same $RREQ$ messages with the same identification number reach that node through different paths.

When the $RREQ$ message originating from a source node S is received by some node u from a node v belonging to the set of immediate neighbors of u, the following actions will be performed:

Step 1: The node u first checks whether it is the destination node. If yes, u prepares itself for sending a route reply message $RREP$ as discussed later. Otherwise, u checks whether it already received an $RREQ$ message from S with the same request identification number. If not, then Step 2(a) as given below is executed; otherwise Step 2(b) will be executed.

Step 2(a): u first creates reverse route information by storing the address of node v in the reverse route (from u to S) along with the hop count information in the received $RREQ$ message, and then executes Step 3 as given below.

Step 2(b): Since u has already processed this $RREQ$ message (received from some other neighbor), u checks whether its stored reverse route information (u to S) has a hop count value smaller than that in the received $RREQ$ message. If yes, then ignores this $RREQ$ message; otherwise it will modify the hop count value in the stored reverse route information with the smaller hop count value in the received $RREQ$ message and execute Step 3 as given below.

Step 3: u checks whether it has route information to the destination node with a sequence number greater than or equal to the destination sequence number in the received $RREQ$ message. If yes, this route information can be used to send a route reply message to the source node S as explained below; otherwise, a fresh route needs to be discovered by executing Step 4 as given below.

Step 4: The hop count field of the $RREQ$ message is incremented by unity and the modified $RREQ$ message is broadcast to the neighboring nodes of u other than v.

In essence, the above route discovery action is based on flooding the whole network with the $RREQ$ message to reach the destination node along different source and destination paths in the network. For an illustration of this route discovery process, consider the network shown in Figure 5.36. The destination sequence number is not shown. The fields of the $RREQ$ message forwarded from a node to its neighbors along any possible route as shown in the figure contains the source id, destination id and hop count in order. The $RREQ$ message forwarded from every node is shown on the corresponding link. Thus, when node F receives the forwarded message with hop count 2, F ignores this message as it has already received an $RREQ$ message with a hop count of 1. Also, node C receives two $RREQ$ messages from A and B, respectively, and only one of them, say the one from A, which reaches C first, is processed.

When the $RREQ$ message is received by the destination node D, D sends a route reply message $RREP$ back to the source node S. The $RREP$ message contains fields for the addresses of the source and the destination nodes, destination sequence number, hop count which is initialized to zero, and the forward route information (from S to D). Initially, the forward route information in the $RREP$ message formed by D is empty. The destination sequence number is incremented from the sequence number used by D in the last $RREP$ message it sent. The destination node D then unicasts this $RREP$ message to its neighbor node from which it received the $RREQ$ message (as known from its corresponding entry for the reverse route information). When an $RREP$ message is received by a node u from one of its neighboring nodes,

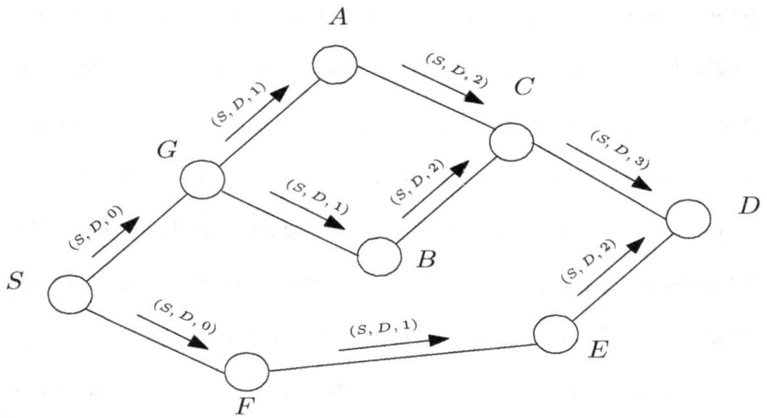

FIGURE 5.36: Request flooding

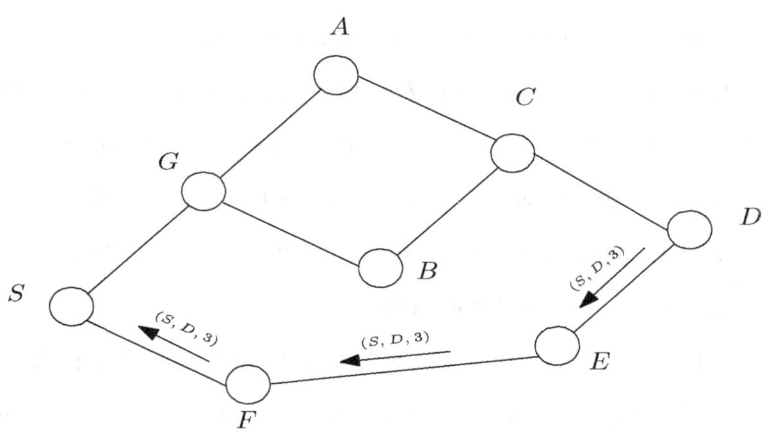

FIGURE 5.37: Request reply

say v, u augments the forward route information of the received $RREP$ message by concatenating the address of the node v to this forward route, and the hop count field is incremented by unity. Thus, referring to Figure 5.37, the destination node D sends back the $RREP$ message with hop count 3 to the source node S backward along the source-destination route (S, F, E, D). Every node has the information about the node id from which it received the corresponding $RREQ$ message, i.e., node D knows that the $RREQ$ message came from E and so it sends the $RREP$ message to E; E knows that it received the $RREQ$ message from F and so it forwarded the $RREP$ message to F; and finally F sent the $RREP$ message to S.

If D later receives another $RREQ$ message originating from S with the same request identification number and a smaller hop count value, D would initiate sending a fresh $RREP$ message back to S with the new route information involving smaller path length.

However, as described above in the route discovery process, it is also possible that the $RREQ$ message is received by an intermediate node u (u is not the destination D) which has a route information from u to D with a higher destination sequence number than that in the received $RREQ$ message. In that case, u sends an $RREP$ message (instead of broadcasting the $RREQ$ message any further) to its neighbor from which it received the $RREQ$ message, by setting the information regarding the route from u to D already stored at u in the forward route information entry of the $RREP$ message, and also setting the hop count field in the $RREP$ message as that of the route from u to D stored in its table.

After forming the route, the source node S starts sending the data packets to the destination D through this route. However, once the route is established, it must be maintained for the duration of sending the packets from S to D. However, if a link breaks due to mobility of nodes or any other reason during this message sending period, similar to the DSR protocol, a route error message $RERR$ is initiated at node u which is on the upstream side of the broken link. This $RERR$ message is sent back to the source node S. Every node that receives this $RERR$ message on the reverse path from u to S invalidates any route stored in its table that might have been using this broken link, and then initiates a fresh route discovery to the corresponding destination nodes. An alternative approach might be that node u first tries to locally repair the broken link problem by finding an alternative route from u to the required destination D which is then unicast to the source node S. If u fails to find such a route, only then a fresh route discovery process will be initiated by S. This latter approach would take less time to repair the link error and would also involve less message communication overhead in the network.

5.10 Temporally Ordered Routing Algorithm (TORA)

The *Temporally-ordered routing algorithm (TORA)* [25] enables us to have a source-initiated on-demand routing protocol based on finding multiple routes from a source node S to a destination node D. This protocol uses three basic activities: *route creation, route maintenance and route erasing*.

The source node S starts route creation by broadcasting a $QUERY$ packet with the destination node's id in it. This packet propagates through the network until it reaches either the destination node D or some other node, say u, which has already found a route to the destination D. When such a node (D or u, as the case may be) receives the $QUERY$ packet, it sends back an $UPDATE$ message noting its height in the network with respect to the destination node D (by initially setting the height of the destination node as zero). Any node v that receives this $UPDATE$ message with a height value of, say, h (which is actually the height of the neighbor from which v received this message), increases its own height to $h+1$ and forwards the $UPDATE$

message to all of its other neighbors indicating this increased height value. If a node receives two or more $UPDATE$ messages corresponding to the same source and destination pair, it sets its height to the minimum of all height values computed from all the $UPDATE$ messages. This process is continued until the $UPDATE$ message reaches the source node S. Thus, a directed acyclic graph is formed connecting S and D if we visualize a directed edge from a node with a height value $h + 1$ to a node having the height value h, $h \geq 0$. The node with the larger height value forming a link is called the *upstream node* and the one with the smaller height value is called the *downstream node* corresponding to that link. If we now start from S and follow the nodes with successively smaller height values along this directed acyclic graph, all possible routes from S to D can be computed. An example is given in Figure 5.38. The $QUERY$ message is broadcast from S to reach D along different paths as shown by the arrows in Figure 5.38(a). If node C has received the $QUERY$ message from node A followed by that from node B, it ignores this duplicate $QUERY$ message received from B and does not forward it again to the node D. In Figure 5.38(b), the destination node is shown to have the height 0 which sends the $UPDATE$ message to nodes C and E with this height value. Nodes C and E then both set their heights as 1 and send the $UPDATE$ message with this height value to their neighbors A, B and F, all of which then set their heights as 2. Note that node A would also later receive an $UPDATE$ message from node B with the height value 2, but then A does not set to its height to $2 + 1 = 3$; rather A sets its height to 2 (minimum of 2 and 3).

The action taken by *TORA* has some resemblance to water flowing downhill from a source along a network of tubes toward a destination. Water always flows from a junction of tubes (a node in the network) to other junctions (nodes in the network) which are at a lower height along the respective tubes (links in the network). In case of a broken link due to mobility of nodes in the network, *route maintenance* is needed to be considered by the *TORA*. This will be similar to the situation in water-

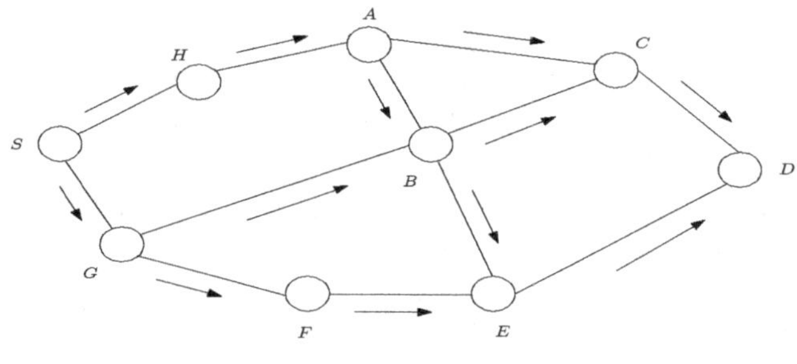

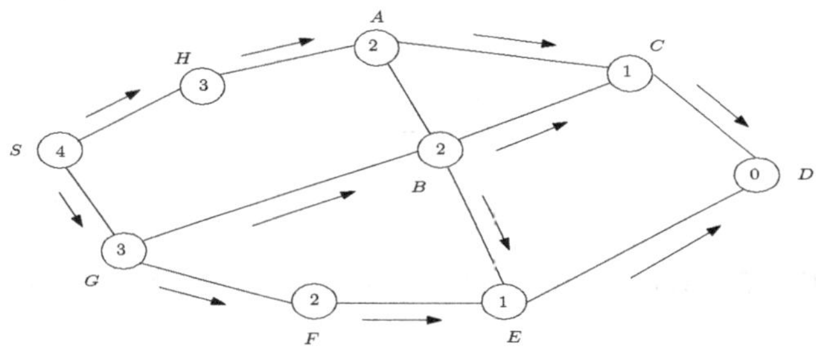

FIGURE 5.38: Route establishment using TORA

flow network when some tube is blocked. In that case, the junctions connecting the blocked tube are placed at some larger height so that water blocked in the tube can eventually flow down to the destination point. Analogously, when a link is broken in *TORA*, a route is found from the upstream node of this broken link toward the source by local changes in the height of the node, e.g., the height of this upstream node is set to one more than the maximum of the heights of all of its neighbors and this process is percolated through all the blocked nodes up to the source node. As an example, referring to Figure 5.39, assume the nodes C and B move away from their neighbors so that the links between C and D, between B and G, and between B and E are all broken. Then the heights of nodes C, B, A and H are finally adjusted to the values 7, 7, 6 and 5, respectively, so that the routes from C, B, A and H toward the destination D are all maintained.

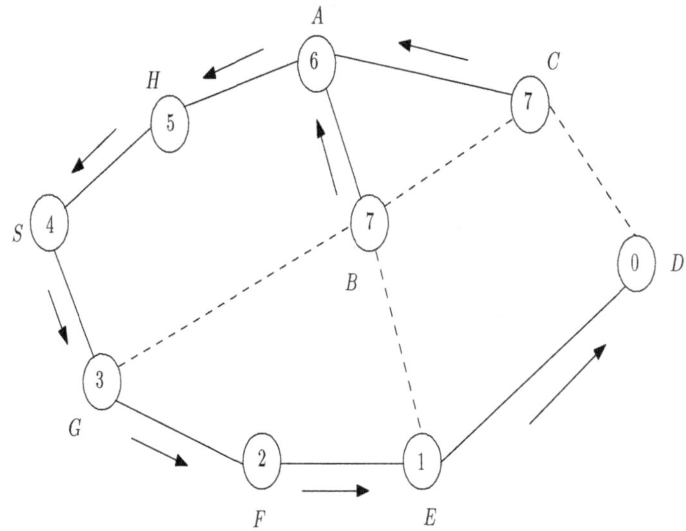

FIGURE 5.39: Route adjustment in TORA due to change in network topology

Finally, when a node u detects a network partition, i.e., a portion of the network containing u becomes separated from the destination so that the destination node is no longer reachable from u, u issues a *route erase* message for removing all invalid routes from u to the destination.

5.11 Zone Routing Protocol (ZRP)

The *Zone routing protocol (ZRP)* proposed in [22] has the combined features of both proactive and reactive protocols. Around each node, a *zone* is defined whose area is measured in terms of the number of hops to cover the entire area. Each node follows a proactive routing protocol within its zone and a reactive routing protocol outside its zone. Thus, referring to Figure 5.40, assume that the zone is defined with two hops. Thus, the source node S has the zone around it containing the nodes A,

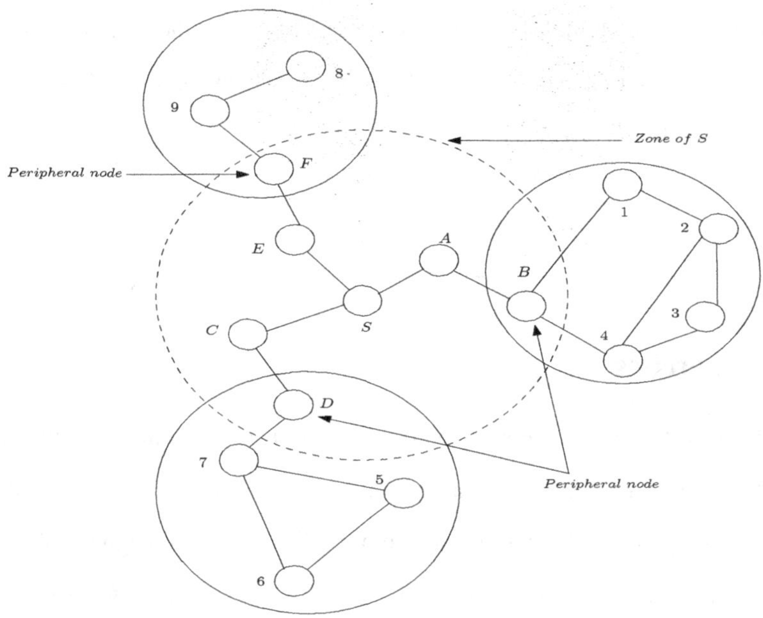

FIGURE 5.40: Zone routing

B, C, D, E and F, with nodes B, D and F at Distance 2 (two hops) from S. B, D and F are the peripheral nodes of the zone around S. Since intrazone routing is proactive in nature, S first checks whether its destination node is within this zone, by checking its routing table. If yes, the routing is readily done. If not, ZRP utilizes an *interzone routing protocol* by discovering the route to the destination outside its zone with the help of these peripheral nodes by a process known as *bordercasting*. Thus, S sends a route discovery message to its peripheral nodes B, D and F. Each peripheral node then broadcasts a route discovery to all nodes in its zone (in this example, within Distance 2 from each of B, D and F) to check whether the destination is found in any of their respective zones. If F finds the destination node within its zone, it proactively finds the route from its table to reach this destination and the overall route from S to the destination is found. If, however, none of the peripheral nodes reports success in locating the destination node, the bordercasting is repeated from the peripheral nodes of all the zones explored at this stage, and this process is continued until the destination is finally found. Thus, if node 2 in Figure 5.40 is the destination node, the possible discovered route is $S - A - B - 1 - 2$, of which the routes $S - A - B$ and $B - 1 - 2$ are found proactively.

5.12 Exercises

1. Consider an ad hoc network with 32 nodes. The node numbers range from 0 to 31. Assume that at some instant of time, nodes 0, 1, 3, 7, 11, 20 and 25 are all trying to broadcast a message to node 10 which is a neighbor of all seven transmitting nodes. Node 10 is yet to receive this message. However, simultaneous transmission from all the transmitting nodes would cause a collision at node 10. Assuming a radix-2 encoding of node numbers, determine a transmission schedule by which node 10 will receive this message from only one of the seven nodes. Find the number of transmission cycles for successful reception of the message by node 10.

2. Repeat the above exercise assuming radix 4 encoding of node numbers. Compare the numbers of transmission schedules required for successful reception of message by node 10 with radix 2 and radix 4 encoding of the node numbers.

3. Assume that more nodes join the network so that it has now 256 nodes, with node numbers ranging from 0 to 255. The set of transmitting nodes remains the same, i.e., 0, 1, 3, 7, 11, 20 and 25 are simultaneously trying to send a message to node 10. Recompute the required transmission schedules assuming radix 2 and radix 4 encoding of the node numbers. Estimate the required numbers of transmission cycles in both cases and then compare these two values.

4. Assume that the mobile terminals in a MANET are spread over a large re-

gion divided into hexagonal cells. The number of mobile terminals lying in a given cell of this overlaying cell structure will indicate the cell demand. Also assume that two mobile terminals lying in two cells Distance 3 or more apart will not interfere with each other during transmission. Thus, two mobile terminals lying in two cells Distance 3 or more apart can use the same time slot for broadcasting their messages. Based on these assumptions, describe an algorithm that will appropriately assign the time slots among the nodes in different cells in an efficient way. Compute the time complexity of your algorithm.

5. Assume a node has an id value of 00011 in binary. This node has three neighbors with binary id values of 11101, 11011 and 01110, all of which want to send a message to node 00011 at the same time. Following the strategy of breadth-first broadcast described in this chapter, explain how the collisions will be resolved and the node 00011 will finally receive the message correctly.

6. Describe a breadth-first broadcast algorithm for deterministic broadcasting in an ad hoc network, clearly explaining the underlying model and the working principle of the algorithm.

7. Prove the correctness of your algorithm for the above problem and also compute its worst case time complexity.

8. What are the basic differences of proactive, reactive and hybrid routing protocols in an ad hoc network? Give two examples of each of these categories of protocols.

9. Explain the AODV routing protocol and indicate in what respect it is different from the DSR protocol.

10. Describe the TORA protocol for routing in an ad hoc network. Illustrate a suitable example to explain the different actions taken by this protocol.

11. Explain the zone routing protocol with the help of an appropriate illustrative example.

Bibliography

[1] Z. J. Haas, J. Y. Halpern and L. Li, "Gossip-based ad hoc routing, *Proc. IEEE Infocom*, 2002.

[2] P. K. Srimani and B. P. Sinha, "Mobility tolerant broadcast in mobile ad hoc networks ," *Proc. Intl. Workshop on Distr. Comput. (IWDC 2004), LNCS 3326*, pp. 435–446, 2004.

[3] S. Basagni, D. Bruschi and I. Chlamtac, "A mobility-transparent deterministic broadcast mechanism for ad hoc networks," *IEEE Trans. Networking*, vol. 7, pp. 799–807, December 1999.

[4] S. Basagni, I. Chlamtac, V. Syrotiuk and B. Woodward, "A distance routing effect algorithm for mobility (DREAM)," *Proc. 4th Annual ACM/IEEE Intl. Conf. on Mobile Computing and Networking (MobiCom)*, pp. 76–84, Dallas, October 1998.

[5] D. Bruschi and M. D. Pinto,"Lower bounds for the broadcast problem in mobile radio networks," *Distributed Computing*, vol. 10, pp. 129–135, 1997.

[6] K. Sinha,"Optimal time slot assignment for mobile ad hoc networks," *Proc. 7th Intl. Workshop on Distributed Computing (IWDC)*, LNCS 3741, pp. 250–261, 2005.

[7] K. Perumal, R. K. Patro and B. Mohan, "Neighbor-based TDMA slot assignment algorithm for WSN," *Proc. IEEE INFOCOM*, 2005.

[8] W.-H. Liao, Y.-C. Tseng and J.-P Sheu, "GRID: a fully location-aware routing protocol for mobile ad hoc networks," *Telecom. Systems*, vol. 18, pp. 37–60, 2001.

[9] Y.-C. Tseng and T.-Y. Hsieh, "Fully power-aware and location-aware protocols for wireless multi-hop ad hoc networks," *Proc. IEEE Int. Conf. on Comp. Comm. Networks*, 2002.

[10] K. Nakano and S. Olariu, "Randomized initialization protocols for ad-hoc networks," *IEEE Trans. Parallel Distributed Systems*, vol. 11, pp. 749–759, 2000.

[11] K. Nakano and S. Olariu, "Uniform leader election protocols for radio networks," *IEEE Trans. Parallel and Distributed Systems*, vol. 13, pp. 516–526, 2002.

[12] K. Nakano and S. Olariu, "A survey on leader election protocols for radio networks," *Proc. Intl. Symp. on Parallel Architectures, Algorithms, and Networks (I-SPAN)*, pp. 71–76, May 2002.

[13] K. Sinha, S. Ghose and P. K. Srimani,"Fast deterministic broadcast and gossiping algorithms for mobile ad hoc networks," *J. Parallel Distributed Computing*, Vol. 68(7), pp. 922–938, July 2008.

[14] K. Sinha and P. K. Srimani, "Broadcast and gossiping algorithms for mobile ad hoc networks based on breadth-first traversal," *Proc. 6th International Workshop on Distributed Computing* (IWDC), *Lecture Notes in Computer Science 3326, Springer*, pp. 459–470, December 2004.

[15] K. Sinha and P. K. Srimani, "Deterministic broadcast and gossiping algorithms for ad hoc networks," *J. Supercomputing* 37(2), pp. 115–144, 2006.

[16] M. Chrobak, L. Gasieniec and W. Rytter, "Fast broadcasting and gossiping in radio networks," *Proc. 41st IEEE Symp. Found. of Computer Science*, pp. 575–581, 2000.

[17] A. E. F. Clementi, A. Monti and R. Silvestri, "Round Robin is optimal for fault-tolerant broadcasting on wireless networks," *Proc. 9th Annual European Symp. on Algorithms*, pp. 452–463, 2001.

[18] The network simulator, http://www.isi.edu/nsnam/ns, 2006.

[19] Bonn Mobility and Networking Suite, http://web.informatik.uni-bonn.de/IV/bomonet/BonnMotion.htm, 2006.

[20] T. Clausen, P. Jacquet, A. Laouiti, P. Muhlethaler, A. Qayyum and L. Viennot, "Optimized link state routing protocol," *Proc. IEEE INMIC*, Lahore, Pakistan, December 2001.

[21] C. E. Perkins and P. Bhagwat, "Highly dynamic destination-sequenced distance-vector routing (DSDV) for mobile computers," *Computer Communications Review*, vol. 24, pp. 234–244, October 1994.

[22] M. R. Pearlman and Z. J. Haas, "Determining the optimal configuration for the zone routing protocol," *IEEE J. Selected Areas in Communications*, vol. 17, pp. 1395–1414, August 1999.

[23] D. B. Johnson and D. A. Maltz, "Truly seamless wireless and mobile host networking. Protocols for adaptive wireless and mobile networking," *IEEE Personal Communications*, vol. 3, pp. 34–42, February 1996.

[24] C. E. Perkins and E. M. Royer "Ad-hoc on-demand distance vector routing," *Proc. 2nd IEEE Workshop on Mobile Computing Systems and Applications*, pp. 90–100, 1999.

[25] V. Park and S. Corson, "Temporally-ordered routing algorithm (TORA) Version 1 Functional Specification," *Internet Draft, draft-ietf-manet-tora-spec-01.txt*, August 7, 1998.

Chapter 6

Wireless Local Area Networks

6.1 Introduction

A wireless local area network ($WLAN$) is used for providing location-independent network access by using radio waves rather than a cable infrastructure.

In $WLAN$ architecture, the region to be covered is subdivided into cells and the network communication follows the IEEE 802.11 protocol. Each cell is controlled by an access point (AP) which coordinates all communications that take place in that cell. The network is formed by several APs connected through a backbone network. The backbone is called the distribution system or DS and is typically an Ethernet. The stations ($STAs$) access the network through these APs. The main benefits of $WLAN$ lie in increased mobility of the nodes and cost-effective setup. In general, $WLAN$ supports different data rates. For example, IEEE 802.11b $WLAN$ supports $1, 2, 5.5$ and 11 $Mbps$ data rates, respectively, for the ISM 2.4 GHz frequency band [19]. It uses the direct sequence spread spectrum ($DSSS$) technique for 1 $Mbps$ and 2 $Mbps$ data rates. In $DSSS$, each information bit is combined via an XOR operation with a pseudo-random numerical (PN) sequence resulting in a high speed digital stream which modulates a carrier frequency using differential binary phase shift keying ($DBPSK$) and differential quadrature phase shift keying ($DQPSK$) for 1 $Mbps$ and 2 $Mbps$ operations, respectively. However, to provide the higher data rates of 5.5 and 11 $Mbps$, it uses the extension of $DSSS$ to a high rate technique known as the direct sequence spread spectrum ($RH/DSSS$) in combination with 4-bit and 8-bit complementary code keying (CCK), respectively [19]. It also uses *dynamic rate shifting* which allows the data rates to be automatically adjusted to changing radio channel conditions. For each rate, there is an optimal range for the successful operation. When $WLAN$ devices move beyond the optimal range for a higher rate operation, they will transmit at the next lower rate. Likewise, if the device moves back within the optimal range of a higher rate transmission, the connection will automatically speed up again. The optimal range for a rate depends on the radio channel condition which, in turn, depends primarily on the Euclidean distance, though other factors are also present. Most IEEE 802.11 vendors provide the optimal range for each data rate [22, 23] in accordance with their supplied devices. The basic access method for $WLAN$ is the distributed coordination function (DCF) which uses carrier sense multiple access with collision avoidance ($CSMA/CA$) . It requires each station to sense the channel before transmitting. If the channel is found to be idle, the station transmits. If the channel is busy, a station defers its transmission until the channel becomes idle and then starts transmitting. If a collision occurs, the colliding stations wait for a random time using a binary exponential backoff algorithm and then try again. In the infrastructure mode, authentication is established between an AP and STA before they start communicating.

6.1.1 Association Problem

The performance of an IEEE 802.11 $WLAN$ depends on how the $STAs$ are associated with the APs. For an infrastructure-based $WLAN$, an STA can send data frames via an AP only when it is associated with that AP. This requires an STA-to-AP association that is used by the distribution system (DS) to accomplish its distribution service. When many APs are deployed in a region of a IEEE 802.11 $WLAN$ network, an STA may be in the range of several APs. A well-known problem is which AP an STA selects. An inappropriate association of the $STAs$ to the APs may result in a degraded throughput for each of the $STAs$ and hence may pull down the aggregate throughput of the network. Thus, finding an appropriate association between $STAs$ and APs is an important design concern. Each AP will be allocated a frequency channel to serve all its associated $STAs$. The APs with common interfering ranges require the allocation of different frequency channels for satisfying the co-channel interference. If STA i is associated with AP j, the throughput obtained by it and the aggregate throughput obtained by all the $STAs$ will depend on the MAC scheduling used. We describe below different association policies under various MAC scheduling strategies.

6.1.2 Different Association Policies

When several APs are available, the AP selection of an STA heavily influences its performance and those of others. An association policy is an algorithm on the basis of which an STA is associated to an AP within its range. Intensive studies have been performed by various researchers to find a feasible association between $STAs$ and APs based on different metrics and objectives. Kumar et al. [2] address an optimal association policy between $STAs$ and APs with the objective of maximizing the throughput of the association, but ignored the co-channel interference by assuming the proper frequency planning of the APs. Many authors advocate load balancing to optimize performance. It has been shown [3, 6, 5] that the traditional association based on the received signal strength indicator ($RSSI$) measurement may lead to poor user experience due to non-uniform load distribution across different APs. To address this issue, some AP access control mechanisms are proposed [5, 6]. Bejerano et al. [3] proposed an association model considering both load balancing and fairness that includes a complicated central AP association mechanism.

Unlike the traditional $RSSI$, a heuristic has been proposed that helps the users to select the AP to be associated on the basis of the number of users already associated with that AP and the mean $RSSI$ [4].

Vasudevan et al. [7] proposed an AP-STA association policy based on the bandwidth that an STA is likely to receive if it is associated with that AP. Before a user is associated to an AP, it estimates the potential bandwidth of neighboring APs by passive measurement and selects the AP with the highest potential bandwidth for connection.

In [12], an AP association strategy is proposed. Before an STA associates to an AP, it calculates the minimal throughput for all the $STAs$. The STA finds its associating AP such that the minimal throughput for all the $STAs$ is maximized.

In [9], the AP selection is modeled as a many-to-one matching problem in a bipartite graph to maximize the total load among all the APs.

Another association policy [10] taken into account the throughput degradation of an STA before it joins an AP. The metric is a weighted sum of the throughput provided to the new STA and the average transmission time of $STAs$ already associated with the concerned AP.

Below we describe a few more association policies under various MAC scheduling strategies.

6.1.2.1 Strongest Signal First (SSF) Association

Strongest signal first(SSF) is one of the simplest association policies based on the metric of strongest signal-to-noise ratio measured by a received signal strength indicator ($RSSI$). According to this policy, an STA is associated to the AP from which it receives the strongest signal. As described in [4], the procedure for finding this metric is simple. An STA scans the available channels of each AP in the region and listens to the beacon or probe response frames. Depending on the country, one to fourteen channels per AP may exist. The STA stores the $RSSI$ of the beacon or the probe response frames and other relevant information (e.g., whether encryption is on or off). After finishing the scanning procedure, the STA selects that AP with the maximum $RSSI$ if the selected AP covers other requirements (usually the extended service set identification ($ESSID$), wired equivalent privacy (WEP) encryption) as well. The STA will leave the AP when the $RSSI$ falls under a predefined threshold. This procedure is based on the premise that the quality of service of the selected AP is the best. However, the result of this procedure is that many stations are connected to a few APs, while neighboring APs remain idle. This overloading of the APs due to non-uniform load distribution will lead to performance degradation [3, 6, 5].

6.1.2.2 Maximum Utility-Based Association

Kumar et al. [2] address an optimal association policy between the $STAs$ and the APs in their ranges in which finding an optimal association policy is considered the maximum utility-based formulation of the problem. The authors note that an STA associates to an AP with a physical bit rate and eventually such an association will yield an STA throughput dependent on the specific MAC scheduling used. They used two MAC scheduling techniques, namely random polling and proportional fair access, to calculate the throughput of an individual association. As per random polling access, the set of $STAs$ associated to a common AP achieves equal individual throughput. If there are m_j $STAs$ $i_1, i_2, \ldots, i_{m_j}$ associated with AP j and these are associated with rate $r_{i_p j}$ where $1 \leq p \leq m_j$, the aggregate uplink throughput obtained by all the $STAs$ served by AP j is of the form:

$$\frac{m_j}{\sum_{p=1}^{m_j} \frac{1}{r_{i_p j}}}.$$

The individual throughput obtained by the STA from AP j can be expressed as

$$t_j = \frac{1}{\sum_{p=1}^{m_j} \frac{1}{r_{i_p j}}}.$$

For fair allocation of user throughput, they used log utility to formulate the aggregate throughput of the network as given by

$$\sum_{j=1}^{n} m_j \ln(t_j).$$

Based on these facts, the association policy will be optimum when the $STAs$ are associated to the APs in range such that the logarithmic sum of the obtained individual throughput is maximized. Basically they used the log utility formulation while calculating the individual throughput. The advantage of the association policy is that it provides a fair throughput allocation to the individuals.

6.1.2.3 Max-Min Fair (MMF) Allocation

One well known association policy [2, 3] is known as the *max-min fair* (MMF) allocation. As stated earlier, the $STAs$ are associated to the APs in their range by obtaining physical bit rates. A rate vector is MMF if there is no other vector for which the rate of some node, say i, is higher without the rate of some other node, say j, that has a smaller than or equal to rate than i, being made lower. $STAs$ associated on the basis of MMF allocation of their individual data rates may obtain uniform or fair throughput allocation. However, such a policy may lead to a reduced network throughput [3]. Furthermore, instead of the obtained data rate, max-min fair allocation can be formulated with the obtained bandwidth. A system provides a fair service if all users have the same allocated bandwidths. Unfortunately, such a degree of fairness may cause significant reduction of the network throughput, since all users have the same bandwidth allocation as bottleneck users. As described in [3], a bandwidth allocation of a weighted system is called max-min fair if there is no way to increase the bandwidth of a user without decreasing the bandwidth of another user with the same or less normalized bandwidth.

6.1.2.4 Least Load First (LLF)

Another association policy known as the *least load first* (LLF) association is a widely used load-balancing heuristic defined in terms of load of an AP [3]. When an STA associates to an AP, it imposes a load on that AP. According to this association policy, an STA checks the load incorporated to an AP before associating it to the AP. A user chooses the least loaded AP, where an AP load is inversely proportional to the current bandwidth that its associated users receive. This policy may lead to unfair bandwidth allocation and poor load balancing.

6.1.2.5 Min-Max Load Balancing Allocation

An association policy aiming at providing fair load balancing across the APs is *min-max load balancing* [3]. The load of an AP needs to reflect its inability to satisfy the requirements of its associated users and it should be inversely proportional to the average bandwidth that they enjoy. A feasible association is min-max load balanced if its corresponding load vector has the same or lower lexicographical value than any other load vector of any other feasible assignment. According to such a policy, $STAs$ are associated to the APs with the objective of min-max load balancing [3]. Load balancing along with mean $RSSI$ can be used as the association metric [4]. According to such a metric, users select the AP on the basis of the number of users already associated with that AP and the mean $RSSI$ value.

6.1.2.6 Link Rate and Throughput-Based Association

Link rate and realized throughput together can be considered to formulate an association policy [13] for a multi-channel multiple AP network. The objective is to maximize the capacity region of the set of APs under a dynamic traffic model. Users arrive at a point within the network coverage area, associate with an access point, realize a download and leave. Network capacity is defined by the limiting load beyond which the number of active users would grow indefinitely as the arrival rate in some region is greater than the rate at which downloads complete. The limitation of such an approach is that using network capacity as a criterion for comparison does not necessarily consider how well policies perform at normal loads.

6.1.2.7 RSSI and Bandwidth-Based Association

According to this association policy [1], users associating to an AP providing the maximum throughput may change to another AP with maximum $RSSI$ depending on the available bandwidth. The drawback is that the change may also lead to an imbalance in the load distribution.

6.1.2.8 Radio Access Technology Association

In [8], a numerical approach based on stochastic dynamic programming proposes to find the optimal association for a network having a small topology consisting of two APs only. Using the insights from the association rule for the two-AP problem, a near-optimal heuristic named RAT, has been proposed. An objective function is defined based on both rate and throughput and then an STA associates with that AP for which is objective function is minimized.

6.2 Placement of APs and Channel Allocation

The AP selection and channel allocation are two fundamental access control problems. For users in the coverage area of one AP, this involves only channel allocation. However, for users that belong in the intersection of coverage areas of more than one AP, the selection of an appropriate AP to establish the connection is also important. Iordanis et al. [16] proposed a joint approach of AP selection and channel assignment with a view to satisfying a given user load vector with a minimum number of channels. The authors formulated this joint problem as a clique load balancing problem and using this as a guideline, a heuristic algorithm has been proposed.

Traditionally, AP placement and channel assignment are considered sequentially. The AP placement is often modeled as a facility location problem and the channel assignment probem is mapped to a graph coloring problem. Authors in [11] argued that treating these two key issues separately may lead to suboptimal designs. Accordingly, they proposed an integrated model based on mathematical programming that addresses both the aspects simultaneously to maximize the net throughput that an STA can enjoy.

In [17], AP placement and channel assignment were considered jointly to maximize the system throughput and fair resource sharing among mobile terminals. However, they allowed co-channel overlapping between cells.

6.3 Exercises

1. Distinguish two key concepts related to WLAN design: ad hoc mode and infrastructure mode.

2. Describe the operating principles of CSMA/CA to resolve collusion in WLANs.

3. State reasons why collusion detection is not a preferred method for a WLAN?

4. Explain the operation of DSSS and FHSS.

5. What is OFDM?

6. Distinguish the operations of DCF and PCF.

7. What are the differences between DBPSK and DQPSK?

Bibliography

[1] G. Athanasiou, T. Korakis, O. Ercetin, and L. Tassiulas, "A cross-layer framework for association control in wireless mesh networks," *IEEE Transactions on Mobile Computing*, vol. 8, no. 1, pp. 65–80, January, 2009.

[2] A. Kumar and V. Kumar, "Optimal association of stations and APs in IEEE 802.11 WLAN," *National Communications Conference (NCC)*, January 2005.

[3] Y. Bejrtano, S-J. Han, and L. Li, "Fairness and load balancing in wireless LANs using association control," *In Proceedings of ACM MobiCom*, Philadelphia, pp. 315–329, Sept.-Oct. 2004.

[4] I. Papanikos and M. Logothetis, "A study on dynamic load balance for IEEE 802.11b wireless LAN," *In Proceedings of COMCON*, Rethymna, Greece, 2001.

[5] G. Judd and P. Steenkiste, "Fixing 801.11 access point selection," *In Poster in Proceedings of ACM MobiCom 2002*, Pittsburgh, Aug. 2002.

[6] A. Balachandran, P. Bahl, and G. Voelker, "Hot-spot congestion relief and service guarantees in public-area wireless networks," *SIGCOMM Computer Communication Review*, vol. 32, no. 1, 2002.

[7] S. Vasudevan, K. Papagiannaki, C. Diot, J. Kurose, and D. Towsley, "Facilitating access point selection in IEEE 802.11 wireless networks," *In Proceedings of 5th ACM SIGCOMM Conference on Internet Measurement*, pp. 293–298, 2005.

[8] G. S. Kasbekar, P. Nuggehalli, and J. Kuri, "Online client-AP association in WLANs," *In Proceedings of RAWNET*, Boston, April 03–04, 2006.

[9] Mingming Lu and Jie Wu, "Localized access point association in wireless LANs with bounded approximation ratio," *In Proceedings of ICCCN*, vol. 1-2, pp. 656–663, 2008.

[10] M. Abusubaih, J. Gross, S. Wiethoelter, and A. Wolisz, "On access point selection in IEEE 802.11 wireless local area networks," *In proceedings of LCN*, pp. 879–886, Nov. 14-16, 2006.

[11] A. Eisenblatter, H.-F. Geerdes, and I. Siomina, "Integrated access point placement and channel assignment for wireless LANs in an indoor office environment," *In Proceedings of WoWMoM 2007*, Espoo, Finland, pp. 1–10, 2007.

[12] F. Xu, C. C. Tan, Q. Li, G. Yan, and Jie Wu, "Designing a practical access point association protocol," *In Proceedings of INFOCOM 2010*, San Diego, pp. 1361–1369, March 15–19, 2010.

[13] T. Bonald, A. Ibrahim, and J. Roberts, "The impact of association on the capacity of WLANs," *In Proceedings of 7th International Symposium on Modeling and Optimization in Mobile, Ad Hoc, and Wireless Networks*, Seoul, pp. 1–10, 2009.

[14] H. Zhou, P. Fan, and D. Guo, "Joint channel probing and proportional fair scheduling in wireless networks," *IEEE Transactions on Wireless Communications*, vol. 10, no. 10, pp. 3496–3505, Oct. 2011.

[15] I. Koukoutsidis and V. A. Siris, "Access point assignment algorithms in WLANs based on throughput objectives," *In Proceedings of WiOPT 2008*, Berlin, pp. 375–383, April 2008.

[16] I. Koutsopoulos and L. Tassiulas, "Joint optimal access point selection and channel assignment in wireless networks," *IEEE/ACM Transaction on Networking*, vol. 15, no. 3, June 2007.

[17] X. Ling and K. L. Yeung, "Joint aceess point placement and channel assignment for 802.11 wireless LANs," *IEEE Transactions on Wireless Communications*, vol. 5, no. 10, pp. 2705–2711, 2006.

[18] Li(Erran) L. Li, M. Pal, and Y. R. Yang, "Proportional fairness in multi-rate wireless LANs," *In Proceedings of the INFOCOM 2008*, Phoenix, pp. 1678–1686, April 2008.

[19] IEEE 802.11 Standard, 1997.

[20] J. Lim, H. G. Myung, K. Oh, and D. J. Goodman, "Proportional fair scheduling of uplink single-carrier FDMA systems," *In Proceedings of the PIMRC*, Helsinki, pp. 1–6, 2006.

[21] H. Kim and Y. Han "A proportional fair scheduling for multicarrier transmission systems," *IEEE Commun. Letters*, vol. 9, no. 3, pp. 210–212, 2005.

[22] http://www.cisco.com/c/en/us/td/docs/solutions/Enterprise/Mobility/emob73dg/emob73.pdf

[23] http://www.zcomax.co.uk/doc/XG-705S%20 Draft%20Product%20Specification_C0_060517.pdf, http://www.zcomax.co.uk/doc/AG-623C.pdf

Chapter 7

Placement of Sensor Nodes in WSN

7.1 Introduction

In this chapter, we consider the problem of deployment of sensor nodes in a wireless sensor network (WSN) to cover a region with a given number of sensor nodes. However, the results we present here will also be applicable for placement of base stations in a cellular mobile network. The overall objective is to place a given number of sensor nodes in a convex region and assign range to each of them so that every point in the region is covered by at least one sensor node and the maximum range assigned is minimized. We assume that the same range is assigned uniformly to all the sensor nodes and that the sensing ranges and communication ranges of the sensor nodes are the same.

In an abstract sense, the above problem basically maps to completely cover a convex region (without leaving any holes) by a given number of circles of equal radius, as illustrated in Figure 7.1, with the objective of minimizing the radius. We describe an elegant algorithm for this problem using a Voronoi diagram, as presented in [8, 9]. Other results for this problem appear in the literature and indicate some special geometric structure of the region to be covered, e.g., a square [2] and an equilateral triangle [1]. For these special cases, the minimum radius obtained by the algorithm presented in [8] compares favorably with the results reported in [1, 2]. Also, the execution time of this algorithm is a fraction of a second, whereas the existing methods may require 2 weeks to yield a reasonable value of the number of circles (≥ 27) as reported in [1, 2], even for special cases.

7.2 Preliminaries

Consider a set of points $P = \{p_1, p_2, \ldots, p_k\}$ inside a convex polygon Π where the i-th base station is located at point $p_i \in \Pi$. We use $VOR(P)$ to refer the Voronoi diagram [3] of the set of points P, and $vor(p_i)$ to denote the Voronoi cell corresponding to the point $p_i \in P$. Since we need to establish communication inside Π, if a part of the region $vor(p_i)$ goes outside Π for some i, the region $vor(p_i) \bigcap \Pi$ is used as $vor(p_i)$.

Note that all the points inside $vor(p_i)$ are closer to p_i than any other point $p_j \in P$, $j \neq i$. Thus, all these points will communicate with p_i. As all the sensor nodes are of equal range, our objective is to arrange the points in P inside the region so that the maximum range required (ρ) among the points in P is as small as possible. The algorithm [8] described below is iterative. At each step, it perturbs the point set P as explained below, and finally attains a local minimum.

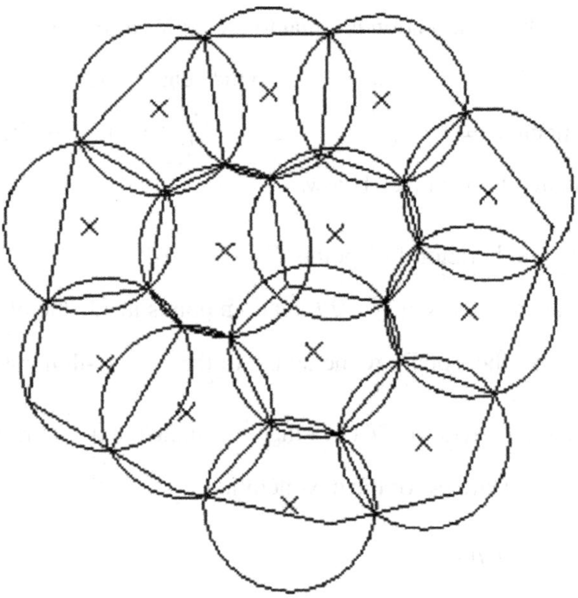

FIGURE 7.1: Illustration of abstract problem

7.3 Deployment Algorithm

In each iteration, we compute $VOR(P)$ [3], and then compute the circumscribing circle C_i of each $vor(p_i)$ using the algorithm proposed in [4] for each $i = 1, 2, \ldots, k$. Let r_i denote the radius of C_i. It can be easily verified that to cover a given convex polygon completely by a sensor node with minimum range, we need to place the sensor node at the center of the circumscribing circle of that convex region, and the range assigned to that base station is equal to the radius of the circumscribing circle. Thus, for each $i = 1, 2, \ldots, k$, we move p_i to the center of C_i and assign range r_i to it. Next, we compute $\rho = max\{r_i, i = 1, 2, \ldots, k\}$. The stepwise description of the algorithm for an iteration is given below.

Input: (* of the j-th iteration *) k points $P = \{p_1, p_2, \ldots, p_k\}$

If $j = 1$, the members of P are k random points inside the polygon Π, and if $j > 1$, the members of P are the output of the $(j-1)$-th iteration.

Output: /* of the j-th iteration */ The range ρ obtained in this iteration and the centers of k circles for the next iteration.

Step 1: Compute $VOR(P)$.

Step 2: Compute the minimum enclosing circle C_i of $vor(p_i)$.

Step 3: Compute $\rho = \max_{i=1}^{k} r_i$, where r_i is the radius of the circle C_i.

Step 4: Output ρ and the centers of C_i for $i = 1, 2, \ldots, k$.

Lemma 24. *At each iteration, i) the newly assigned position of each point p_i lies inside the corresponding $vor(p_i)$, and ii) the value of ρ decreases.*

Proof. (i) The smallest enclosing circle of a convex polygon passes through the farthest pair of vertices of the polygon and the line segment joining that pair of vertices

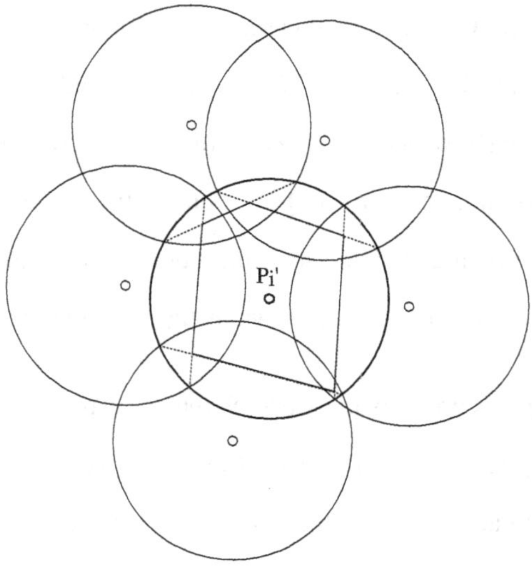

FIGURE 7.2: Illustration of $vor(p_i')$

defines the diameter of the smallest enclosing circle, or it passes through three or more vertices of the polygon. In the first case, the lemma obviously follows. In the second case, if the center lies outside the convex polygon, there exists a point-free arc of the enclosing circle having length greater than half of the perimeter of the circle. Thus it is not the minimum enclosing circle of that convex polygon (see Chapter 16 of [5]).

(ii) Let $\{p_1, p_2, \ldots, p_k\}$ be the position of the base stations prior to an iteration and ρ_{old} be the corresponding value of ρ. We have drawn the Voronoi diagram of $\{p_1, p_2, \ldots, p_k\}$ and then computed the smallest enclosing circle of each $vor(p_i)$. Let $\{p_1', p_2', \ldots, p_k'\}$ be the center of these circles, and C^* be the largest one among these circles. In other words, ρ_{old} is equal to the radius of C^*. In this iteration, the positions of the base stations are revised to $\{p_1', p_2', \ldots, p_k'\}$. The Voronoi polygon around each p_i' in the next iteration is obtained as follows:

Draw copies of C^* with centers at $\{p_1', p_2', \ldots, p_k'\}$. Let C_i be the copy of C^*

with center at p'_i. Consider the line segment defining the chords of intersection of each C_j with C_i for all $j \neq i, j = 1, 2, \ldots, k$, and the edges of the convex polygon Π. Next, compute the *envelope* of these line segments around p'_i. This defines the Voronoi polygon $vor(p'_i)$ around p'_i for the next iteration. In Figure 7.2, the region bounded by the solid lines demonstrates the Voronoi region of p'_i; the cords of intersection are shown using the dotted line segments.

Note that, $vor(p'_i)$ is properly inscribed by the corresponding circle C_i with center at p'_i, and having radius ρ_{old}. In the next iteration, we compute the smallest enclosing circle C'_i of each $vor(p'_i)$ which is completely enclosed in C_i. This proves that if ρ_{new} is the revised value of ρ in the next iteration, $\rho_{new} \leq \rho_{old}$. □

Remark 21. *The iteration terminates when the value of ρ reaches a local minimum, or in other words, $\rho_{new} = \rho_{old}$ is attained.*

To get rid of the local minimum, we apply a refinement step during the iterative process. Note that if a point (base station) p_i is on the boundary of Π, at least 50% of the area of C_i lies outside Π, and hence this region need not be covered. This indicates the scope of further reduction in the area of C_i. Thus, if a point goes very close to the boundary of Π, we move it to the centroid of Π, which is computed as follows:

Let Π be a m vertex convex polygon, and (x_j, y_j) denote the j-th vertex of Π, $j = 1, 2, \ldots, m$. The centroid of Π is the point having the coordinates $(\frac{1}{m}\sum_{j=1}^{m} x_j, \frac{1}{m}\sum_{j=1}^{m} y_j)$. It can be shown that the centroid of a convex region is always inside that region.

Such a major perturbation brings the solution away from local minima, and it leads to a further reduction in ρ. We continue the iteration with this initial placement until it reaches more local minima.

The following theorem analyzes the time complexity of each iteration of the above heuristic algorithm.

Theorem 24. *The worst case time complexity of an iteration is* $O(n + k\log k)$.

Proof. The factors involved in this analysis are as follows:

• Computing $VOR(P)$ can be done in $O(k\log k)$ time [3].

• $VOR(P)$ splits the convex polygonal region Π into k closed cells. Each edge of $VOR(P)$ appears in at most two cells. As the number of edges of the region Π is n, identifying these k cells requires $O(n + k)$ time.

• Computing the minimum enclosing circle of a convex polygon needs time linear in its number of edges [4]. Thus, computing $\{C_i, i = 1, 2, \ldots, k\}$, needs $O(n+k)$ time.

□

The number of iterations needed to reach local optima from an initial configuration is reasonably small. The overall time complexity depends on the number of times we apply the refinement step.

7.4 Experimental Results

An exhaustive experiment has been performed with several convex shapes of region Π and with different values of k. For a given initial placement of P the value of ρ is decreased at each iteration. As the process reaches local minima, the quality of the result depends on the initial choice of the positions of P. Studies have been made of random distribution of P. In an optimal solution, the distribution of points is very regular. To get the best possible results, the initial distribution of the points was carefully chosen. Thus, while performing the experiment with a unit square region, the initial placement of the points in P was chosen as follows: compute $m = \lfloor\sqrt{k}\rfloor$. If $m^2 = k$, we split the region into $m \times m$ cells, and in each cell placed a point of P randomly. If $(k - m^2) < m$, we split the region into m rows of equal width, then arbitrarily chose $(k - m^2)$ rows and split each of these rows into $(m + 1)$ cells; the other rows were split into m cells. We then placed one point in each cell. If $(k - m^2) > m$, we split the square region into $(m + 1)$ rows, and each row was split into m or $(m + 1)$ rows to accommodate all the points in P.

For each k, authors in [8] considered 1000 initial instances. For each instance, they computed the value of ρ by executing the above iterative algorithm. Finally, they reported the value of ρ^*_{opt} = minimum value of ρ over all 1000 instances. Table 7.1 shows a comparison ρ^*_{opt} with the value of ρ_{opt} obtained by the algorithm in [2] for different values of k.

Authors in [8] compared their method with that of [1] when the region to be covered was an equilateral triangle. The experimental results for different values of k are presented in Table 7.2. Figure 7.1 demonstrates the output of the deployment algorithm described above for covering a convex polygon with 13 circles.

To test the performance of the heuristic used in designing the algorithm, we need an idea of the minimum, average and standard deviation of the values of ρ over all the 1000 instances for different values of k and the unit square region as Π (see Table 7.3). Thus, column 3 of Table 7.1 is identical to column 2 of Table 7.3. Das et al. [8] performed the entire experiment in Sun Blade 1000 workstations with 750 MHz CPU speed, and used LEDA [6] for computing the Voronoi diagram. The average time for processing each instance is shown. Similar results were observed with equilateral triangle.

TABLE 7.1: Covering unit square

k	ρ_{opt} using method in [2]	ρ^*_{opt} using the above method
4	0.35355339059327376220	0.353553
5	0.32616054400398728086	0.326165
6	0.29872706223691915876	0.298730
7	0.27429188517743176508	0.274295
8	0.26030010588652494367	0.260317
9	0.23063692781954790734	0.230672
10	0.21823351279308384300	0.218239
11	0.21251601649318384587	0.212533
12	0.20227588920818008037	0.202395
13	0.19431237143171902878	0.194339
14	0.18551054726041864107	0.185527
15	0.17966175993333219846	0.180208
16	0.16942705159811602395	0.169611
17	0.16568092957077472538	0.165754
k	ρ_{opt} using method in [2]	ρ^*_{opt} using the above method
18	0.16063966359715453523	0.160682
19	0.15784198174667375675	0.158345
20	0.15224681123338031005	0.152524
21	0.14895378955109932188	0.149080
22	0.14369317712168800049	0.143711
23	0.14124482238793135951	0.141278
24	0.13830288328269767697	0.138715
25	0.13354870656077049693	0.134397
26	0.13176487561482596463	0.132050
27	0.12863353450309966807	0.128660
28	0.12731755346561372147	0.127426
29	0.12555350796411353317	0.126526
30	0.12203686881944873607	0.123214

TABLE 7.2: Covering equilateral triangle

k	ρ_{opt} using method in [1]	ρ^*_{opt} using the above method	k	ρ_{opt} using method in [2]	ρ^*_{opt} using the above method
4	0.2679491924311227065	0.267972	21	0.0962250448649376274	0.099165
5	0.2500000000000000000	0.250006	22	0.0951772351261450917	0.095877
6	0.1924500897298752548	0.192493	23	0.0937742911094478264	0.094625
7	0.1852510855786008545	0.185345	24	0.0923541375945022204	0.093982
8	0.1769926664029649641	0.177045	25	0.0906182448311340175	0.091688
9	0.1666666666666666667	0.166701	26	0.0887829248953373781	0.090231
10	0.1443375672974064411	0.144681	27	0.0868913397937031505	0.088238
11	0.1410544578570137366	0.141252	28	0.0824786098842322521	0.086795
12	0.1373236156889236662	0.137633	29	0.0818048133956910115	0.084545
13	0.1326643857765088351	0.133379	30	0.0808828500258641436	0.082246
14	0.1275163863998600644	0.127829	31	0.0798972448089536737	0.081665
15	0.1154700538379251529	0.115811	32	0.0788506226168764215	0.080457
16	0.1137125784440782042	0.114574	33	0.0776371221483728244	0.079604
17	0.1113943099632405880	0.112141	34	0.0763874538343494465	0.078827
18	0.1091089451179961906	0.109890	35	0.0751604548962267707	0.076918
19	0.1061737927289732618	0.107288	36	0.0721687836487032206	0.075950
20	0.1032272183417310354	0.104049			

TABLE 7.3: Performance evaluation of algorithm

k	ρ^*_{opt}	$\rho_{average}$	std. devn.	Time in sec.	k	ρ^*_{opt}	$\rho_{average}$	std. devn.	Time in sec.
4	0.353553	0.395284	0.040423	0.052	18	0.160682	0.164347	0.001092	0.351
5	0.326165	0.326247	0.000201	0.073	19	0.158345	0.160797	0.000885	0.377
6	0.298730	0.309837	0.008433	0.090	20	0.152524	0.156772	0.000877	0.405
7	0.274295	0.27603	0.001668	0.107	21	0.149080	0.153131	0.001253	0.436
8	0.260317	0.26131	0.003079	0.124	22	0.143711	0.148640	0.000582	0.465
9	0.230672	0.231119	0.000540	0.143	23	0.141278	0.145498	0.001738	0.499
10	0.218239	0.218244	0.000004	0.164	24	0.138715	0.142105	0.001507	0.531
11	0.212533	0.213855	0.000894	0.184	25	0.134397	0.139549	0.001572	0.557
12	0.202395	0.205567	0.000908	0.206	26	0.132050	0.136489	0.001618	0.587
13	0.194339	0.194960	0.000645	0.228	27	0.128660	0.133725	0.001298	0.623
14	0.185527	0.189217	0.001722	0.258	28	0.127426	0.131589	0.001357	0.655
15	0.180208	0.182782	0.001883	0.279	29	0.126526	0.129241	0.000964	0.688
16	0.169611	0.174669	0.003178	0.303	30	0.123214	0.127069	0.000881	0.719
17	0.165754	0.168231	0.002336	0.327					

7.5 Exercises

1. Assume that there are six points on a two-dimensional plane with coordinate values $(\frac{-2}{\sqrt{3}}, 2)$, $(\frac{2}{\sqrt{3}}, 2)$, $(\frac{4}{\sqrt{3}}, 0)$, $(\frac{2}{\sqrt{3}}, -2)$, $(\frac{-2}{\sqrt{3}}, -2)$, and $(\frac{-4}{\sqrt{3}}, 0)$. Construct the Voronoi diagram for this set of six points.

2. Assume that 10 base stations, all of the same transmitting power, must be placed at suitable locations to cover an entire city which is bounded by a circle of 10 km radius. Describe a suitable algorithm for finding the best possible locations of the base stations which will completely cover the whole city region (without leaving holes or uncovered regions).

3. Write a program implementing an algorithm you designed for the problem above and find the corresponding locations of the base stations and range of a base station, i.e., the maximum distance up to which the signal from a base station should be available.

*4. Assume you have identical transceivers for all base stations, each of which has the omnidirectional maximum transmission distance of 1 km. The problem is to again cover the whole city region which is bounded by a circle of 10 km radius. Devise an algorithm to find the minimum number of base stations required for this cover and compute the corresponding locations where the base stations must be placed.

*5. Repeat the above problem if the city area is bounded by a square with each side of length 10 km. Generalize your algorithm, if possible, for covering a city that may be bounded by any arbitrary convex polygon.

Bibliography

[1] K. J. Nurmela, "Conjecturally optimal coverings of an equilateral triangle with up to 36 equal circles," *Experimental Mathematics*, vol. 9, pp. 241–250, 2000.

[2] K. J. Nurmela and P. R. J. Ostergard, "Covering a square with up to 30 Equal Circles," Research Report HUT-TCS-A62, Laboratory for Theoretical Computer Science, Helsinky University of Technology, 2000.

[3] M. de Berg, M. Van Kreveld, M. Overmarse and O. Schwarzkopf, *Computational geometry algorithms and applications*. Springer, Heidelberg, 1997.

[4] N. Megiddo, "Linear-time algorithms for linear programming in R^3 and related problems," *SIAM Journal on Computing*, vol. 12, pp. 759–776, 1983.

[5] H. Rademacher and O. Toeplitz, *The Enjoyment of Mathematics*. Princeton University Press, Princeton, NJ, 1957.

[6] K. Mehlhorn and S. Naher, *The LEDA platform of combinatorial and geometric computing*. Cambridge University Press, Cambridge, UK, 1999.

[7] J. B. M. Melissen and P. C. Schuur, "Improved coverings of a square with six and eight circles," *Electronic Journal on Combinatorics*, vol. 3, R32, 1996.

[8] G. K. Das, S. Das, S. C. Nandy and B. P. Sinha, "Efficient algorithm for placing a given number of base stations to cover a convex region," *J. Parallel and Distributed Computing*, vol. 66, pp. 1353–1358, 2006.

[9] G. K. Das, S. Das, S. C. Nandy and B. P. Sinha, "Placing a given number of base stations to cover a convex polygon," *Proc. Intl. Workshop on Distributed Computing (IWDC)*, Lecture Notes in Computer Science, No. 3741, pp. 57–62, 2005.

Chapter 8

Energy-Efficient Communication

8.1 Introduction

Energy has become a *buzz* word in recent times. Worldwide resources for generating energy in conventional ways are limited. Scientists and technologists continually search for innovative techniques to produce energy from unconventional, but safe sources including sunlight, tides, and winds. Other researchers are working on techniques to reduce the energy consumed by real-life domestic and industrial applications.

The Internet, social networks, mobile telephony, sensor networks and pervasive computing require huge amounts of information to be communicated through wired and wireless media. Although the energy required for transmitting and receiving a single bit of information may be very small, the total energy requirement for incessant communication of the amount of information for global applications is no longer negligible. It constitutes a substantial fraction of the world's total energy consumption for all activities. As a result, research focusing on energy-efficient communication has become very important, particularly in the context of the modern civilization which is so dependent on seamless information dissemination. In this chapter, we focus on energy-efficient methodologies for only wireless communication systems.

In a wireless communication system, energy can be conserved in various layers of the communication protocol stack, e.g., in the physical, MAC, network, and application layers, as outlined below.

8.1.1 Physical Layer-Centric Approach

In the physical layer, we think of designing efficient modulation and demodulation schemes so that energy spent for transmitting and/or receiving every bit of a message is reduced as much as possible without compromising required quality of services (QoS) in areas such as bit error rate, bandwidth requirement, and delay. Also, we can think of designing low cost hardware using minimum numbers of low power logic gates and improved VLSI chip design for energy-efficiency in communication.

8.1.2 MAC Layer-Centric Approach

In the MAC layer, we need improved techniques for avoiding collision in grabbing a common channel for communication, reducing collisions of messages and reducing the frequency of message retransmissions. Note that collision leads to retransmission of a collided message which, in turn, involves additional energy that is in a sense wasted. Thus, collision of messages should be avoided or minimized in the designs of energy-efficient communication schemes. Also, in the MAC layer, there is another possible way of conserving energy. Usually, a transmitter may operate in *transmit* mode in which some bits can be transmitted, or *idle* mode. The power required to keep the transmitter in the *idle* mode is usually negligible compared to that used in its *transmit* mode. Because of this difference in power requirements, an intelligent way of saving energy would be to keep the transmitter in *idle* mode when the communication channel is sensed to be idle (i.e., no message to be transmitted), rather than keeping it always in the *transmit* mode. The decision for keeping the transmitter in the *transmit* or *idle* mode is made by the MAC layer of the communication protocol. Thus, a judicious selection (tweaking) of the idle and transmit periods of the transmitter may result in substantial energy savings in the long run. For further study of optimizing data transmissions by reducing collisions and retransmissions, the reader is referred to [7, 15, 13, 8, 14, 20].

8.1.3 Network Layer-Centric Approach

In the network layer, energy required for communication can be reduced if the path for routing a message from a source node to a destination node is so chosen to minimize the overall energy spent by all the nodes in that communication path. It may seem logical to seek the shortest path between a source and destination pair to route the messages; but this may not be the best solution as explained below.

Let u_1 and v_1 be the source and destination nodes. Assume that the shortest path between u_1 and v_1 passes through another node w. Repeated transmission of messages from u_1 to v_1 through this node w would eventually deplete the energy of w to such a low level that it will no longer be able to communicate to other nodes of the network. Thus, if w has some message to be transmitted to another node of the network, it can no longer do so. In the same way, it would not be possible for some other node to send its data to w. Thus, effectively, the node w is altogether eliminated from the network. The situation can worsen if node w falls in the shortest paths between more than one source-destination pair, e.g., the shortest path between source node u_2 and destination node v_2 also passes through w. In such a situation, the node w will have a reduced lifetime serving as an intermediate node for message communication between several source-destination pairs. In essence, blind shortest path routing does not yield the best possible result considering the operational lifetime of a whole network. This is particularly true for low power wireless networks, e.g., wireless sensor networks in which each sensor node is battery powered. Despite the limited energy source, we want to extend the lifetime of each sensor node as much as possible.

To mitigate the above situation, several alternative message routing techniques have been proposed and consider the residual energy of the constituent nodes. Essentially, the idea is i) to maintain the residual energy level of each node above some chosen threshold value so that no node goes out of network connectivity, ii) the individual messages are transmitted with minimum communication latency (delay) and iii) the total energy required for communication of all messages throughout the network is also minimized. Note that if the communication ranges (which are dependent on the transmitting power of a node) of all nodes are same, the last two conditions are not mutually conflicting. Communicating each individual message via the shortest path will lead to the minimum energy expenditure for the total message communication between all source-destination pairs (we assume that no congestion or queuing delay at any intermediate node). The first requirement may be mutually conflicting with the last two as explained above. Thus, a tradeoff is needed between these conflicting requirements to maintain as many nodes with residual energy above the threshold value as possible while keeping the total energy requirement for communication of all messages as low as possible.

One possibility for meeting the requirement is clustering of nodes with a cluster head in each cluster so that the heads form the backbone network. Usually, the node in a cluster having the maximum residual energy is chosen as the cluster head. Thus, a node in a specific cluster sends its message first to its cluster head, then the cluster head routes this message directly to the destination node (if it lies in the same cluster) or to another cluster head that may function as an intermediate node for a shortest path routing in the backbone network. Note that a node may have a direct (single hop communication) link to its destination, but this would not be allowed by the above technique if the source node is not a cluster head. A variant of this technique may be that when the destination node is in the same cluster, the source node may directly send its message to the destination node, relieving the cluster head of functioning as the intermediate node for this communication. This would reduce the time required for communication and the total energy for routing this message, with the negative point that the source node (which is a non-cluster head node) may be required to increase its peak transmission power, i.e., spend more energy to directly reach the farthest node of the same cluster. This, in turn, needs a careful clustering algorithm that considers the residual energy and transmission range of each node to be included in the cluster along with the selection of the appropriate nodes as the cluster heads. Thus, if a node in the cluster does not have enough residual energy to increase the peak transmission power to reach another node in the same cluster at the farthest distance, this node can not be included in the cluster. Basically, this requirement should reduce the cluster size and hence the maximum delay in the backbone network. Again there is a question of tradeoff to get the best possible result.

The technique of clustering and routing requires changing the clusters and the cluster heads dynamically. After certain time, a cluster head may be depleted of its energy and will no longer have the maximum residual energy among all nodes in the cluster. Also, the network connectivity may be changed if the nodes are mobile, energy of a node is so much depleted that it is disconnected from the network, or new nodes may have joined since the last computation of the clusters (and corresponding cluster heads). These situations should invoke computing the clusters afresh, depending on the current status of the network, e.g., network connectivity, residual energy of nodes, and transmission range of nodes.

An alternative approach for energy-efficient message routing is the use of a connected dominating set (CDS) as the backbone network, instead of the cluster heads forming the backbone network in the earlier technique. In this approach, we first construct a network graph $G(V, E)$ with V as the set of vertices and E as the set of edges from the network. Corresponding to each node in the network, there is a vertex in the set V, and two vertices will have an undirected edge between them if the corresponding two nodes of the network are within the transmission range of each other. We assume, for simplicity, that the transmission range of each node is the same. A dominating set DS of the graph $G(V, E)$ consists of a subset of V such that every vertex of the graph is in DS or adjacent (at Distance 1) from at least one vertex in DS. As an example, consider the graph of Figure 8.1 with 10 vertices. One possible dominating set of this graph is $\{2, 5, 9\}$ as shown in Figure 8.2. However, the induced graph on this dominating set is not connected and accordingly the nodes corresponding to this dominating set cannot constitute a backbone network. To form a backbone network, we include a minimal set of other vertices in this dominating set to make it a connected dominating set CDS. Thus, a possible connected dominating set of the graph in Figure 8.3 is $\{2, 4, 5, 7, 9\}$. Note that the additional vertices 4 and 7 added

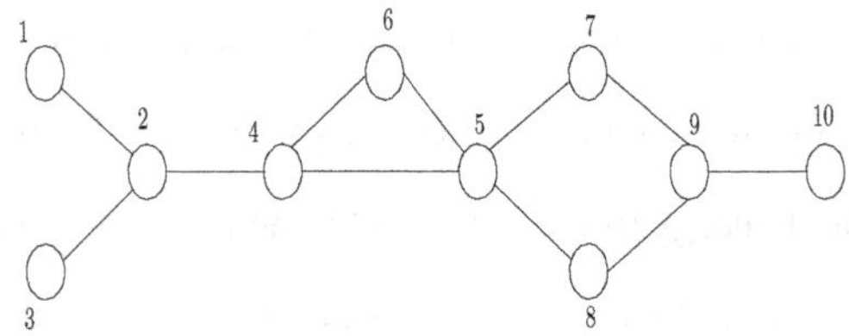

FIGURE 8.1: Network graph G with 10 nodes

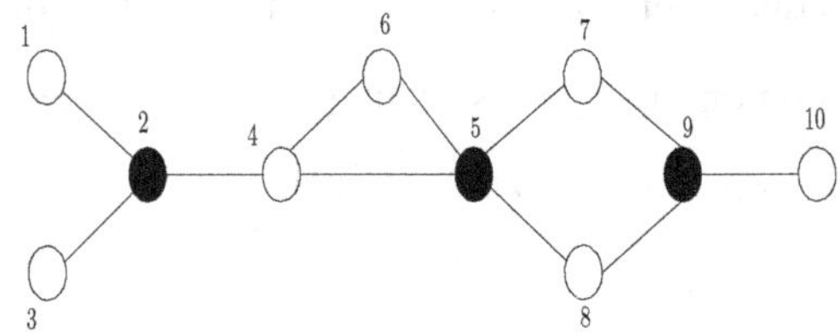

FIGURE 8.2: One possible dominating set of G

to the dominating set $\{2, 5, 9\}$ to make it a connected dominating set is minimal, i.e., not including any of the nodes 4 and 7 would not make the dominating set connected.

Once we form the backbone network with the induced graph on a connected dominating set of the network graph, routing of messages can be done in the same technique used in the approach based on clustering, with cluster heads forming a backbone network. A message from any node (not included in the chosen connected dominating set) will be sent first to an adjacent node in the connected dominating set (CDS) and from there it will be routed to a node in the (CDS) adjacent to the destination node, following the shortest route along the backbone network.There may be a number of possible dominating sets in a graph and corresponding to a given

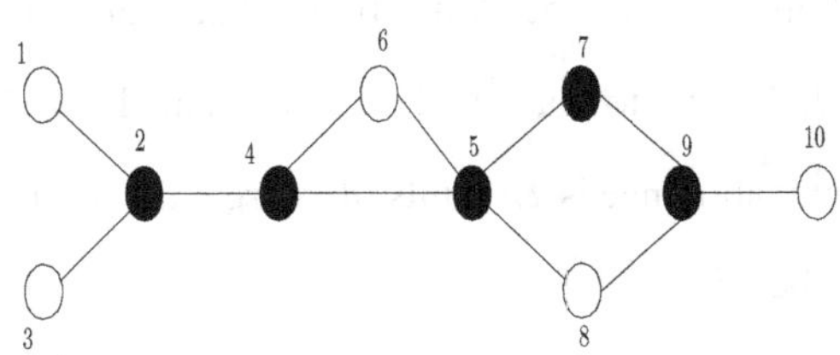

FIGURE 8.3: One possible connected dominating set of G

dominating set there may be a number of possible connected dominating sets. Thus, while selecting a connected dominating set, we should look for the nodes with maximum possible residual energy to form the backbone network to increase the lifetime of the network. Also, periodic recomputation of such a connected dominating set is necessary to take care of changed network topology and any reduction in the residual energy of the nodes due to ongoing message communication through the network. For further information on intelligent selection of paths or special architectures for sending data, the reader is referred to [12, 16, 19].

8.1.4 Application Layer-Centric Approach

Another possible approach for reducing energy required for message communication is to reduce the message size or the number of messages to be transmitted, depending on the specific application. As an example, imagine that the network graph shown in Figure 8.1 represents a sensor network with each node representing a temperature sensing device. Suppose the specific application for which this sensor network will be used is to measure the average temperature of a big auditorium with sensor devices placed at different locations inside the auditorium. Assume that node 5 collects the data for temperature sensed by all other nodes as shown in Figure 8.4. Thus, as shown in Figure 8.4, to collect the temperature data from node 1, three hops of communication are needed. Similarly, node 3 also needs three hops of communication to send its data to node 5, and the corresponding number of hops for the data from nodes 2 and 4 are two and one, respectively. The total number of hops for the data from nodes 1, 2, 3 and 4 to reach node 5 is thus equal to $3 + 3 + 2 + 1 = 9$. If each hop of communication needs E_h units of energy, the total energy spent in this communication will be $9E_h$.

In contrast, let us try another scheme. Since our overall objective is to compute the average temperature sensed by all the nodes, we know that eventually collector

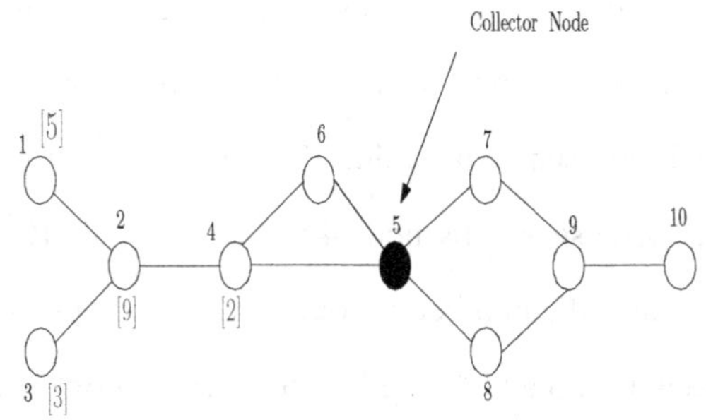

FIGURE 8.4: Typical application on a sensor network

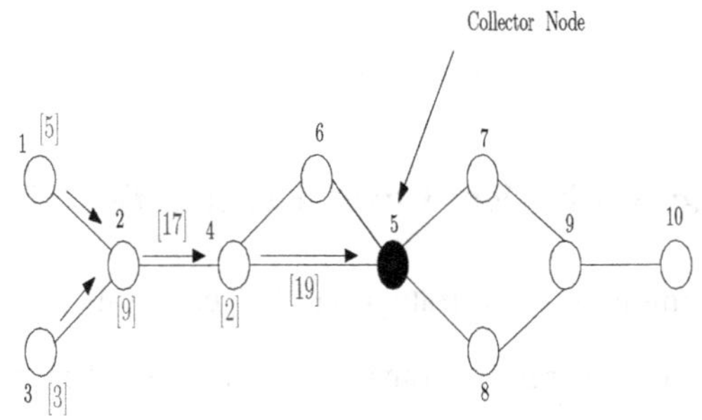

FIGURE 8.5: Reduced number of communication hops in sensor network

node 5 has to sum all the temperature values received from all nodes. Assume that the temperature values sensed by the nodes $1, 2, 3$ and 4 are $5, 9, 3$ and 2, respectively. Node 1 sends its data value 5 to node 2 which adds this value to its own sensed data 9 to store the partial sum 14 in its memory. Then node 3 sends its data value 3 to node 2 which is added to the previous partial sum 14 to generate the new partial sum 17 at node 2. Node 2 later sends this value 17 to node 4 (refer to Figure 8.5) which is added to the data value sensed by node 4 to get the result 19, which is then sent to collector node 5. Thus, the number of communication hops used to send the data values from the nodes 1, 2, 3 and 4 is only four, spending only $4E_h$ units of energy for these steps. This saves $5E_h$ units of energy. Instead of the average temperature, we could have computed the maximum or minimum temperature among all points of observation inside the auditorium and this scheme would be equally effective in saving energy. However, it must be noted that such a method of energy saving in data communication through data aggregation will be effective only if the required application allows such aggregation. That is, energy saving at the application layer is not a general technique and very application specific.

8.2 Some Elegant Physical Layer-Centric Techniques

Conventional communication strategies utilize energy-based transmission (EbT) schemes, which require energy for transmitting each bit of information, whether a 0 or a 1. Depending on the transmission strategy, both 0 and 1 are represented by non-zero voltage levels or one of the bit values is represented by a zero voltage level while a non-zero voltage level is used to distinguish the other bit value. An example of the latter is the polar return-to-zero (polar-RZ) transmission scheme, where a 0 corresponds to a zero voltage level, while a 1 represents a non-zero voltage level.

However, most existing transmission schemes utilize non-zero voltage levels for both 0 and 1 to distinguish between silent and busy channels. Communication strategies that require energy expenditure for transmitting both 0 and 1 bit values are known as *energy-based transmission* (EbT) schemes. For example, to communicate a value of 278, a node will transmit the bit sequence $< 1, 0, 0, 0, 1, 0, 1, 1, 0 >$, consuming energy for every bit it transmits. Thus, if the energy required per bit transmitted is e_b, the total energy consumed to transmit the value 278 would be $9e_b$.

In 2005, Zhu and Sivakumar [9] proposed a communication strategy called *communication through silence* (CtS), that involved using silent periods instead of energy-based transmissions. CtS, however, suffers from the disadvantage of being exponential in time. It introduces long delays and adversely affects the network throughput. Another problem is maintaining synchronization between sender and receiver over long periods of silence. Chen et al. proposed [10] an alternative strategy called variable-base tacit communication (VarBaTaC) that uses a variable radix-based information coding coupled with CtS for communication. The throughput with VarBaTac decreases exponentially with an exponential increase of the radix. Also the hardware circuit complexity increases significantly for representing higher radix numbers and handling dynamic changes of radix.

We now describe elegant techniques for energy-efficient communication in low power wireless networks, e.g., in wireless sensor networks, mobile ad hoc networks and applications which typically involve low cost devices and demand low power operations with simple modulation techniques like ASK, OOK and FSK [22]. All these techniques are based on some type of source coding along with selective silent communication of the dominant symbol (most frequently occurring) after encoding the message.

With some appropriate source coding, it may be possible to convert a given binary bit stream (message) into a stream of m possible symbols ($m > 2$), with one of these m symbols having a higher frequency of occurrence than the remaining $m - 1$ symbols. Most modern radio devices have multiple states that consume varying amounts of power. In general, a radio in *transmit* or *receive* state consumes the maximum power while in *active* and *idle* states, it consumes very little power. Typically, the radio consumes the least power in idle state but takes longer to switch to transmit or receive state from idle state than from the active state. Thus, given a biased statistics of the symbols after encoding the message, one might choose to transmit the stream of m symbols through an appropriate digital modulation scheme in which the transmitter is kept silent during the occurrence of the dominant symbol. During these silence intervals, the transmitter can save significant energy. The family of phase and frequency modulations such as phase-shift keying (PSK), frequency-shift keying (FSK), m-ary PSK, m-ary FSK and their variants are superior to the amplitude-shift keying (ASK) and m-ary ASK techniques. When one chooses to employ m-ary PSK/FSK modulation along with silent intervals for the dominant symbol which corresponds to an ASK transmission, the instantaneous power levels of the $m - 1$ non-silent symbols may have to be increased appropriately due to the higher SNR requirements of ASK for the same bit error rate (BER) value compared to m-ary PSK/FSK. Thus, for a successful energy-efficient communication, the difference between the energy saved due to silence and the additional energy needed for the increased instantaneous power for m-ary PSK/FSK needs to be positive.

Several successful approaches implementing the appropriate source coding technique and silent communication of the dominant symbol, along with suitable modulation and demodulation techniques have already been proposed by various researchers [39, 3, 4, 6, 5, 40]. We discuss some of these schemes in the following sections.

8.3 Redundant Binary Number Encoding with Silent Zero Communication (RBNSiZeComm)

This communication scheme is based on encoding the messages to be transmitted in a redundant binary number system ($RBNS$). We first discuss $RBNS$ basics and the frequency distribution of the symbols in all possible messages (assuming that every binary string is equally likely to occur as a message to be transmitted) in this number system. After that, we describe the communication protocol using the concept of silent communication on encoded messages and then evaluate the performance of the protocol in terms of energy savings in the transceiver as described in [39].

8.3.1 Preliminaries on Redundant Binary Number System

The redundant binary number system (RBNS) [21] utilizes the digits from the set $\{-1, 0, 1\}$ for representing numbers using radix 2. In the rest of the chapter, for convenience, we denote the digit -1 by $\bar{1}$. In RBNS, there can be more than one possible representation of a given number. For example, the decimal number 7 can be represented as either 111 or $100\bar{1}$ in RBNS. We utilize this property to recode a message string to reduce the number of 1's in the string while transmitting the message. The original binary message can, however, be obtained at the receiver end by reconverting the received message from RBN to the binary number system.

The basic idea of the recoding scheme is as follows: Consider a run of k 1's, $k > 1$. Let i be the bit position for the first 1 in this run, $i \geq 0$ (bit position 0 refers to the least significant bit at the rightmost end). Let v represent the value of this run of k 1's. Then,

$$v = 2^i + 2^{i+1} + 2^{i+2} + \ldots + 2^{k+i-1}. \quad (8.1)$$

Alternatively, we can rewrite Equation (8.1) as,

$$v = 2^{k+i} - 2^i. \quad (8.2)$$

Equation (8.2) can be represented in RBNS by a 1 at bit position $(k+i)$ and a $\bar{1}$ at bit position i, while all the intermediate 1's between them are converted to 0's. We formalize this observation in the form of the following reduction rule.

Reduction 1. *A run of k 1's ($k > 1$) starting from bit position i, is replaced by an equivalent representation consisting of a 1 at bit position $k+i$ and a $\bar{1}$ at bit position i, with 0's in all intermediate bit positions.*

Observe that for a run of k 1's, $k > 1$, the savings in terms of the number of non-zero digits is $k-2$. Note that the number of non-zero digits remains unchanged for $k = 2$. Also, if we consider a string, say 110111, with only one 0 trapped between runs of 1's, after applying reduction rule 1, we would get the string $10\bar{1}100\bar{1}$. Note the presence of the pattern $\bar{1}1$ for this trapped 0. A second reduction can now be applied to the bit pattern $\bar{1}1$ to replace it with $0\bar{1}$. Thus, from $10\bar{1}100\bar{1}$, we get the string $100\bar{1}00\bar{1}$ with fewer non-zero digits. We thus have a second reduction rule.

Reduction 2. *Every occurrence of the bit pattern $\bar{1}1$ in a string obtained after applying reduction rule 1 is replaced by the equivalent bit pattern $0\bar{1}$.*

The application of reduction rules 1 and 2 warrants a two-pass algorithm. However, it is easy to combine the two rules to obtain a single-pass algorithm. as stated below.

```
procedure RBNSIZECOMM(IN bit vector data_frame, IN integer len)
    /* The data frame is converted to RBNS */
```

```
    /* by applying reduction rules 1 and 2 */

    bit vector out_buf : array [0..len]                    ▷ size of len + 1 bits;
    boolean runflag : false;
    integer runlengthcount, lowbit : 0;

    Call ReductionRule1(data_frame, len, out_buf);         ▷ Reduction Rule 1
    Call ReductionRule2(out_buf, len);                     ▷ Reduction Rule 2
    for i = 0 to len do
        if out_buf[i] ≠ 0 then
            transmit bit out_buf[i];                       ▷ Either 1 or 1̄
        else
            remain silent;                                 ▷ No transmission of bit value '0'
        end if
    end for
end procedure
```

```
procedure REDUCTIONRULE1(IN data_frame, IN len, OUT out_buf)
    for i = 0 to len − 1 do                                ▷ Test each bit of data frame
        if data_frame[i] = 1 then
            if runlengthcount = 0 then
                lowbit ← i;                                ▷ Start index of run of 1's
            end if
            runlengthcount ← runlengthcount + 1;           ▷ Run length
            if runlengthcount > 1 then
                runflag ← true;                            ▷ Found a run of ones
            end if
        else                                               ▷ data_frame[i] = 0
```

```
            if runlengthcount > 1 then                      ▷ Apply Reduction Rule 1
                out_buf[lowbit] ← 1̄;
                for j = lowbit + 1 to i − 1 do
                    out_buf[j] ← 0;
                end for
                out_buf[i] ← 1;
                runflag ← false;
            else                                            ▷ Single 1, just copy to out buffer
                out_buf[lowbit] ← 1;
                out_buf[i] ← 0;
            end if
            runlengthcount ← 0;                             ▷ Reset run length count
        end if
    end for

    if runflag = true then                                  ▷ run of 1 ends at msb
        out_buf[lowbit] ← 1̄;
        for j = lowbit + 1 to len − 1 do
            out_buf[j] ← 0;
        end for
        out_buf[len] ← 1;                                   ▷ out_buf of length len + 1
        runflag ← false;
    end if
end procedure
```

```
procedure REDUCTIONRULE2(INOUT out_buf, IN len)
    for i = 1 to len do
        if (out_buf[i] = 1̄) and (out_buf[i − 1] = 1) then   ▷ Reduction Rule 2
```

```
            out_buf[i − 1] ← 1̄;
            out_buf[i] ← 0;
        end if
    end for
end procedure
```

To convert *protocol RBNSiZeComm* to a single-pass routine, we replace procedures *ReductionRule1* and *ReductionRule2* by a single procedure with a change in the lines 13 and 27 (where we write a $\bar{1}$ in the output buffer) of the procedure *ReductionRule1*. Each of these lines 13 and 27 of the procedure *ReductionRule1* will be replaced with the following:

```
if (lowbit ≠ 0) and (out_buf[lowbit − 1] = 1) then
    out_buf[lowbit − 1] ← 1̄;
    out_buf[lowbit] ← 0;
else
    out_buf[lowbit] ← 1̄;
end if
```

On the receiver side, the reverse process must be done to correctly receive the data and then convert it into its binary equivalent representation for further processing by the different layers of the network protocol stack. The application of the reduction rules 1 and 2 on the RBN coded data during transmission prevents occurrences of the bit patterns $1\bar{1}$ and $\bar{1}1$ in the transmitted data. Also, if the original data was an n-bit binary data frame, RBN encoding can result in a frame of size $n + 1$ RBN bits. Protocol *RBNSizeRecv* outlines the steps involved in correctly receiving and reconverting an RBNS encoded data frame to its binary equivalent.

```
procedure RBNSIZERECV(IN RBN vector integer len, OUT bit vector
    out_buf)
```

```
    /* First receive an RBNS data frame of length len + 1 */
    /* Then converted it to its equivalent binary value */

    RBN vector recv_buf : array [0..len];          ▷ received RBNS data frame
    bit vector out_buf : array [0..len];            ▷ Output binary frame

    /* Receive the data frame */
    Call ReceiveRBNDataFrame(recv_buf, len);
    Call ConvertRBNtoBinary(recv_buf, len, out_buf);

    /* The binary data in out_buf is len + 1 bits. The msb */
    /* can be ignored as it will always be 0, and only the */
    /* lower order len bits needs to be returned */
end procedure

procedure RECEIVERBNDATAFRAME(INOUT recv_buf, IN len)
    /* null indicated no signal in channel */
    /* time_slot[i] refers to the i^th time slot */

    for i = 0 to len do
        if time_slot[i] = null then
            recv_buf[i] ← 0;                ▷ channel status null during i^th time slot
        else
            recv_buf[i] ← time_slot[i];     ▷ received 1 or 1̄ in i^th time slot
        end if
    end for
end procedure

procedure CONVERTRBNTOBINARY(IN recv_buf, IN len, INOUT out_buf)
```

```
/* Convert received RBNS data frame to binary data */
/* The final output binary data will be in out_buf */
boolean runflag : false;
runflag ← false;                          ▷ To decode the run of 1's
for i = 0 to len do
    repeat
        if recv_buf = 1̄ then
            if runflag = false then
                runflag ← true;
                out_buf[i] ← 1;           ▷ output i^th bit = 1
            else
                out_buf[i] ← 0;           ▷ result of reduction rule 2
            end if
        end if

        if recv_buf = 1 then
            if runflag = true then
                out_buf[i] ← 0;           ▷ output i^th bit = 0
                runflag ← false;          ▷ end of a run of 1's
            else
                out_buf[i] ← 1;           ▷ single 1 in data
            end if
        end if

        if recv_buf = 0 then
            if runflag = true then
                out_buf[i] ← 1;           ▷ inside a run of 1's
```

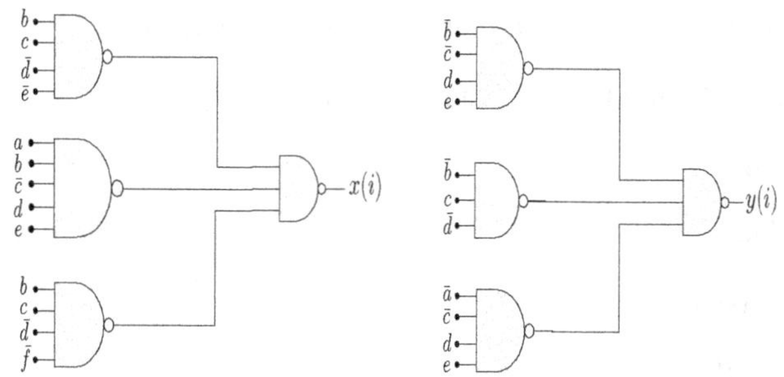

FIGURE 8.6: Hardware circuit for output bits $x(i)$ and $y(i)$ of RBNS symbol b_i^{rbn} generated from b_i at transmitter

```
                else
                    out_buf[i] ← 0;
                end if
            end if
        until runflag = false;
    end for
end procedure
```

As an alternative to the above software version, it is also possible to design a simple hardware circuit to convert from binary to RBNS serially. Let the RBN digit $b_i^{rbn}, 0 \leq i \leq n$, corresponding to the bit b_i of the given binary number be represented by two bits $x(i)$ and $y(i)$ with $x(i)$ as the most significant bit. Assuming encoding of the RBN digits as 0 by 00, 1 by 01 and $\bar{1}$ by 10 and denoting the bits $b_{i+2}, b_{i+1}, b_i, b_{i-1}, b_{i-2}$ and b_{i-3} by a, b, c, d, e and f, respectively, Figure 8.6 depicts a representative hardware circuit for evaluating $x(i)$ and $y(i)$ with b_{-3}, b_{-2}, b_{-1}, b_n and b_{n+1} initialized to 0. The circuit in Figure 8.6 can be used for a *pipelined* transmission scheme where the conversion of the $(i+1)$-th bit is overlapped with the transmission of the i-th RBNS symbol.

Similarly, the received RBNS digits can also be converted serially by a hardware circuit as shown in Figure 8.7. The lines R_0, R_1 and $R_{\bar{1}}$ assume the logic level 1

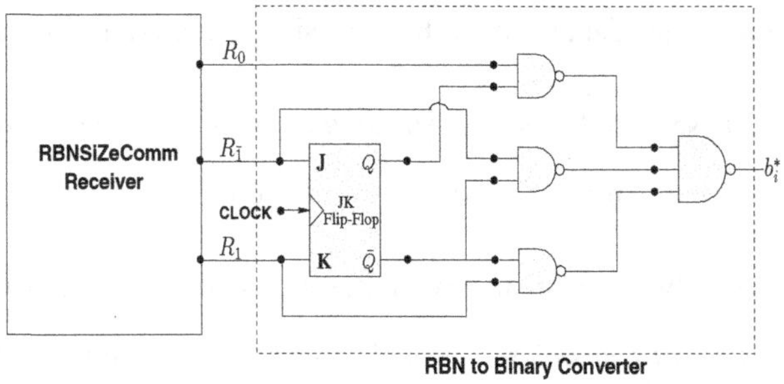

FIGURE 8.7: Circuit diagram for converting received i-th RBN symbol to its equivalent binary symbol b_i^* at receiver

when the received i-th symbol is a 0, 1 or $\bar{1}$, respectively, and b_i^* is its equivalent binary form.

We now make the following observations:

Observation 1. *The application of the reduction rules 1 and 2 on the binary data during transmission ensures that the digit patterns* $\bar{1}\bar{1}$, $1\bar{1}$ *and* $\bar{1}1$ *do not occur in the transmitted data.*

Observation 2. *If the original data was an* n*-bit binary data frame, RBNS encoding can result in a frame of size* $n+1$ *RBNS digits.*

Observation 3. *The encoding and decoding processes need to scan from the least significant to the most significant digit position, and these can be conveniently overlapped with a pipelined serial transmission and reception of the digits.*

It is thus clear that the logic complexities for hardware conversion from binary to RBNS and vice versa are minimal. Thus, the increase in delay and energy consumption due to the encoding and decoding logic would be too small to considerably affect the energy savings offered by this scheme.

8.3.2 RBNSiZeComm Communication Protocol

The RBNSiZeComm communication protocol involves the execution of only two steps:

1. Recode the binary data frame in RBNS using reduction rules 1 and 2.

2. Send the RBNS data frame, transmitting a non-zero voltage level only for the 1 and $\bar{1}$ bits, while remaining silent for the 0 bits.

To distinguish between a channel with no signal and a zero bit inside an ongoing transmission, a *start of transmission* signal is sent at the beginning of transmission of the data and an *end of transmission* signal at the end of the data frame.

8.3.3 Analysis of Energy Savings

Let us first compute the average number of non-zero digits after applying reduction rule 1 on a binary string of length n. We first find the total number of occurrences of runs of 1's of length $k, 1 \leq k \leq n$ in all possible binary strings of 0's and 1's of length n.

We denote a run of 1's of length k by R_k. Clearly, two consecutive runs of 1's of length k_1 and k_2, $1 \leq k_1, k_2 \leq n$ in a bit string will be separated by at least one zero. Let us append a zero on left of each such $R_k, 1 \leq k \leq n$ and denote the symbol 0 R_k by y_k. We also denote a single zero by the symbol y_0. Each such $y_k, 0 \leq k \leq n$, will be a string of length $k+1$. We describe two techniques for solving this counting problem: one using generating functions and the other using recurrence relations.

8.3.3.1 Generating Functions Approach

To find the total number of occurrences of R_k's, $1 \leq k \leq n$, in all possible 2^n strings of length n, we first compute the total number of occurrences of exactly i_k number of y_k's. Let this number be denoted by the symbol $N_n^{i_k,k}$. Thus, to compute $N_n^{i_k,k}$, we consider only those bit strings of length $n+1$ which contain exactly i_k number of y_k's. In effect, we want to partition the integer $n+1$ in all possible ways where *order counts*, such that there are exactly i_k blocks, each of size $k+1$. Each ordered partition will correspond to a distinct string of length n (because of the appended zero on the left of each $y_k, 1 \leq k \leq n$), in which there are exactly i_k number of R_k's. We find this number of possible partitions in three steps [2]:

Step 1: We take out i_k blocks of size $k+1$ each, $1 \leq k \leq n$, from the integer $n+1$ and find the total number of ordered partitioning of $m = (n+1) - (k+1)i_k$ in exactly r blocks ($r \geq 1$ for $k \leq n$ and $r = 0$ for $k = n$), such that there is no block of size $k+1$ in the partition.

Step 2: On each such ordered partition obtained in step 1 above, we insert the i_k blocks of size $(k+1)$ each and find the total number of distinct possible ways. This step is similar to putting i_k identical balls into $r+1$ boxes, where some boxes may remain empty.

Step 3: Finally, we sum over the expression obtained in step 2 above for all possible values of r and multiply it by i_k to get $N_n^{i_k,k}$.

Note that for $k = n$, there is only one such run in the whole string and hence, $N_n^{1,n} = 1$. For $k \leq n$, for Step 1 above, if we scan the bit string of length $n+1$ from left to right, the occurrence of a $y_i, 0 \leq i \leq n$, would correspond to a generating function $(x + x^2 + x^3 + \cdots)$. To exclude the occurrence of y_k's, we subtract x^{k+1} from it. Hence, for r such blocks, we consider the generating function

$$G(x) = (x + x^2 + x^3 + \cdots - x^{k+1})^r,$$

such that the co-efficient of x^m in $G(x)$ will be the required number of partitions for step 1.

Now, $G(x)$ can be written as

$$\begin{aligned}
G(x) &= x^r(1+x+x^2+\ldots-x^k)^r \\
&= x^r\left[\frac{1-x^k(1-x)}{(1-x)}\right]^r \\
&= x^r[1-x^k(1-x)]^r(1-x)^{-r} \\
&= x^r\left[\binom{r}{0}-\binom{r}{1}x^k(1-x)+\binom{r}{2}x^{2k}(1-x)^2-\ldots\right. \\
&\quad \left.+(-1)^q\binom{r}{q}x^{kq}(1-x)^q+\ldots\right]\left[1+\binom{r}{1}x+\binom{r+1}{2}x^2+\ldots\right] \\
&= x^r\left[\binom{r}{0}-\binom{r}{1}x^k(1-x)+\binom{r}{2}x^{2k}(1-x)^2-\cdots\right. \\
&\quad \left.+(-1)^q\binom{r}{q}x^{kq}(1-x)^q+\ldots\right]\left[1+\binom{r}{1}x+\binom{r+1}{2}x^2+\ldots\right] \\
&= x^r\sum_{q=0}^{r}(-1)^q\binom{r}{q}x^{kq}\sum_{j=0}^{q}(-1)^j\binom{q}{j}x^j\sum_{p=0}^{\infty}\binom{r+p-1}{p}x^p. \qquad (8.3)
\end{aligned}$$

Let $m = n+1-(k+1)i_k - r$. Hence the coefficient of $x^{n+1-(k+1)i_k}$ in $G(x)$ is the coefficient of x^m in Equation (8.3), which is equal to

$$N_m = \sum_{q=0}^{r}\sum_{j=0}^{q}(-1)^{q+j}\binom{r+m-1-kq-j}{m-kq-j}\binom{r}{q}\binom{q}{j}.$$

Hence, the required number $N_n^{i_k,k}$ for $k \geq 1$ is given by

$$i_k\sum_{r=1}^{n+1-(k+1)i_k}\binom{(r+1)+i_k-1}{i_k}N_m$$

$$= i_k\sum_{r=1}^{n+1-(k+1)i_k}\binom{r+i_k}{i_k}\sum_{q=0}^{r}\sum_{j=0}^{q}(-1)^{q+j}\binom{r+m-1-kq-j}{m-kq-j}\binom{r}{q}\binom{q}{j}.$$

Example 29. *For $n = 8$, $k = 2$ and $i_k = 2$, we get:*

TABLE 8.1: Number of occurrences of runlengths for $n = 8$

Runlength size (k)	Number of possible values of i_k	Number of occurrences
1	4	320
2	3	144
3	2	64
4	1	28
5	1	12
6	1	5
7	1	2
8	1	1

$$
\begin{aligned}
N_8^{2,2} &= 2\sum_{r=1}^{9-6}\binom{r+2}{2}\sum_{q=0}^{r}\sum_{j=0}^{q}(-1)^{q+j}\binom{2-2q-j}{3-r-2q-j}\binom{r}{q}\binom{q}{j} \\
&= 2\binom{3}{2}\left[\binom{2}{2}\binom{1}{0}\binom{0}{0}-\binom{0}{0}\binom{1}{1}\binom{1}{0}+\binom{-1}{-1}\binom{1}{1}\binom{1}{1}\right] \\
&\quad +2\binom{4}{2}\left[\binom{2}{1}\binom{2}{0}\binom{0}{0}-0+0\right]+2\binom{5}{2}\left[\binom{2}{0}\binom{3}{0}\binom{0}{0}-0+0\right] \\
&= 2(0+12+10)=44.
\end{aligned}
$$

For a given $k \geq 1$, if we now sum the expression $N_n^{i_k,k}$ for all possible values of i_k, $1 \leq i_k \leq \lfloor (n+1)/(k+1) \rfloor$, we get the total occurrences of R_k in all possible strings of length n. Tables 8.1 and 8.2 show the numbers of occurrences of all possible runlengths of 1's in binary strings of length 8 and 16 bits, respectively. The tables along with Figure 8.8 demonstrate that the number of occurrences of run of 1's of length k in a binary string of length n, $1 \leq k \leq n$, decreases exponentially as k increases.

8.3.3.2 Recurrence Relations Approach

Our objective is to successively generate all possible strings containing R_{n-k}'s, $1 \leq k \leq n-1$, starting from all possible strings containing R_{n-k+1}. First we

TABLE 8.2: Number of occurrences of runlengths for $n = 16$

Runlength size (k)	Number of possible values of i_k	Number of occurrences
1	8	147456
2	5	69632
3	4	32768
4	3	15360
5	2	7168
6	2	3328
7	2	1536
8	1	704
9	1	320
10	1	144
11	1	64
12	1	28
13	1	12
14	1	5
15	1	2
16	1	1

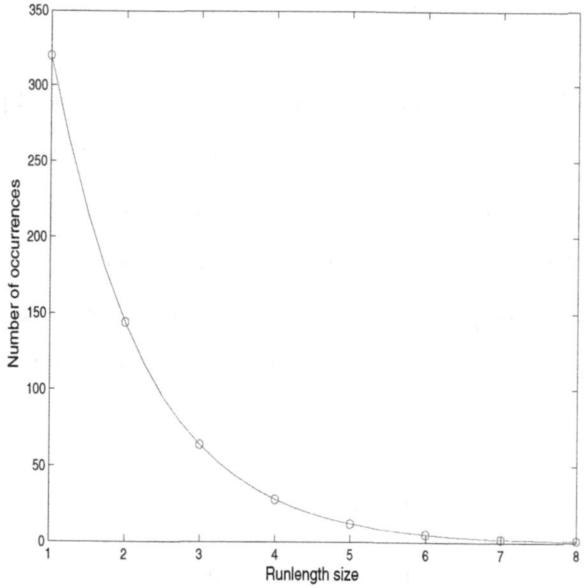

FIGURE 8.8: Number of occurrences of different runlengths for $n = 8$

note that there is only one run of 1's of length n, i.e., the number of R_n's in all possible strings is equal to 1 for any value of n. Also, there are only two possible strings containing R_{n-1}'s. Consider $n = 8$. The only string containing R_8 is 1111 1111, and there are only two possible strings containing R_7 given by 0111 1111 and 1111 1110. For R_6, there are five possible strings: 0011 1111, 1011 1111, 1111 1100, 1111 1101 and 0111 1110. We now try to find a systematic way to generate the strings containing R_6 from the two strings containing R_7. For this we define the following rules.

Left Replacement Rule (*LRR*): If there is a string containing R_{n-k+1}, $k \geq 2$, by applying this rule, we can generate two possible strings containing R_{n-k} from R_{n-k+1} by replacing the leftmost 1 in R_{n-k+1} by a 0 and its immediate preceding 0 (if it exists) by either a 0 or a 1.

Remark 22. *If there is no 0 preceding R_{n-k+1}, then only one string containing R_{n-k} is generated by replacing the leftmost 1 bit.*

Right Replacement Rule (*RRR*): If there is a string containing R_{n-k+1}, $k \geq 2$, by applying this rule, we can generate two possible strings containing R_{n-k} from R_{n-k+1} by replacing the rightmost 1 in R_{n-k+1} by a 0 and its immediate succeeding 0 (if it exists) by either a 0 or a 1.

Remark 23. *If there is no 0 following R_{n-k+1}, then only one string containing R_{n-k} is generated by replacing the rightmost 1 bit.*

Example 30. *Thus, from R_7 in 0111 1111, we can generate two strings containing R_6 by left replacement rule (LRR) as 0011 1111 and 1011 1111. Also, by right replacement rule (RRR) on R_7 in 1111 1110, we can generate the two strings containing R_6 as 1111 1100 and 1111 1101. By LRR on R_7 in 1111 1110 or by RRR on R_7 in 0111 1111, we get the same string 0111 11110, i.e., $0R_60$ (duplicate generation).*

From all these five strings containing R_6 and applying LRR and RRR, we can

TABLE 8.3: Runlength 5 strings from strings containing runs of length 6

Source string	Generated string by LRR	Generated string by RRR
0011 1111	0001 1111 0101 1111	0011 1110
1011 1111	1001 1111 1101 1111	1011 1110
1111 1100	0111 1100	1111 1000 1111 1010
1111 1101	0111 1101	1111 1001 1111 1011
0111 1110	0011 1110 1011 1110	0111 1100 0111 1101

generate all possible strings containing R_5's as shown in Table 8.3. However, by this process, duplicate strings are generated; we see that the strings generated from 0011 1111 and 1011 1111 by RRR are the same as those generated from 0111 1110 by LRR. Also, the strings generated from 1111 1100 and 1111 1101 by LRR are the same as those generated from 0111 1110 by RRR. To avoid duplicate counting of runs of 1's, we follow the biased replacement strategy discussed below.

In the string 0111 1110, the substring R_6 starts at the second bit position from left. By applying RRR on this string, we get two strings $0R_500$ and $0R_501$ in which the substring R_5 also starts at the second bit position from left. Hence, neither of these two strings can be generated by successively applying RRR on any string in which a run of 1's starts at the leftmost bit position. In general, if we apply RRR on a string $0R_k0 * * \cdots *$, where a * may take any value (0 or 1), these strings cannot be duplicated by successively applying RRR on any string of the form $R_j0 * * \cdots *$, $j \geq k$. Also, successive applications of LRR on a string of the form $* * \cdots * 0R_j$ will never be identical with any string generated from $0R_k0 * * \cdots *$ by RRR only. Thus to avoid duplicate string generation, we formulate the biased replacement rule as follows:

Biased Replacement Rule: To generate new strings containing R_{n-k} from a string S containing R_{n-k+1}, $2 \leq k \leq n-1$,

1. Apply LRR on the substring R_{n-k+1} unless R_{n-k+1} starts at the leftmost bit position of the string S.

2. If R_{n-k+1} starts at the leftmost bit position of S, then apply RRR on S.

3. If R_{n-k+1} starts at the second bit position from left in S, then apply both LRR and RRR.

We now show that by using the biased replacement rule, we can avoid duplicate counting of R_{n-k}'s generated from R_{n-k+1}'s in all possible strings.

In the following discussions, unless otherwise mentioned, the substring R_k in a string S always refers to R_k at a given position in S.

The effect of LRR on a substring R_k in S starting at a bit position, say b, will be denoted by the function $\mathcal{L}(R_k, S)$, whose value is the substring R_{k-1} (starting at the bit position $b+1$) obtained from the R_k by LRR. Similarly, the effect of RRR on a substring R_k in S ending at a bit position, say b, will be denoted by the function $\mathcal{R}(R_k, S)$, whose value is the substring R_{k-1} (ending at the bit position $b-1$) obtained from the R_k by RRR.

The result of successive m number of LRR operations on a substring R_k in a string S will be denoted by $\mathcal{L}^m(R_k, S)$. Similarly, the result of successive m number of RRR operations on a substring R_k will be denoted by $\mathcal{R}^m(R_k, S)$. The result of an application of LRR on a substring R_k in S followed by the application of a RRR on the resulting substring will be denoted by $\mathcal{LR}(R_k, S)$, while application of RRR followed by LRR will be denoted by $\mathcal{RL}(R_k, S)$.

Hence, we note that

$$\mathcal{L}^m(R_{n-1}, R_{n-1}0) = *^m 0 R_{n-m-1},$$

and

$$\mathcal{R}^m(R_{n-1}, 0R_{n-1}) = R_{n-m-1}0*^m,$$

where the symbol $*$ stands for a 0 or 1.

We now have the following results:

Lemma 25. $\mathcal{L}^m\mathcal{R}(R_{n-1}, 0R_{n-1}) = \mathcal{L}^m(R_{n-2}, 0R_{n-2}0)$.

Proof. Follows from the definitions of $\mathcal{L}$ and $\mathcal{R}$ functions. □

Lemma 26. $\mathcal{R}^m\mathcal{L}(R_{n-1}, R_{n-1}0) = \mathcal{R}^m(R_{n-2}, 0R_{n-2}0)$.

Proof. Follows from the definitions of $\mathcal{L}$ and $\mathcal{R}$ functions. □

In view of Lemma 25, $\mathcal{L}^m\mathcal{R}$ on R_{n-1} in $0R_{n-1}$ can be equivalently replaced by m successive $\mathcal{L}$ functions on R_{n-2} in $0R_{n-2}0$. Also, due to Lemma 26, $\mathcal{R}^m\mathcal{L}$ on R_{n-1} in $R_{n-1}0$ can be equivalently replaced by m successive $\mathcal{R}$ functions on R_{n-2} in $0R_{n-2}0$.

Lemma 27. $\mathcal{L}\mathcal{R}(R_{n-2}, 0R_{n-2}0) = \mathcal{R}\mathcal{L}(R_{n-2}, 0R_{n-2}0)$.

Proof. Follows from the definitions of $\mathcal{L}$ and $\mathcal{R}$ functions. □

Because of Lemma 27, we can reverse the order of application of LRR and RRR on the string $0R_{n-2}0$, and repeated application of this property may result in applying all LRR (or RRR) first, followed by applications of all RRR (or LRR) on $0R_{n-2}0$. For example, we can write, $\mathcal{L}^2\mathcal{R}^3 = \mathcal{L}(\mathcal{L}\mathcal{R})\mathcal{R}^2 = \mathcal{L}\mathcal{R}\mathcal{L}\mathcal{R}\mathcal{R} = \mathcal{L}\mathcal{R}(\mathcal{L}\mathcal{R})\mathcal{R} = \mathcal{L}\mathcal{R}\mathcal{R}\mathcal{L}\mathcal{R} = \mathcal{L}\mathcal{R}\mathcal{R}\mathcal{R}\mathcal{L} = \cdots = \mathcal{R}\mathcal{R}\mathcal{R}\mathcal{L}\mathcal{L}$. Hence, we get the following result.

Lemma 28. $\mathcal{L}^p\mathcal{R}^q(R_{n-2}, 0R_{n-2}0) = \mathcal{R}^q\mathcal{L}^p(R_{n-2}, 0R_{n-2}0)$.

In view of Lemma 28, $\mathcal{L}^m\mathcal{R}$ on R_{n-2} in $0R_{n-2}0$ may be equivalently replaced by only $\mathcal{L}$ functions on strings generated by $\mathcal{R}^m$ on R_{n-2} in $0R_{n-2}0$.

By repeated applications of LRR on $R_n - 1$ in $0R_{n-1}$, we may get a resulting string S given by $S = ** \cdots * 0R_k0 ** \cdots * 0R_{n-m-1}$, for some value of k. We claim that application of LRR or RRR on this resulting R_k in S is not needed to generate any new substring R_{k-1}, as this R_{k-1} will also be generated by LRR and RRR on R_{n-2} in $0R_{n-2}0$. This follows from the following lemma.

Lemma 29. *Given that* $S = *^{k_1}0R_k0 *^{k_2} 0R_{n-k-k_1-k_2-3}$, *we have*

$$\mathcal{L}(R_k, S) = \mathcal{R}^{n-k-k_1-2}\mathcal{L}^{(k_1+1)}(R_{n-2}, 0R_{n-2}0).$$

Proof. $\mathcal{L}(R_k, S)$ is the substring R_{k-1} in $*^{k_1+1}0R_{k-1}0 *^{k_2} 0R_{k-k_1-k_2-3}$.

On the other hand, the string $\mathcal{R}^{n-k-k_1-2}(R_{n-2},\ 0R_{n-2}0)$ is the substring $R_{n-2-n+k+k_1+2}$ $(= R_{k+k_1})$ in the string $0R_{n-2-n+k+k_1+2}0*^{n-k-k_1-2}$, which is $0R_{k+k_1}0*^{n-k-k_1-2}$.

Hence, $\mathcal{R}^{n-k-k_1-2}\mathcal{L}^{k_1+1}(R_{n-2}, 0R_{n-2}0) = *^{k_1+1}0R_{k-1}0*^{n-k-k_1-2}$, which contains the string $*^{k_1+1}0R_{k-1}0 *^{k_2} 0R_{n-k-k_1-k_2-3}$.

Note that we have $\mathcal{L}^m(R_{n-1}, 0R_{n-1}) =$ the substring R_{n-m-1} in $*^m0R_{n-m-1}$, while the string $\mathcal{R}^m(R_{n-1}, R_{n-1}0)$ = the substring R_{n-m-1} in $R_{n-m-1}0*^m$. Hence, although $*^m0R_{n-m-1}$ may be identical with $R_{n-m-1}0*^m$ when $m = \frac{(n+1)}{2}$ for odd n, these two appearances of R_{n-m-1} (one at the leftmost position and the other at the rightmost position) in the same string are different and should be counted twice, as obtained through the above two operations. Hence we get the following result. □

Lemma 30. $\mathcal{L}^m(R_{n-1}, 0R_{n-1})$ *and* $\mathcal{R}^j(R_{n-1}, R_{n-1}0)$ *always generate two distinct substrings* R_{n-m-1} *and* R_{n-j-1} *for any value of* m *and* j.

Theorem 25. *The biased replacement rule always produces all possible distinct appearances of* R_{n-k}*'s in all possible strings without any duplicate counting.*

Proof. Follows from all the above lemmas. □

Now we see that, by LRR on a substring R_{n-k+1}, we generate two possible distinct substrings R_{n-k}. By RRR on a substring R_{n-k+1} in each string of the form $0R_{n-k+1}0*^{k-3}$, we always generate two different substrings of the form

$0R_{n-k}0{*}^{k-2}$, whose number is thus doubled on each application of RRR. Since for $k = 2$, the number of such strings of the form $0R_{n-k}0{*}^{k-2}$ = 1, we get the following theorems about the total number $\mathcal{N}_{n-k}$ of R_{n-k}'s in all possible strings of length n.

Theorem 26. $\mathcal{N}_n = 1$, $\mathcal{N}_{n-1} = 2$, *and for* $k \geq 2$,

$$\mathcal{N}_{n-k} = 2\mathcal{N}_{n-k+1} + 2^{k-2}.$$

Theorem 27. $\mathcal{N}_n = 1$, $\mathcal{N}_{n-1} = 2$, *and for* $k \geq 2$,

$$\mathcal{N}_{n-k} = (k+3)2^{k-2}.$$

Proof. Follows by solving the recurrence relation in Theorem 26. □

Hence, the total number of runs of 1's of all possible lengths $(n-k), 0 \leq k \leq n-2$, is given by

$$\begin{aligned} S &= \sum_{k=2}^{n-2}(k+3)2^{k-2} + 3 \\ &= 5 \cdot 2^0 + 6 \cdot 2^1 + 7 \cdot 2^2 + \cdots + (n+1) \cdot 2^{n-4} + 3. \end{aligned}$$

Hence, $2S = 5 \cdot 2^1 + 6 \cdot 2^2 + \cdots + n \cdot 2^{n-4} + (n+1) \cdot 2^{n-3} + 6$, from which we get, $S = n \cdot 2^{n-3}$.

The total number of 1's and $\bar{1}$'s in the RBN coded message obtained after applying the reduction rule 1 (considering also the presence of R_1's) is equal to $2S + (n+2) \cdot 2^{n-3} = (3n+2) \cdot 2^{n-3}$.

Hence the fraction of energy savings over EbT schemes obtained by applying reduction rule 1 is given by

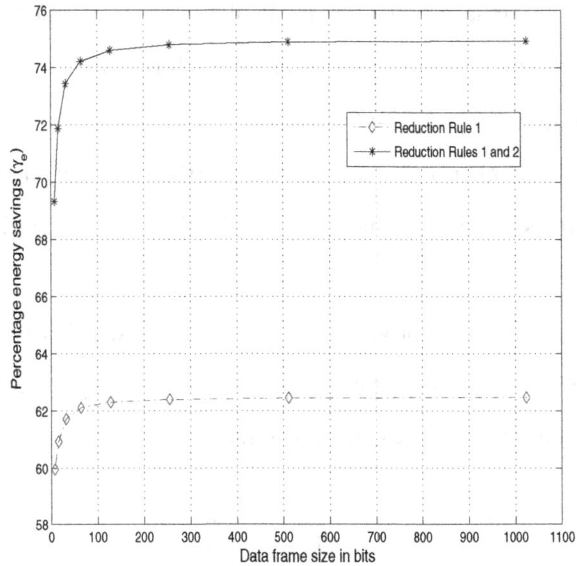

FIGURE 8.9: Plot of theoretical savings versus data frame size

$$\begin{aligned}\gamma_{e_1} &= 1 - \frac{(3n+2)\cdot 2^{n-3}}{n \cdot 2^n} \\ &= 1 - \frac{(3n+2)}{8n}. \end{aligned} \tag{8.4}$$

Typically, for $n = 8$, γ_{e_1} is nearly equal to 60% and for $n = 1024$, $\gamma_{e_1} \simeq 63\%$.

Let us now consider the effect of applying reduction rule 2 on the RBN coded string (after applying reduction rule 1). Every appearance of the pattern $\bar{1}1$ in the RBN coded string will be replaced by $0\bar{1}$ after applying reduction rule 2, and this appearance corresponds to every single 0 in between two runs of 1's in the original binary string. To compute the total number of such appearances of singleton 0's trapped between two runs of 1's, we proceed as follows.

Note that the total number of singleton 0's in all possible binary strings of length n is same as the total number of appearances of R_1's, and is equal to $\mathcal{N}_1 = (n +$

$2)2^{n-3}$. From this, we have to subtract the total number of appearances of singleton 0's at either end of the string, i.e., in strings of the form $01*^{n-2}$ or $*^{n-2}10$, whose number is equal to $2^{n-2} + 2^{n-2} = 2^{n-1}$.

Thus, the total number of 1's and $\bar{1}$'s in the RBN coded message after applying both reduction rules 1 and 2 is equal to $(3n+2) \cdot 2^{n-3} - (n+2) \cdot 2^{n-3} + 2^{n-1} = (n+2) \cdot 2^{n-2}$.

If we assume that in the original message each bit, whether 0 or 1, consumes e_b units of energy for communication, the fraction of energy saving obtained by applying reduction rule 1 followed by reduction rule 2 is given by,

$$\gamma_{e_2} = 1 - \frac{(n+2)}{4n}. \tag{8.5}$$

Typically, for $n = 8$, γ_{e_2} is nearly equal to 69% and for $n = 1024$, $\gamma_{e_2} \simeq 75\%$. Figure 8.9 shows the plot of theoretical energy savings (γ_{e_1} and γ_{e_2}) in data transmission against the data frame size n. From the graph we see that if we apply only reduction rule 1 on binary encoded data, the maximum possible energy savings in transmitting data is 63%. Application of both reduction rules 1 and 2, increases the maximum possible theoretical savings to 75%.

8.3.4 Physical Implementation Issues

The above results of the performance of the *RBNSiZeComm* protocol are valid for a transmitter with ideal device characteristics. We now consider various issues arising from the practical implementation of this protocol.

8.3.4.1 Representation of RBNS Encoded Numbers

To represent the three symbols ($\{\bar{1}, 0, 1\}$) at the MAC layer, we can use two bits to encode each digit of the redundant binary number (RBN). As an example, the

RBN digit 0 may be encoded using the bit pair 00, while 1 and $\bar{1}$ use the bit pairs 01 and 10, respectively.

8.3.4.2 Synchronization Issues

A requirement for the correct detection of silent symbols in the RBNSiZeComm protocol is clock synchronization between the sender and the receiver. We assume that the receiver and the transmitter are i) synchronized for every packet on the MAC packet header as common in most wireless and wired communication protocols (e.g., in the IEEE 802.3 protocol the *preamble* field in the MAC frame header is utilized for synchronization of the sender and the receiver) or, ii) there exists a global clock source, for example, the *network time protocol* (NTP), which disseminates *International Atomic Time* (TAI) in the form of *Coordinated Universal Time* (UTC) to provide network-wide clock synchronization. For WSNs, a number of low overhead clock synchronization protocols are described in the literature [30, 31]. The transmission time of the payload of the packet is assumed to be small enough not to lose synchronization between the receiver and the transmitter during the transmission of consecutive silent symbols. The exact implementation of maintaining synchronization of nodes is left to the data link layer designer.

8.3.4.3 Collision Avoidance Issues

For contention-based medium access protocols, an issue that arises with the use of silence-based symbols in the transmitted data is how neighboring nodes determine whether a channel is free for transmission or whether there is an ongoing transmission containing a run of 0's. An erroneous interpretation of the absence of any signal on air (carrier and/or baseband) can lead to corruption of an ongoing transmission (thus leading to retransmission and an increase in energy consumption by the communicating node pair) or unnecessary increase in communication delay due to waiting when a channel is actually free.

One way to avoid this would be to utilize only designated time slots in a predetermined transmission schedule for data communication. For example, in the 802.15.4 protocol, one may choose to employ RBNSiZeComm-based data transmission only during the *contention-free period* (CFP).

For contention-based medium access, three possible MAC layer schemes can detect and resolve collisions for acquiring the channel.

The first is based on a longer carrier sensing time, equal to the duration of the maximum allowable number of bits in the message as determined by the protocol, by a neighboring node (that wishes to transmit over the same channel) so that collision is avoided.

The second scheme is based on inserting a 1 or $\bar{1}$ (similar to *bit stuffing*) after every run of 0's of a predefined length. This avoids collision, but the overhead would be a small increase in transmission time and energy. This is in line with similar techniques used in protocols like IEEE 802.15.4.

A third possible scheme is based on waiting for a few, say w, symbol periods to get a non-zero symbol and then trying to retransmit which may, however, cause a collision with a probability as a function of w. If a collision is detected, standard backoff techniques similar to those used in IEEE 802.3 and IEEE 802.15.4 may be used. As an example, in CSMA-CA-based IEEE 802.15.4, before a node attempts to send a frame in the *contention access period* (CAP), it waits for a random backoff period chosen from a specific integer range. After completing a backoff, a node performs a *clear channel assessment* (CCA) for eight symbol periods to assess whether the channel is busy. If after eight symbol periods the channel is assessed to be free, the node attempts a transmission.

To understand the effects of the second and the third scheme on energy savings using the above results on run distribution [1], we note that for a binary string of length $n \geq 8$, the sum of the numbers of all possible runs of length less than or equal to 6 is given by

$$S_6 = \sum_{k=7}^{n} \mathcal{N}_k = (63n + 69)2^{n-8}. \tag{8.6}$$

Thus, the fraction f of all possible numbers of runs of length greater than 6 is given by

$$f = \frac{S - S_6}{S} = \frac{n - 5}{2^6(n + 1)}. \tag{8.7}$$

For $n = 8$, $f = 0.005208$, while the asymptotic value of f is 0.015625. Thus, if we choose w greater than 6, the collision probability is only about 1.5%, which is quite low. Thus, depending on the typical application environment, one may choose any of the above schemes to handle collisions while using RBNSiZeComm in contention-based medium access protocols.

8.3.4.4 Effect of Device Characteristics

For most commercially available radio devices, the power drawn in the *transmit* or *receive* state is considerably more than the power drawn in the *idle* or *active* state - i.e., when the radio is in *low power* operation mode [22]. The penalty paid for this low power operational state is the switching time from the active to transmit state and vice versa. However, this does not limit the applicability of the *RBNSiZeComm* protocol in a WSN, which essentially is classified as a low data rate one. With chips having faster switching time, the data rate can be increased. We now investigate the effect of the device characteristics on the energy saving aspect of the *RBNSiZeComm* protocol. We will use the terms *idle* and *active* interchangeably to refer to this low power operation state.

We assume that the device moves from idle to transmit (TX) state in time t_{rise} and during this transition period, it draws current I_{tran}. Similarly, when the device moves from a TX state to the idle state, it takes a finite amount of time, say t_{fall}. Usually t_{fall} is small and we neglect the energy consumed by the device during t_{fall} time. For example, for the CC1100 or CC2420 chips, the device moves from TX to idle state quickly, drawing insignificant extra energy over that used in the idle state. Let V_{cc} be the supply voltage, and I_{high} and I_{low} denote the currents drawn in the TX and *active* states, respectively. Let $I_0 = I_{high} - I_{low}$ and t_p denote the width of the symbol pulse. The energy expenditure for transmitting the equivalent RBNS data frame can be represented as the sum of the following three terms.

1. A base energy value P_{base} needed throughout the transmission, which is equal to the energy required when the device is in the *active* state. For an n-digit RBN data frame, considering all possible 2^n strings of length n each, $P_{base} = n2^n t_p I_{low} V_{cc}$.

2. Extra energy $P_{1\bar{1}}$ in addition to the base energy required to transmit a 1 or $\bar{1}$ symbol in TX state. For an n-digit RBN data frame, considering all possible strings, for $(n+2)2^{n-2}$ non-zero symbols (1 and $\bar{1}$) in the TX state, $P_{1\bar{1}} = (n+2)2^{n-2}(t_p - t_{rise})I_0 V_{cc}$.

3. Extra energy P_{rise} in addition to the base energy required during the edge transitions from a 0 to 1 or $\bar{1}$. This corresponds to switching from the active to TX state and the number of such transitions is equal to the number of non-zero symbols. Thus, we have, $P_{rise} = (n+2)2^{n-2} t_{rise}(I_{tran} - I_{low})V_{cc}$.

Hence, the total energy required to transmit an n-digit RBN data frame is

$$\begin{aligned} E_{rbn} &= P_{base} + P_{1\bar{1}} + P_{rise} \\ &= V_{cc}\Big[n2^n t_p I_{low} + (n+2)2^{n-2}(t_p - t_{rise})I_0 \\ &\quad + (n+2)2^{n-2} t_{rise}(I_{tran} - I_{low})\Big]. \end{aligned} \tag{8.8}$$

In contrast, the total energy consumed by an EbT transmission scheme for transmitting an n-bit binary data frame is $E_{ebt} = n2^n t_p(I_{low} + I_0)V_{cc}$. Considering the device characteristics, the fractional energy savings generated by the *RBNSiZeComm* protocol over EbT transmission schemes is

$$\eta_e^{'} = \frac{E_{ebt} - E_{rbn}}{E_{ebt}} = \frac{(3n-2)I_0 + (n+2)(I_{high} - I_{tran})\frac{t_{rise}}{t_p}}{4n \cdot I_{high}}. \tag{8.9}$$

From Equation (8.9), we see that as $n \rightarrow \infty$, $\eta_e^{'} \approx \frac{3}{4}\left(\frac{I_0}{I_{high}}\right) + \frac{1}{4}\frac{(I_{high} - I_{tran})}{I_{high}}\left(\frac{t_{rise}}{t_p}\right)$. To estimate the effect of device characteristics on energy savings, we consider two examples using two commercially available popular chips:

Example 31. *For the CC1100 chip supporting O-QPSK modulation, we have* $I_{low} = 1.9mA$ *and* $I_{tran} = 8.7mA$. *At 868/915 MHz, values of* I_{high} *are 30.3 mA, 19.7 mA, 16.6 mA and 14.0 mA for 10 dBm, 5 dBm, 0 dBm and -5 dBm output power respectively. Also,* $t_{rise} = 88.4\mu s$ *and for a data rate of 2.5 Kbps, we have* $t_p = 400\mu s$. *Assuming* $n = 1024$, *we get from Equation(8.9),* $\eta_e^{'} = 74.20\%, 70.81\%, 69.01\%$ *and* 66.87% *for 10 dBm, 5 dBm, 0 dBm and* -5 *dBm output power, respectively.*

Example 32. *For the CC2420 chip,* $I_{low} = 0.426$ *mA,* $I_{tran} \approx I_{low}$ *and values of* I_{high} *are 17.4 mA, 14 mA, 11 mA, 9.9 mA and 8.5 mA at 0 dBm,* -5 *dBm,* -10 *dBm,* -15 *dBm and* -25 *dBm output power, respectively.* $t_{rise} = 0.1\mu s$. *As in the case of CC1100, we assume that* $t_p = 400\mu s$ *and* $n = 1024$. *Hence, from Equation(8.9), we get* $\eta_e^{'} = 73.12\%, 72.68\%, 72.05\%, 71.73\%$ *and* 71.20% *for 0 dBm,* -5 *dBm,* -10 *dBm,* -15 *dBm and* -25 *dBm output power, respectively.*

Observation 4. $\eta_e^{'}$ *approaches the asymptotic value of* $\eta_e = 75\%$ *for radios where* $I_{low} \ll I_{high}$. *Two examples of such radios include the RFM TR1000 (*$I_{high} =$

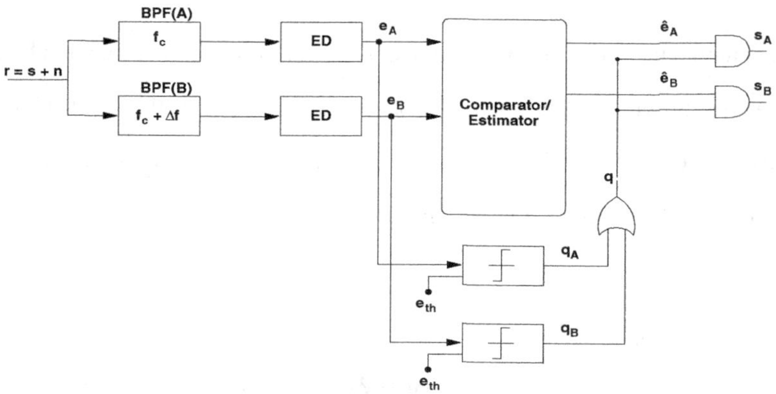

FIGURE 8.10: Representative non-coherent FSK receiver for RBNS scheme

$12mA$, $I_{low} = 7.0 \times 10^{-4}\ mA$) *and the Maxim 1479* ($I_{high} = 7.3mA$, $I_{low} = 0.2 \times 10^{-6}\ mA$).

8.3.5 Transceiver Design and Analysis for Noisy Channels

We now turn our attention to a realistic situation where a channel is noisy. We assume that the channel has *additive white Gaussian noise* (AWGN) noise. Assume that the transmitter uses FSK modulation with two frequencies, f_c and $f_c + \Delta f$ corresponding to 1 and $\bar{1}$, respectively, and is switched off during 0's. Effectively this will be a hybrid modulation scheme involving FSK and ASK. To demonstrate energy savings, we use a non-coherent detection-based receiver as shown in Figure 8.10.

In the figure, r is the signal received at the input which consists of the transmitted signal s and the channel noise n, i.e, $r = s + n$. The received signal r travels through two bandpass filters (BPFs) as shown in Figure 8.10, with center frequencies at f_c and $f_c + \Delta f$ (corresponding to 1 and $\bar{1}$, respectively). The outputs q_A and q_B of the threshold detectors are applied to the inputs of an OR gate to generate $q = q_A + q_B$. If the transmitted symbol is 0 and e_{th} is correctly chosen, the outputs of the threshold detectors will be 0, making q equal to 0. On the other hand, if the transmitted symbol is 1 or $\bar{1}$, and the threshold value e_{th} is correctly chosen, it will generate the desired output symbol 1 or $\bar{1}$ at the output $\hat{s}$ of the receiver. Thus, the desired output $\hat{s}$ is as follows:

i) If $q = 0$, $\hat{s} = 0$, corresponding to $s_A = s_B = 0$.

ii) If $q = 1$,

$$\hat{s} = \begin{cases} 1 & \text{if } e_A > e_B, \text{ (i.e., } s_A = 1,\ s_B = 0) \\ \bar{1} & \text{if } e_B > e_A, \text{ (i.e., } s_B = 1,\ s_A = 0). \end{cases} \tag{8.10}$$

8.3.5.1 Error Analysis

We consider a message frame containing n binary bits. Corresponding to each such frame, we consider the RBNS frame after converting the binary message to RBNS form containing symbols 0, 1 and $\bar{1}$. Instead of the conventional bit error rate (BER) encountered while receiving the transmitted message, we consider a frame error rate (FER) for both the original binary frame (communicated through non-coherent binary FSK (BFSK)) and the RBNS frame communicated through the non-coherent hybrid FSK-ASK modulation scheme as in Figure 8.10. This is because RBNS symbols have some interdependency due to the nature of $\bar{1}$. An erroneous reception of $\bar{1}$ may affect a number of higher order bits in the decoded binary data. Hence, FER becomes a more convenient performance metric for comparing typical binary and RBNS-based number systems.

In the hybrid FSK-ASK modulation scheme, let α be the probability that a transmitted RBNS symbol was 0 and it is received correctly as 0. Similarly, let β (or, δ) be the probabilities that a transmitted symbol was 1 (or, $\bar{1}$) and the received symbol is also 1 (or $\bar{1}$). From symmetry, $\beta = \delta$. Note that $\alpha = P\{q_A = 0|0\} \cdot P\{q_B = 0|0\}$, where $P\{q_A = 0|0\}$ and $P\{q_B = 0|0\}$ are the probabilities of correctly receiving 0 at q_A and q_B respectively, given that 0 was transmitted. Hence,

$$\alpha = P\{q_A = 0|0\} \cdot P\{q_B = 0|0\}$$
$$= [P\{q_A = 0|0\}]^2 \text{ (by symmetry).} \tag{8.11}$$

Also, note that q_A for 0 transmission follows Rayleigh distribution. Assuming a zero mean and standard deviation of σ for AWGN channel, we can write

$$P\{q_A = 0|0\} = \int_0^{e_{th}} \frac{e_A}{\sigma^2} e^{-e_A^2/2\sigma^2} \cdot de_A = (1 - e^{-e_{th}^2/2\sigma^2}).$$

Hence, we have $\alpha = (1 - e^{-e_{th}^2/2\sigma^2})^2$. Next, we define β as

$$\beta = P\{q = 1|1\} P\{e_A > e_B, \forall e_A|1\}) = X_1 X_2 \qquad \text{(say)},$$

where $X_1 = P\{q = 1|1\}$ and $X_2 = P\{e_A > e_B|1, \forall e_A|1\}$. From Figure 8.10, we can write:

$$X_1 = \int_0^{\infty} [1 - P\{q_A = 0|1\} P\{q_B = 0|1\}] p_1(e_A)\, de_A,$$

where $P\{q_B = 0|1\}$ is governed by a Rayleigh distribution, while $P\{q_A = 0|1\}$ follows a Rician distribution. Thus,

$$P\{q_A = 0|1\} = \int_0^{e_{th}} \frac{e_A}{\sigma^2} e^{-\frac{e_A^2 + s_A^2}{2\sigma^2}} I_0\left(\frac{s_A e_A}{\sigma^2}\right) de_A \quad \text{and}$$
$$P\{q_B = 0|1\} = \int_0^{e_{th}} \frac{e_B}{\sigma^2} e^{-e_B^2/2\sigma^2}\, de_B = 1 - e^{-e_{th}^2/2\sigma^2}.$$

Using the above expressions, we obtain X_1 and X_2 as

$$X_1 = \int_0^\infty \left[1 - \left(1 - e^{-e_{th}^2/2\sigma^2}\right)\left(\int_0^{e_{th}} \frac{e_A}{\sigma^2} e^{-\frac{e_A^2+s_A^2}{2\sigma^2}} \right.\right.$$
$$\left.\left. \times\, I_0\left(\frac{s_A e_A}{\sigma^2}\right) de_A\right) \frac{e_A}{\sigma^2} e^{-\frac{e_A^2+s_A^2}{2\sigma^2}} I_0\left(\frac{s_A e_A}{\sigma^2}\right)\right] de_A$$
$$X_2 = \int_0^\infty \left[\left(1 - e^{-e_A^2/2\sigma^2}\right) \frac{e_A}{\sigma^2} e^{-\frac{e_A^2+s_A^2}{2\sigma^2}} \right.$$
$$\left. \times\, I_0\left(\frac{s_A e_A}{\sigma^2}\right) u(e_A)\right] de_A.$$

Let P_0, P_1 and $P_{\bar{1}}$ be the probabilities of occurrences of 0, 1 and $\bar{1}$, respectively, in the transmitted message. The probability that the transmitted RBNS frame with $n+1$ RBNS digits will be received correctly by the above hybrid FSK-ASK scheme can be expressed as *frame-correct rate* (FCR):

$$FCR = \alpha^{nP_0} * \beta^{nP_1} * \delta^{nP_{\bar{1}}}. \tag{8.12}$$

Assuming that all possible binary strings with n bits are equally likely to appear in the message, we found in Section 8.6.3.2 that with the application of both reduction rules 1 and 2, the average number of 0's in the RBNS frame will be asymptotically (for large n) 75% and those of all non-zero symbols, i.e., 1 and $\bar{1}$ taken together will be 25%. Thus, $P_0 = 0.75$ and $P_1 + P_{\bar{1}} = 0.25$. The probability that a message frame in RBNS will be received correctly (using the hybrid FSK-ASK modulation scheme), is:

$$FCR = \alpha^{0.75n} * \beta^{0.25n}. \tag{8.13}$$

Note that $\log(FCR) = n(0.75 \log \alpha + 0.25 \log \beta)$, from which we can easily evaluate FCR. Next we obtain the frame error rate (FER) for the RBNS message as $FER_{rbns} = 1 - FCR$.

Let us consider the situation that we transmit the original binary message of n bits using non-coherent FSK with a BER of ρ. The probability that the whole frame will be received correctly is $(1-\rho)^n$. Thus, the FER for binary FSK will be given by $FER_{bfsk} = 1-(1-\rho)^n$, which may be approximated as $n\rho$, since ρ is usually very small. For a given frame size n, we can compare FER_{rbns} with $FER_{bfsk}(= n\rho)$, or in other words, FER_{rbns}/n with ρ. To compute FER_{rbns}/n, we set $s_A = 1$, i.e., signal power = $s_A^2/2 = 1/2$ and a signal-to-noise ratio (SNR) = $1/2\sigma^2$. The computations to find the optimum threshold voltage e_{th} that gives the minimum FER_{rbns}/n for a given SNR were implemented in $MATLAB^{®}$[1]. Note that from Equation(8.13), we get the required peak transmitter power with RBNSiZeComm for a given value of FER_{rbns}.

The SNR values for binary FSK modulation are determined from the relation, $\rho = \frac{1}{2}e^{-\gamma/2}$ where SNR in dB = $10 \log_{10} \gamma$. For different values of frame size in the range $8 \leq n \leq 1024$, the SNR values of the hybrid modulation scheme with RBNS message and the corresponding SNR values for binary FSK to meet the same FER values are summarized in Table 8.4. We see that for the same FER, the required SNR values with RBNSiZeComm are on an average 2.76 dB higher than those with binary non-coherent FSK. However, in the RBNSiZeComm scheme, the transmitter will be in the transmit state only during the non-zero symbols (1 or $\bar{1}$), and in the idle state during the 0's in the RBNS encoded message. Hence, the required average transmitter power for a given FER will be less than the value demanded by Equation(8.13). Taking the percentage of non-zero symbols in the RBNS string as 25%, the required *average* transmitter power will be 0.25 times the values as may

[1] $MATLAB^{®}$ is a registered trademark

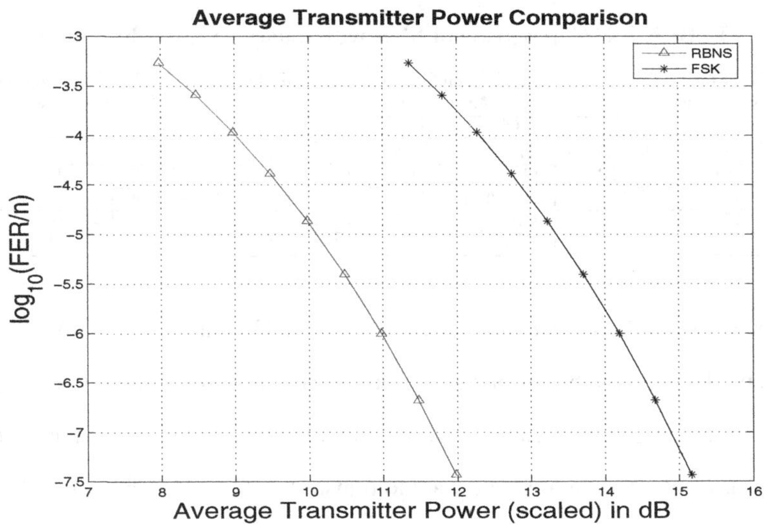

FIGURE 8.11: Comparison of average transmitter power (scaled) for given BER

be obtained from Equation(8.13). This implies that the required average transmitter power for RBNSiZeComm will be reduced from the peak by $10\log_{10}(0.25)$ dB = 6.0206 dB. A plot of the scaled transmitter power averaged over $8 \leq n \leq 1024$ for RBNSiZeComm and for binary FSK (with non-coherent detection) is presented in Figure 8.11.

From Figure 8.11, we see that for a given $\rho = FER/n$ in the range 10^{-4} to 10^{-8}, RBNSiZeComm needs approximately 3.26 dB less average power than binary FSK. Let P_b and P_r be the required average transmitter power levels for binary FSK and RBNSiZeComm, respectively. Noting that $10^{-3.26/10} = 0.472$, we have $P_r/P_b = 0.472$. Compared to binary FSK, this amounts to a saving of 52.8% transmitter energy on average.

8.3.6 Performance Comparison with CtS and VarBaTaC

We now compare the performance of RBNSiZeComm scheme with those of CtS [9] and VarBaTac [10], which are also based on communication through silence.

TABLE 8.4: Comparison of SNR with binary FSK for different n

SNR for RBNSiZeComm	Corresponding SNR with binary FSK for different frame sizes (n)						Average SNR for binary FSK
	16	64	128	256	512	1024	
14.0	11.3121	11.321	11.3328	11.356	11.4012	11.4869	11.3545
15.0	12.2674	12.2688	12.2706	12.2742	12.2813	12.2955	12.2741
16.0	13.2278	13.2279	13.2281	13.2285	13.2292	13.2306	13.2285
17.0	14.1921	14.1922	14.1922	14.1922	14.1922	14.1923	14.1922
18.0	15.1649	15.1649	15.1649	15.1649	15.1649	15.1649	15.1649

We first consider the transmission length duration. The transmission time in CtS is exponential in the number of bits to be sent. VarBaTaC uses a variable coding base and controls the communication time by suitably tuning the base value. Suppose, without any loss of generality, we want to transmit a value M, where $0 \leq M < 2^k$, $k \geq 0$. Then the representation of M in base r will have $m = \left\lceil \frac{k}{\log r} \right\rceil$ digits. Each of these m digits is then transmitted using CtS with a pulse separating each digit. Let t_{avg} be the average number of time slots required in the VarBaTaC scheme. We need to transmit $m + 1$ pulses for the m digits to be transmitted. In addition, we need on average, $\frac{m(r-1)}{2}$ time slots for the digits (assuming equal likelihood for all possible values of the message). Hence, t_{avg} is given by

$$t_{avg} = (m+1) + \frac{m(r-1)}{2} = \frac{m(r+1)}{2} + 1. \tag{8.14}$$

Let $r = 2^p$. Hence, $m = \left\lceil \frac{k}{p} \right\rceil$, and from Equation (8.36) we have

$$\frac{k}{p}\left(\frac{2^p+1}{2}\right) + 1 \leq t_{avg} < \left(\frac{k}{p}+1\right)\left(\frac{2^p+1}{2}\right) + 1.$$

Thus, for a given k, t_{avg} monotonically increases with p due to the function $\frac{2^p+1}{p}$. Table 8.5 shows t_{avg} for different values of $r = 2^p$. We see that the transmission time of the RBNSiZeComm scheme will be less than those of CtS and VarBaTaC. Like RBNSiZeComm, neither CtS nor VarBaTaC generates energy savings at the receiver. Thus, in CtS and VarBaTaC, the energy savings generated during transmis-

TABLE 8.5: Number of slots for VarBaTaC with radix as power of 2 and $k = 64$

p	radix r	$t_{avg} = \lceil \frac{64}{p} \rceil (\frac{2^p+1}{2}) + 1$
2	4	81
3	8	100
4	16	137
5	32	215.5
6	64	357.5
7	128	646
8	256	1029

sions would be substantially outweighed by the excessive receiver energy due to far longer duration of communication than observed with RBNSiZeComm. While for single hop networks this is not a problem, with multi-hop WSNs, this may become an issue for CtS and VarBaTaC.

8.3.7 Performance Comparison with QPSK

We now present a comparison with QPSK, using both non-coherent and coherent receivers for the RBNSiZeComm scheme.

8.3.7.1 Using Non-coherent Receiver Design

We compute the required SNR values with QPSK for a given $\rho = FER/n$ from the following relation:

$$\rho = Q(\sqrt{(2\gamma)}) \approx \frac{1}{2\sqrt{\pi\gamma}} e^{-\gamma}. \tag{8.15}$$

The results are summarized in Table 8.6. From a plot of the SNR values with the data from this table against different values of ρ, we see that the peak transmitter power with RBNSiZeComm employing non-coherent FSK-ASK detection is about 6.537 dB more than that with QPSK modulation. Noting that the average transmitter power with RBNSiZeComm will be 6.0206 dB less than the peak power, it appears that the average power requirement with the above implementation of RB-

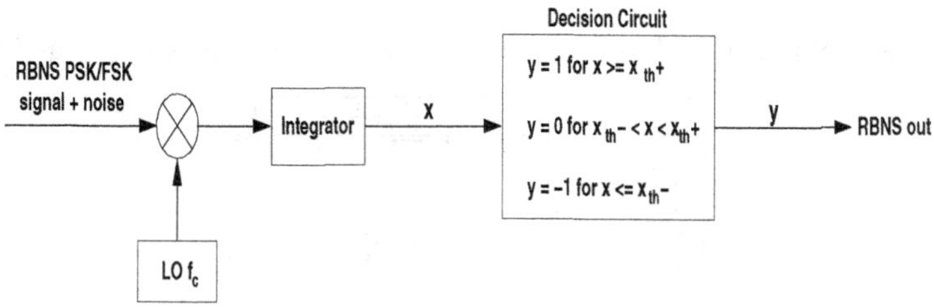

FIGURE 8.12: Representative coherent FSK receiver for RBNS scheme

NSiZeComm is about 0.514 dB more than that with QPSK, which translates into 1.12564 times more energy i.e., about 12.56% more energy than QPSK.

TABLE 8.6: Comparison of peak transmitter powers between QPSK and non-coherent RBNSiZeComm

$\rho = FER/n$	Required SNR in dB	
	RBNSiZeComm Protocol	QPSK
5.39173×10^{-4}	14.0	7.33626
1.18339×10^{-4}	15.0	8.39008
1.493×10^{-5}	16.0	9.47063
9.94673×10^{-7}	17.0	10.5466
4.17347×10^{-8}	18.0	11.5815

8.3.7.2 Using Coherent Receiver Design

If, instead of using the non-coherent FSK-ASK detection with the RBNSiZeComm technique, we used coherent detection, a possible receiver diagram appears in Figure 8.12. Assuming the threshold voltage levels as $|x_{th}^{+}| = |x_{th}^{-}| = x_{th}, 0 \leq x_{th} \leq 1$, due to the presence of Gaussian noise, the expressions for α, β and δ would have been modified as

$$\alpha = 1 - 2Q(\sqrt{\gamma/2}x_{th}) \tag{8.16}$$

$$\beta = \delta = 1 - Q(\sqrt{\gamma/2}(1 - x_{th})). \tag{8.17}$$

Using these values of α, β and δ in Equation(8.13), the required SNR values for a given FER may be re-computed and compared with the corresponding SNR values required with QPSK modulation. The results are summarized in Table 8.7 and we can see that the peak transmitter power with RBNSiZeComm implemented by coherent FSK-ASK is about 6.077 dB higher than that with QPSK modulation. Noting that in RBNSiZeComm the average transmitter power will be 6.0206 dB less than the peak power, it appears that the average transmitter power with RBNS will be about 0.057 dB higher than that with QPSK, which translates into 1.01321 times or about 1.32% more energy than the QPSK scheme. Thus, with coherent FSK-ASK detection, performance of RBNSiZeComm is comparable to that of coherent binary FSK, so far as energy consumption is concerned.

TABLE 8.7: Comparison of peak transmitter powers between coherent RBNSiZeComm and QPSK

$\rho = FER/n$	Required SNR in dB		
	Coherent RBNSiZeComm		QPSK
	x_{th} (volts)	SNR	
2.29217×10^{-4}	0.54	14.0	7.9256
4.17229×10^{-5}	0.53	15.0	8.9167
4.91399×10^{-6}	0.52	16.0	9.91931
3.37337×10^{-7}	0.52	17.0	10.9267
1.22031×10^{-8}	0.51	18.0	11.9265

However, the coherent detection scheme requires more complex electronics and consequently consumes additional power. Thus, in typical applications where we prefer to keep the circuit complexity simple and the cost low, it would be preferable to opt for non-coherent transmission supported by the RBNSiZeComm-based scheme. In non-coherent systems, RBNSiZeComm would indeed offer significant advantage (about 53% savings) over the receiver implementation of FSK-ASK modulation using envelope detectors.

8.3.8 Simulation Results

We evaluate the energy savings generated by RBNSiZeComm for two typical applications in WSNs: remote healthcare and agricultural sensor networks. For the healthcare application, we considered the transmission of electrocardiogram (ECG) signals, JPEG images of blood samples and ASCII text data representing various physiological parameters like blood pressure, blood sugar level and temperature. [23]. For evaluating the savings for ECG signals we used 50 real-life samples obtained from the QT database in [27]. JPEG images of blood, tissue and bone samples were obtained from [28]. For agricultural sensor networks we assumed that the data generated by the sensor devices were transmitted as ASCII text messages [25, 26, 24].

In Table 8.8, the savings for noiseless channel are obtained using Equation(8.9) for the CC1100 chip, transmitting at 10dBm. These savings are nearly equal to the percentage of 0's in the RBNS encoded version of the respective data types. The table entries for the AWGN noisy channel are obtained using Equation(8.12) with P_0 equal to the values in the second column of Table 8.8. Figures 8.13 and 8.14 show the energy savings on different data types for the three application domains. Figure 8.13 depicts the average transmitter power (scaled with respect to noise power) in dB for given values of $\rho = FER/n$, averaged over $8 \leq n \leq 1024$, while Figure 8.14 shows the relative average energy savings generated by RBNSiZeComm at the transmitter compared to binary FSK, using Equation(8.12) and the receiver shown in Figure 8.10.

TABLE 8.8: Energy savings for different application file types

Signal Type	Energy Savings by RBNSiZeComm	
	Noiseless Channel	Noisy Channel
ECG Signals	79.70%	62%
JPEG Blood/Tissue Sample Images	65.11%	33.4%
ASCII Text data	69.44%	41.8%

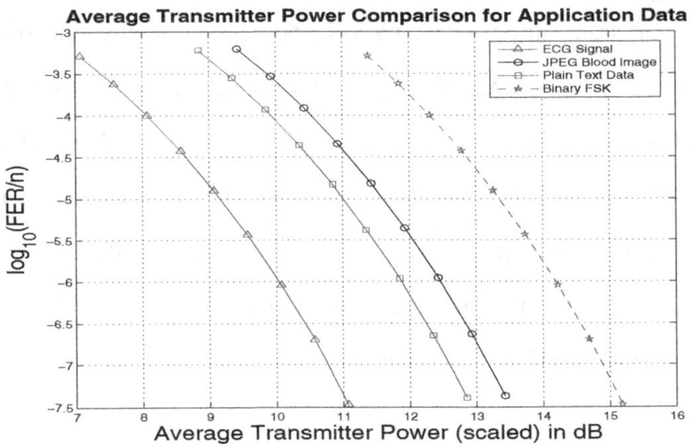

FIGURE 8.13: Average transmitter power (scaled) for various application data types

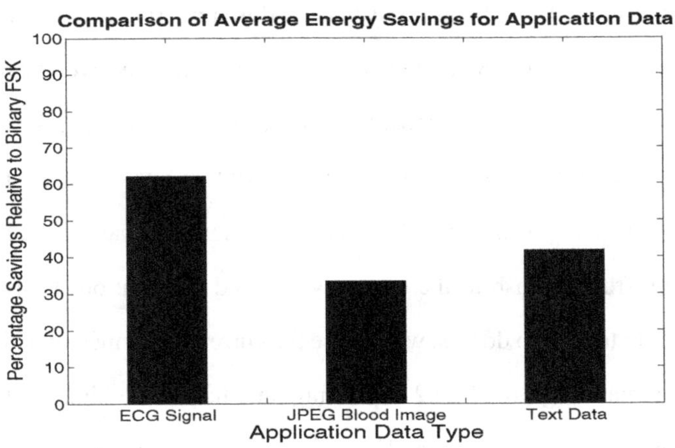

FIGURE 8.14: Relative energy savings for various wireless application data types

8.4 Ternary with Silent Symbol (TSS) Communication Protocol

We now describe another energy-efficient communication protocol called *ternary with silent symbol (TSS)* [3, 4] which is based on a source coding scheme using the ternary number system, along with silent communication. Of the three symbols 0, 1 and 2 in the encoded message, one symbol (say 0) is made silent while the other two (say 1 and 2) are transmitted using a hybrid modulation scheme. In a noiseless channel there will be energy savings of 33% in the transmitter due to the non-transmission of carrier during the periods of 0. However, in a channel with additive white Gaussian noise (AWGN) and a receiver based on non-coherent detection, the amount of transmitter energy savings is expected to be less than this value as the peak signal power needs to be increased in the presence of the noise to distinguish the silent symbol from the no-signal situation of the channel.

8.4.1 Basic Idea and Algorithm Descriptions

Let a given binary message B be represented by an n-bit binary string $b_{n-1}b_{n-2}\ldots b_2b_1b_0$. Let T be its equivalent m-digit ternary representation given by $T = t_{m-1}t_{m-2}\ldots t_2t_1t_0$. We assume that n is even, otherwise we pad the message with a 0 bit at the leftmost, i.e., most significant bit (msb) position. For conversion from binary to ternary, we successively scan every two bits of the binary message starting from its msb position and then convert the leftmost part of the binary message (starting from its msb to the currently scanned bit position) to its equivalent ternary representation. To do this, we replace the currently scanned two bits (bit pair) by a single ternary digit (0, 1 or 2 depending on whether the binary bit pair is 00, 01 or 10), or by two ternary digits $(10)_3$ (if the binary bit pair is 11), and add it to four times the obtained ternary number T from all the previous bit positions (on the left of this bit pair). Note that this addition will be in the ternary number system, and the multiplying factor four is to adjust the weight of T with respect to the currently scanned bit pair. Also, since this addition will be in the ternary number system, the weight of four can be assigned to T by adding the two ternary numbers $T0$ and $0T$. Thus, the equivalent ternary representation of the part of the binary message from its msb to the currently scanned bit pair will then be obtained as $T0 \oplus 0T \oplus x$, where $\oplus$ represents a ternary addition operator and x is the ternary number equal to 0, 1, or 2 (when the scanned bit pair is 00, 01 or 10, respectively), or equal to $(10)_3$ (when the scanned bit pair is 11).

Example 33. *Consider the binary number* 10011100. *First, we get* $T = 2$ *from the leftmost two bits* 10. *Then, for the next two bits, the equivalent ternary digit is 1. We add (in ternary) the three numbers* $(20)_3$, $(02)_3$ *and* $(1)_3$ *to get the changed value of* $T = (100)_3$ *(which is equal to 9 in decimal corresponding to the binary digits* 1001 *scanned so far). For the next two bits, we get the equivalent ternary representation as* $(10)_3$, *which is added to* $(1000)_3(= T0)$ *and* $(0100)_3(= 0T)$ *in ternary to get the changed value of* $T = (1110)_3 = (39)_{10}$. *Finally, for the bit pair* 00, *we add* $(11100)_3$ *to* $(01110)_3$ *to get the equivalent ternary number as* $(12210)_3 = (156)_{10}$.

The *Binary2Ternary* algorithm outlines the steps required to convert a number from binary to ternary.

procedure BINARY2TERNARY(IN **bit vector** B, **integer** n, OUT **ternary number** T)
    /* Initialization */
    if $b_{n-1}b_{n-2} = 00$ **then**
        $T \leftarrow 0$;
    end if
    if $b_{n-1}b_{n-2} = 01$ **then**
        $T \leftarrow 1$;

```
    end if
    if b_{n-1}b_{n-2} = 10 then
        T ← 2;
    end if
    if b_{n-1}b_{n-2} = 11 then
        T ← 10;
    end if

    for i = n − 3 downto 1 step -2 do
        if b_i b_{i-1} = 00 then
            T ← T0 ⊕ 0T;                    ▷ this is a ternary addition
        end if
        if b_i b_{i-1} = 01 then
            T ← T0 ⊕ 0T ⊕ 1;                ▷ ternary addition
        end if
        if b_i b_{i-1} = 10 then
            T ← T0 ⊕ 0T ⊕ 2;                ▷ ternary addition
        end if
        if b_i b_{i-1} = 11 then
            T ← T0 ⊕ 0T ⊕ 10;               ▷ ternary addition
        end if
    end for
end procedure
```

In a similar manner, we can reconvert the received m-digit ternary message to its equivalent binary form by using the *Ternary2Binary* algorithm, and again we scan the given ternary number from its most significant (leftmost) position in a digit-by-digit manner, and convert the scanned ternary digits to their equivalent binary representation. In the algorithm, we denote a ternary digit by t_i and the converted binary number by B. As in the *Binary2Ternary* algorithm, we need to adjust the weight of the binary representation of the previously scanned ternary digits by attaching a weight of three with respect to the currently scanned ternary digit, and this is achieved by adding $B0$, $0B$ and the equivalent binary representation of the currently scanned ternary digit (note that this time it is a binary addition).

Example 34. *Let the given ternary number* $T = 2102$. *The leftmost digit is 2 whose binary representation is* 10. *We initially form the binary number as* $B = 10$. *The next ternary digit is 1. We add the three binary numbers* $100(= B0)$, $010(= 0B)$ *and 1 to get the changed value of* $B = 111$. *The next scanned ternary digit is 0 and we add* 1110 *to* 0111 *to get the changed value of B as* 10101. *Finally, for the rightmost ternary digit 2, we add* 101010, 010101 *and* 10 *to get* 1000001 *as the equivalent binary representation B of the given ternary number T.*

```
procedure TERNARY2BINARY(IN ternary number T, integer m, OUT binary number B)
    /* Initialization */
    if t_{m-1} = 0 then
        B ← 00;
    end if
    if t_{m-1} = 1 then
        B ← 01;
    end if
    if t_{m-1} = 2 then
        B ← 10;
    end if

    for i = m − 2 to 0 step -1 do
```

```
        if t_i = 0 then
            B ← B0 + 0B;                      ▷ this is a binary addition
        end if
        if t_i = 1 then
            B ← B0 + 0B + 1;                  ▷ binary addition
        end if
        if t_i = 2 then
            B ← B0 + 0B + 10;                 ▷ binary addition
        end if
    end for
end procedure
```

Binary to ternary conversion of data provides the advantage of reduction in the total number of symbols to be transmitted and reduces total transmission duration. As both the transmitter and the receiver consume energy proportional to the duration for which they are switched on, regardless of the symbol transmitted, recoding to a higher radix will lead to a savings in energy at both the transmitter and the receiver. Additional energy savings can be obtained at the transmitter by adopting the silent symbol strategy as explained in the preceding sections for one of the three possible ternary digits. The *ternary with silent symbol* (TSS) combines these two strategies to derive an energy-efficient communication scheme that saves energy at both the transmitter and the receiver. Assuming equal likelihood of all the three symbols, we assume 0 to be the silent symbol. In practice, however, the relative frequencies of the three symbols will depend on the nature of the data to be transmitted. A better strategy would be to assign the most frequent symbol as the *silent symbol* based on the relative frequencies of the symbols, determined statistically or assigned for each message to be transmitted.

The transmission protocol of TSS thus consists of two steps:

Protocol TransmitTSS

Step 1: Recode the binary data to ternary using the *Binary2Ternary* protocol.

Step 2: Transmit the ternary data symbol by symbol, obtained by the recoding process from step 1 using the following rules:

Step 2.1: If the symbol to be transmitted is 0, then switch off the transmitter for the symbol period.

Step 2.2: Otherwise, switch on the transmitter and transmit the symbol (1 or 2) in the the time slot.

The receiver side protocol would be the reverse of that for the transmitter:

Protocol ReceiveTSS

Step 1: Receive the ternary encoded data symbol by symbol as follows:

Step 1.1: If a signal is received in the given time slot (symbol period), the received symbol is either a 1 or a 2, depending on the predetermined received signal-to-symbol mapping.

Step 1.2: Otherwise, if the channel is quiet, the received symbol is a 0.

Step 2: Convert the received ternary message to binary using the *Ternary2Binary* protocol.

8.4.2 Performance Analysis

The TSS transmission scheme involves the recoding of an outgoing binary message to its equivalent ternary form and transmitting it. The receiver then reconverts the received ternary data to binary and passes it onto the higher layers of the communication stack for further processing.The transmitter uses FSK modulation with two frequencies, such as f_c and $f_c + \Delta f$ corresponding to the symbols 1 and 2, respectively, and is switched to *idle* state during the 0's, when it consumes very little power compared to the *transmit* or *receive* state. Effectively, this is a hybrid modulation scheme involving FSK and ASK. The receiver diagram as shown in Figure 8.10 for RBNSiZeComm can also be used for the TSS scheme based on non-coherent detection and the frequencies f_c and $f_c + \Delta f$ will correspond to the symbols 1 and 2, respectively. Also, q_A and q_B will now be the binary outputs of the threshold detectors corresponding to the symbols 1 and 2, respectively. Thus, in this case, the desired output $\hat{s}$ of the receiver can be expressed as follows:

i) If $q = 0$, $\hat{s} = 0$, corresponding to $s_A = s_B = 0$.

ii) If $q = 1$,

$$\hat{s} = \begin{cases} 1 & \text{if } e_A > e_B, \text{(when } s_A = 1, s_B = 0) \\ 2 & \text{if } e_B > e_A, \text{(when } s_B = 1, s_A = 0). \end{cases} \tag{8.18}$$

We consider a message frame containing n binary bits. Corresponding to each such frame, we consider the message frame after converting the binary to ternary form containing symbols 0, 1 and 2, and having a length of m digits, where n and m are related as $m = \left\lceil n \log 2 / \log 3 \right\rceil = \lceil 0.6309n \rceil$.

For performing the energy saving analysis of the TSS scheme, again the conventional bit error rate (BER) will not be suitable. Instead, the frame error rate (FER) will be used as in RBNSiZeComm scheme. BER has the limitation that the propagation of errors over the higher significant digits due to an erroneously received ternary symbol cannot be captured appropriately since it considers the transmission of each bit independent of the other bits. It may be noted that due to the nature of ternary (and also other radix encodings that are not powers of 2), the number of higher significant bits in the original binary number through which an error at a received digit position (in the higher radix) will propagate is difficult to predict. This is because the conversion from binary to such a radix involves *carry propagation* from lower order digits to higher order digits. When using a radix that is a power of 2, say 2^k for $k > 1$, such carry propagations do not occur as the equivalent digit in the new radix of 2^k can be obtained by simply grouping successive k bits from the least significant bit position, independent of the other bits.

Consider as an example the transmission of the binary number $(0111111)_2 = (63)_{10}$ using ternary encoding. The ternary representation of $(63)_{10}$ is $(2100)_3$. If the least significant digit of this ternary number transmission, i.e., 0 is wrongly interpreted at the receiver as 1, the received binary number after decoding will be $(1000000)_2$, representing a propagation of error from the least significant bit to the most significant bit position. A BER-based error analysis would not be able to capture such *ripple effects* (on the original binary number) of error in interpreting a single digit (in radix other than power of 2) at the receiver.

The FER-based error analysis overcomes the limitation of BER-based error analysis in capturing the transitive nature of error propagation to the higher significant bit positions of the decoded binary number, caused by a single digit error in the received ternary data. Although we show the error analysis of only the TSS scheme based on FER, this concept of FER-based error analysis is generally applicable to any transmission scheme that uses recoding of binary data to another radix which is not a power of 2 for the reason mentioned above. We now describe the error analysis of the TSS scheme using both non-coherent and coherent receiver designs.

8.4.3 Non-coherent Detection

In this section, we compare the FER of communicating a binary data frame using non-coherent BFSK, with the FER of communicating the equivalent ternary data in the TSS scheme using a receiver based on non-coherent detection as shown in Figure 8.10.

In the hybrid FSK-ASK modulation scheme, let α be the probability that a transmitted ternary symbol was a 0 and was received correctly as a 0. Similarly, let β (or δ) be the probabilities that a transmitted symbol was a 1 (or 2) and the received symbol is also a 1 (or 2). From symmetry, $\beta = \delta$. Note that $\alpha = P\{q_A = 0|0\} \cdot P\{q_B = 0|0\}$, where $P\{q_A = 0|0\}$ and $P\{q_B = 0|0\}$ are the probabilities of correctly receiving a 0 at q_A and q_B, respectively, given that a 0 was transmitted. Hence,

$$\alpha = P\{q_A = 0|0\} \cdot P\{q_B = 0|0\} = [P\{q_A = 0|0\}]^2 \text{ (by symmetry).} \qquad (8.19)$$

Also, note that q_A for 0 transmission follows Rayleigh distribution. Assuming a zero mean and a standard deviation of σ for AWGN channel, we can write

$$P\{q_A = 0|0\} = \int_0^{e_{th}} \frac{e_A}{\sigma^2} e^{-e_A^2/2\sigma^2}\, de_A = (1 - e^{-e_{th}^2/2\sigma^2}).$$

Hence, we have $\alpha = (1 - e^{-e_{th}^2/2\sigma^2})^2$. Next, we define β as

$$\beta = P\{q = 1|1\} P\{e_A > e_B, \forall e_A|1\}) = X_1 X_2 \qquad \text{(say)},$$

where $X_1 = P\{q = 1|1\}$ and $X_2 = P\{e_A > e_B|1, \forall e_A|1\}$. From Figure 8.10, we write:

$$X_1 = \int_0^{\infty} [1 - P\{q_A = 0|1\}P\{q_B = 0|1\}]p_1(e_A)\, de_A,$$

where $P\{q_B = 0|1\}$ is governed by a Rayleigh distribution, while $P\{q_A = 0|1\}$ represents a Rician distribution. Thus,

$$P\{q_A = 0|1\} = \int_0^{e_{th}} \frac{e_A}{\sigma^2} e^{-\frac{e_A^2+s_A^2}{2\sigma^2}} I_0\left(\frac{s_A e_A}{\sigma^2}\right) de_A \text{ and,}$$
$$P\{q_B = 0|1\} = \int_0^{e_{th}} \frac{e_B}{\sigma^2} e^{-e_B^2/2\sigma^2}\, de_B = 1 - e^{-e_{th}^2/2\sigma^2}.$$

Using the above expressions, we obtain X_1 and X_2 as

$$X_1 = \int_0^{\infty} \left[1 - \left(1 - e^{-e_{th}^2/2\sigma^2}\right)\left(\int_0^{e_{th}} \frac{e_A}{\sigma^2} e^{-\frac{e_A^2+s_A^2}{2\sigma^2}} I_0\left(\frac{s_A e_A}{\sigma^2}\right) de_A\right) \frac{e_A}{\sigma^2} e^{-\frac{e_A^2+s_A^2}{2\sigma^2}} \times I_0\left(\frac{s_A e_A}{\sigma^2}\right)\right] de_A. \tag{8.20}$$

$$X_2 = \int_0^{\infty} \left[\left(1 - e^{-e_A^2/2\sigma^2}\right) \frac{e_A}{\sigma^2} e^{-\frac{e_A^2+s_A^2}{2\sigma^2}} \times I_0\left(\frac{s_A e_A}{\sigma^2}\right) u(e_A)\right] de_A. \tag{8.21}$$

Let P_0, P_1 and P_2 be the probabilities of occurrences of the symbols 0, 1 and 2, respectively, in a transmitted message. The probability that the transmitted ternary frame with m ternary digits will be received correctly by the above hybrid FSK-ASK scheme can be expressed as the *frame-correct rate* (FCR), given by:

$$FCR = \alpha^{mP_0} \times \beta^{mP_1} \times \delta^{mP_2}. \tag{8.22}$$

Assuming that all possible binary strings with n bits are equally likely to appear in the binary message, the average number of 0's in the ternary frame will be asymptotically (for large n) 33% and those of all non-zero symbols, i.e., 1 and 2 taken together, will be 67%. Thus, $P_0 = 0.33$ and $P_1 + P_2 = 0.67$. The probability that a frame in ternary will be received correctly (using the above hybrid FSK-ASK modulation scheme) is given by:

$$FCR = \alpha^{0.33m} \times \beta^{0.67m}. \tag{8.23}$$

From this FCR, we obtain the FER for a ternary frame as $FER_{ter} = 1 - FCR$.

As already discussed in the context of RBNSiZeComm protocol, the FER for BFSK will be given by $FER_{bfsk} = 1 - (1 - \rho)^n$, where ρ is the BER in BFSK. Thus, FER_{bfsk} can be approximated as $n\rho$, since ρ is usually very small. Hence, for a given frame size n, we can compare FER_{ter} with $FER_{bfsk}(= n\rho)$. In other words, we can compare FER_{ter}/n with ρ. To compute FER_{ter}/n, we set $s_A = 1$, i.e., signal power = $s_A^2/2 = 1/2$ and a signal-to-noise ratio (SNR) = $1/2\sigma^2$. The computations to find the optimum value of threshold voltage e_{th} that gives the minimum FER_{ter}/n for a given SNR value can be done using MATLAB. Note that from Equation (8.23), we get the required peak transmitter power with TSS for a given value of FER_{ter}.

The SNR values for non-coherent BFSK are determined from the relation, $\rho = \frac{1}{2}e^{-\gamma/2}$, where SNR in dB = $10\log_{10}\gamma$. For different values of frame size in the range $8 \leq n \leq 1024$ and using the expressions for X_1 and X_2, the SNR values for the TSS scheme with the non-coherent receiver and the corresponding SNR values for non-coherent BFSK to meet the same FER values are summarized in Table 8.9. The results in Table 8.9 indicate that on average, for a given FER/n, the SNR values with the TSS are nearly 2.64 dB higher than the value with binary non-coherent FSK.

However, in the TSS scheme, the transmitter will be on only during the non-zero symbols (1 or 2), and switched off during the zero symbols in the ternary encoded message. Hence the required average transmitter power for a given BER will be less than the value as demanded by Equation (8.23). Using $P_0 = 1/3$, the average transmitter power for TSS will be reduced from peak transmitter power shown in

TABLE 8.9: Comparison of SNR of binary FSK with TSS for different values of n

SNR with TSS (non-coherent)	Corresponding SNR with non-coherent BFSK for different frame sizes (n)						Average SNR with non-coherent BFSK
	16	64	128	256	512	1024	
14.0	11.4526	11.5025	11.5183	11.5347	11.5671	11.6299	11.5107
15.0	12.3851	12.4214	12.4287	12.4312	12.4363	12.4470	12.4124
16.0	13.3209	13.3494	13.3545	13.3547	13.3552	13.3569	13.3396
17.0	14.2776	14.3005	14.3044	14.3044	14.3045	14.305	14.2923
18.0	15.2270	15.2454	15.2486	15.2486	15.2486	15.2490	15.2388

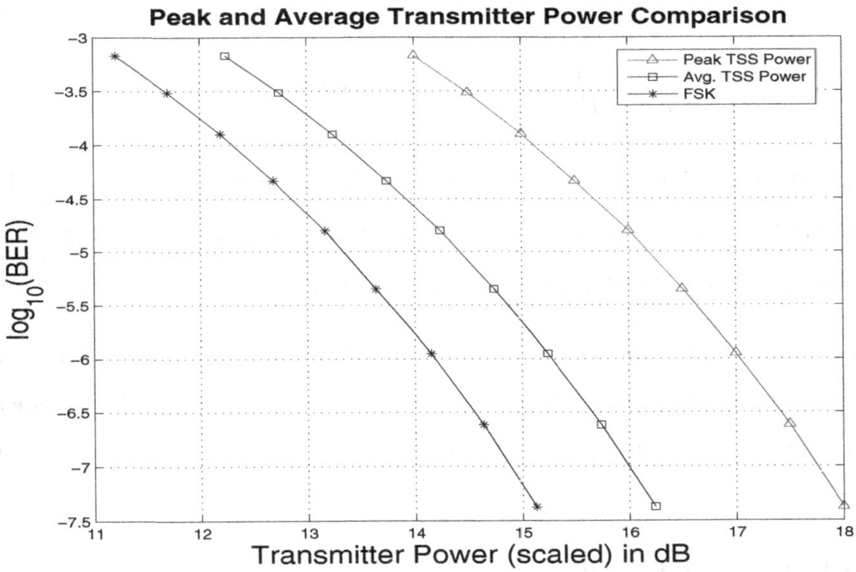

FIGURE 8.15: Peak and average transmitter power comparison

Table 8.9 by $10\log_{10}(1 - 1/3)$ dB $= 1.76$ dB. Thus, the average transmitter power P_1 for TSS communication will be higher than the corresponding average power P_b for non-coherent BFSK by $(2.64 - 1.76)$ dB $= 0.88$ dB, thereby making $\frac{P_1}{P_b} = 1.2246$. A plot of the scaled average transmitter power for TSS and BFSK (with non-coherent detection) is presented in Figure 8.15.

8.4.3.1 Energy Savings Analysis

Let us assume that the symbol duration of the ternary encoded message in TSS is same as that in BFSK. Thus, the transmission duration of the ternary message will be smaller than that of the equivalent binary message by virtue of encoding. If T_b and T_1 are the transmission duration of the binary message and its equivalent ternary message, respectively, $\frac{T_1}{T_b} = \frac{m}{n}$ which is asymptotically (for large n) equal to $\frac{\log 2}{\log 3}$ = 0.6309. Thus, the total (scaled) energy E_1 in TSS with a non-coherent receiver is related to the energy E_b in non-coherent BFSK as

$$\frac{E_1}{E_b} = \frac{P_1 \times T_1}{P_b \times T_b} = 1.2246 \times 0.6309 = 0.7726. \tag{8.24}$$

There is an energy saving at the front-end amplifier section of the receiver too because of the reduced value of m in the ternary encoded message, which is given by $1 - \frac{\log 2}{\log 3} = 0.3691$. We thus have the following result.

Result 5. *The above implementation scheme of TSS will result in energy savings of approximately* 23% *at the transmitter and* 37% *at the front-end amplifier section of the receiver, compared to non-coherent BFSK.* □

8.4.4 Coherent Detection

In this section, as we did in the case of RBNSiZeComm, we consider a coherent receiver design for the hybrid FSK-ASK modulation, with a receiver schematic as shown in Figure 8.16. To compute FCR we can still use Equation (8.23). However, $P\{q_A = 0|0\}, P\{q_B = 0|0\}$ and α are modified to $P'\{q_A = 0|0\}, P'\{q_B = 0|0\}$ and α_c, respectively:

$$\begin{aligned}\alpha_c &= P'\{q_A = 0|0\}.P'\{q_B = 0|0\} = [P'\{q_A = 0|0\}]^2 \\ &= \frac{1}{2\pi\sigma^2}\left[\int_{-\infty}^{e_{th}} e^{-\frac{e_A^2}{2\sigma^2}} de_A\right]^2. \end{aligned} \tag{8.25}$$

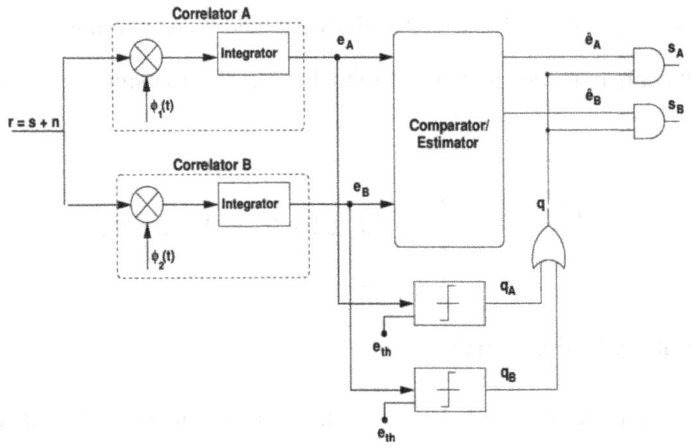

FIGURE 8.16: Representative coherent FSK receiver for TSS scheme

Also, the expressions for X_1 and X_2 in [5] will be modified to X_1' and X_2', respectively, where

$$X_1' = \frac{1}{\sqrt{2\pi}\sigma}\int_{-\infty}^{\infty}\left[1 - \frac{1}{\sqrt{2\pi}\sigma}\int_{-\infty}^{e_{th}} e^{-\frac{(e_A-1)^2}{2\sigma^2}} de_A \frac{1}{\sqrt{2\pi}\sigma}\int_{-\infty}^{e_{th}} e^{-\frac{e_B^2}{2\sigma^2}} de_B\right] \times e^{-\frac{(e_A-1)^2}{2\sigma^2}} de_A. \tag{8.26}$$

$$X_2' = \frac{1}{\sqrt{2\pi}\sigma}\int_{-\infty}^{\infty}\left[\frac{1}{\sqrt{2\pi}\sigma}\int_{-\infty}^{e_A} e^{-\frac{e_x^2}{2\sigma^2}} de_x\right] e^{-\frac{(e_A-1)^2}{2\sigma^2}} de_A. \tag{8.27}$$

Thus, the value of β will now be modified to $\beta_c = X_1'X_2'$ and we get

$$FCR_c = \alpha_c^{0.33m} \times \beta_c^{0.67m}. \tag{8.28}$$

The FER value for coherent detection is given by $FER_c = 1 - FCR_c$. The average SNR values of TSS with coherent detection for $8 \leq n \leq 1024$ are found to be more or less the same as those for $n = 128$. The SNR values for a given FER/n in a BFSK receiver with coherent detection can be computed using

$FER/n = \frac{1}{2} erfc\sqrt{\gamma/2}$. These SNR values and their corresponding FER/n values of TSS with coherent detection are presented in Table 8.10 using $n = 128$.

$$\frac{E_t}{E'_b} = \frac{P_t \times T_t}{P'_b \times T_b} = 1.3122 \times 0.6309 = 0.828. \tag{8.29}$$

8.4.4.1 Energy Savings Analysis

Note from Table 8.10 that for different FER/n values in the range 10^{-5} to 10^{-7}, the peak transmitter power for TSS scheme with coherent detection is on average 2.94 dB higher than that with coherent BFSK. Noting that $10\log_{10}(1 - 1/3)$ dB = -1.76 dB, the average transmitter power for TSS with coherent detection will be $(2.94 - 1.76)$ dB = 1.18 dB higher than that with coherent BFSK. Denoting the average transmitter powers for coherent TSS and coherent BFSK by P_t and P'_b, respectively, we have $\frac{P_t}{P'_b} = 10^{1.18/10} = 1.3122$. Using E_t and E'_b to denote the transmitter energy requirements for the coherent TSS and BFSK schemes, respectively, and considering the reduction in transmission time as in Section 8.6.3.5, we get:

$$\frac{E_t}{E'_b} = \frac{P_t \times T_t}{P'_b \times T_b} = 1.3122 \times 0.6309 = 0.828. \tag{8.30}$$

For a given FER/n in the range 10^{-4} to 0^{-8}, the comparative figures for relative energy requirements with non-coherent TSS, coherent TSS, coherent BFSK and non-coherent BFSK are shown in Table 8.10. We can see that the above TSS implementation scheme with coherent detection will require 0.70 times the energy needed for non-coherent BFSK. We summarize these results as follows.

Result 6. *The above implementation scheme of TSS with a coherent detection-based receiver will result in saving transmitter energy by approximately* 30% *and* 17%

TABLE 8.10: Comparison of peak and relative transmitter energy requirements in TSS

FER/n **value**	**Required peak SNR in db**		**Relative energy units required for**			
	TSS (Coherent)	BFSK (Coherent)	TSS (Coherent)	TSS (Non-Coherent)	BFSK (Coherent)	BFSK (Non-Coherent)
1.64525×10^{-4}	14.0	11.105	10.24	10.834	11.105	12.0517
7.29469×10^{-5}	14.5	11.591	10.74	11.2855	11.591	12.4712
2.93585×10^{-5}	15.0	12.08	11.24	11.744	12.08	12.8973
1.06484×10^{-5}	15.5	12.5695	11.74	12.211	12.5695	13.3272
3.42302×10^{-6}	16.0	13.0612	12.24	12.668	13.0612	13.763
9.64132×10^{-7}	16.5	13.5536	12.74	13.1318	13.5536	14.2025
2.33905×10^{-7}	17.0	14.047	13.24	13.605	14.047	14.6468
4.80461×10^{-8}	17.5	14.541	13.74	14.086	14.541	15.0939
8.24905×10^{-9}	18.0	15.0336	14.24	14.5456	15.0336	15.5437

relative to non-coherent BFSK and coherent BFSK schemes, respectively, along with about 37% *energy saving at the front-end amplifier section of the receiver.* □

In addition to RBNSiZeComm [39], another protocol called RZE [5] was proposed by Sinha and Sinha. Comparing the performances of transmitter and receiver (at the front-end amplifier section) energy savings for non-coherent TSS, coherent TSS, RBNSizeComm and RZE protocols relative to non-coherent BFSK, we can see that while the transmitter side saving for TSS is less than that of RBNSiZeComm and RZE, it saves more energy at the front-end amplifier of the receiver section than both these schemes. Thus, considering the combined energy savings at the transmitter and the receiver, TSS is more suitable than RBNS or RZE for data communication in multi-hop wireless networks.

8.5 Compression with Null Symbol (CNS) Communication Protocol

8.5.1 Preliminaries and Basic Idea

In an *energy-based transmission* (EbT) scheme, the communication of each bit requires energy. Both the transmitter and the receiver spend energy as they must be kept powered on during the entire duration of the transmission. The receiver usually draws substantial power (typically of the same order as that drawn by the transmitter) when message bits are received. By reducing the duration of communication, we save energy at the transmitting and receiving nodes by switching off the radio circuit earlier. Additionally, by using silent periods for communicating symbols in a data frame, we can further reduce the energy consumption required for transmission.

To reduce the duration of transmission, the *compression with null symbol (CNS)* communication protocol [6] proposes a fixed length encoding of the data. If k bits are encoded using a single symbol, the duration of transmission is reduced k times. Unfortunately, this is associated with an increase in the complexity of the hardware by 2^k times. This is because, for a k-bit encoding, there are 2^k different possible values, each of which needs to be uniquely identifiable. For example, if we set $k = 4$, the hardware has to be able to distinguish between $2^4 = 16$ different symbols compared to two symbols in a device operating on the binary number system. The increase in hardware complexity results in a corresponding increase in energy consumption by the encoding and decoding circuits, and increases the cost of the device.

Let c_{sym} denote the complexity of the hardware for each distinct symbol it needs to handle. Hence, for a k bit encoding scheme, the total complexity C_{hw} of the hardware is $C_{hw} = c_{sym}2^k$. Looking at the energy consumption side in conventional communication strategies relying on energy-based transmissions, energy is spent for each bit to be transmitted, whether 0 or 1. The total energy spent in EbT during transmission or reception is directly proportional to the duration of transmission of the data frame. Thus, by using a k-bit encoding scheme, the amount of energy spent for transmitting an n-bit data frame in an ideal channel is given by $E_t = e_b \left\lceil \frac{n}{k} \right\rceil$, where e_b represents the energy required for each transmitted symbol. Note that E_t decreases with increasing k.

Many WSN applications typically require low cost hardware and low energy for communication. We know that hardware cost C_{hw} increases with k while communication energy E_t decreases with k. We may choose an appropriate composite function to be optimized depending on the weights assigned to cost and energy. However, for simplicity, we consider the product of C_{hw} and E_t as the function to be minimized over k. Our objective is to determine the value(s) of k for which the following equation yields a minimum:

$$y = 2^k \left\lceil \frac{n}{k} \right\rceil. \tag{8.31}$$

It is easy to see that for all real $k > 0$, the function $f(k) = \frac{2^k}{k}$ has the minimum value at $k = \log_2 e$. As we are interested in only integral values of k, $\frac{2^k}{k}$ can be shown to attain its minimum value of 2 at both $k = 1$ and $k = 2$. Figure 8.17 shows the plot of $f(k)$ against k for $k > 0$. When k divides n, $y = nf(k)$; otherwise $y > nf(k)$. For simplicity, let us assume that n is always even (which, for all practical purposes, is not a restriction). Hence, it follows that the minimum of y still occurs at $k = 1$ and $k = 2$ only.

Since $k = 1$ degenerates to the binary encoding scheme, we concentrate on the $k = 2$ case, where we reduce the duration of transmission by 50% compared to the binary encoding scheme, due to halving of the transmitted frame length. By encoding each successive pair of bits using one of the $2^2 = 4$ different possible symbols (namely for the bit patterns 00, 01, 10 and 11), we can save 50% energy at both the

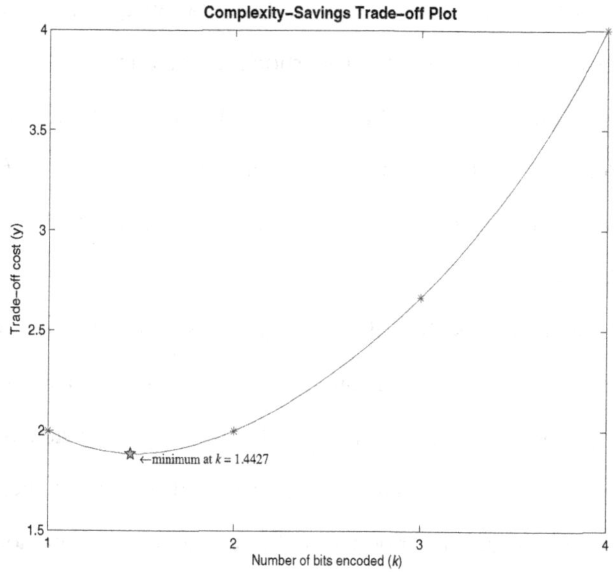

FIGURE 8.17: Plot of Equation ((8.31)) against k

transmitting and receiving nodes. This idea of grouping bits is similar to what is done in modulation schemes such as PSK and QAM [29]. However, our focus here is on arriving at an optimal value of k to strike a balance between hardware complexity and transmission energy. It is possible to arrive at a different value for optimal k by using an optimization function different from Equation (8.31).

While increasing the number of permissible symbols increases the energy consumption by the encoding and decoding logic, typically the energy requirement for communication is significantly larger than that for encoding and decoding [22, 41]. However, we use only four symbols corresponding to $k = 2$, so that the effect of this increased encoding and decoding logic on energy consumption will be nominal. Moreover, we need to use only three frequencies for communication in the corresponding hybrid FSK-ASK modulation scheme for implementing this idea with $k = 2$. Higher values of k would imply more frequencies for transmission, which in turn would lead to an increase in transceiver cost, complexity and energy consumption. Due to these two factors, the choice of $k = 2$ is attractive for communication in WSNs.

Considering only the transmitter side, if we use silence to communicate one of the four symbols and put the transmitter in the *active* or *idle* state for the duration of each such occurrence of that symbol in the data (similar to the RBNSiZeComm and TSS protocols described above), we can further increase the energy saved during transmission. For an n-bit data frame, considering all possible 2^n strings of equal likelihood, the percentage occurrence of each of the bit patterns 00, 01, 10 and 11 is 25%.

Observation 5. *Using the above communication strategy, theoretically it is possible to have an energy saving of* $\gamma = (50.0+0.25\times50.0)\% = 62.5\%$ *at the transmitter for noiseless channels, while simultaneously generating a savings of 50% at the receiver (as the number of symbols to be transmitted is halved and thus the communication time is reduced by factor of two), when compared to EbT schemes.*

For the rest of our discussions on CNS communication protocol, we use the terms *bit pair* and *codeword* interchangeably.

8.5.2 Implementation Issues

The implementation issues of the CNS scheme concern the i) representation of the CNS symbols in the internal buffers at the MAC layer, ii) hardware consideration of communicating the encoded CNS symbols and, iii) maintenance of receiver-transmitter synchronization for the duration of transmission of a data packet. We also present the transmitter and receiver side algorithms for the CNS communication scheme.

8.5.2.1 Representation of Encoded Numbers

From Section 8.5.1, we see that the optimal encoding scheme corresponds to $k = 2$. Therefore, starting from the least significant bit (lsb) position of the binary data frame, every bit pair (00, 01, 10 or 11) can be assumed to represent a CNS symbol. We shall refer to the four CNS symbols as A, B, C and N and the mapping of the bit pairs to the symbols as *bit pair to symbol mapping* or *BpS Mapping*. N stands for *null* to indicate communication through silent time periods or null signal on the channel. As an example, for a particular scenario, the bit pairs 00, 01, 10 and 11 may be mapped to the symbols N, A, B and C, respectively. Starting from the lsb of the binary data frame to be transmitted, the system would then read every successive bit pair to determine whether the corresponding transmitted CNS symbol is to be A, B, C or N and do the reverse mapping at the receiver to reconstruct the original binary string.

8.5.2.2 Mapping to Codewords

For a binary string of length n, considering all 2^n possibilities, the theoretical average percentage of occurrence of each of the 4 bit pairs 00, 01, 10 and 11 is 25%. However, in practice, the distribution of these four bit patterns in a data set depends on the nature of the data transmitted, as can be seen from Figure 8.18 which depicts the frequency of the four codewords in different file types obtained from statistics in 691 files. Table 8.11 lists the number of files of each type used to derive these statistics. File type *text* represents a set of ASCII plain text files (e.g., C/C++ or Java source codes, LaTex source files, etc.) other than *html* source files. The file set *bin* consists of binary files for both Unix and Windows operating systems: executables, object codes and library files.

Two schemes can be adopted to choose the BpS mapping:

- **Scheme 1**: Fix one of the four possible bit patterns as the N symbol and use

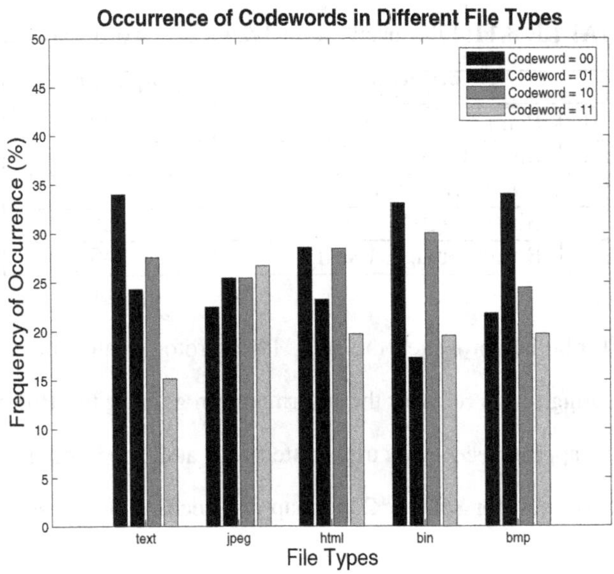

FIGURE 8.18: Occurrence of codewords in different file types

that, regardless of the actual distribution of the four symbols in the data to be transmitted. One way of implementing this could be performing a prior analysis of the application scenario or data set to determine the overall distribution of the four symbols in the data and choosing the one with the highest frequency. For example, from Figure 8.18 we see that for *text* data, the bit pair 00 has the highest frequency of occurrence (nearly 35%) and we can choose that as the N symbol to maximize the generated energy savings at the transmitter. On the other hand, the bit pair 01 has the highest frequency of occurrence for *bmp* files.

- **Scheme 2**: For every data frame to be transmitted, scan the frame to determine the symbol with highest frequency in that frame and set that to be the N symbol. This approach has a slightly higher overhead in terms of frequency distribution computation of the four symbols for each frame but may, however, generate more savings in energy at the transmitter than Scheme 1.

TABLE 8.11: Files used for analyzing codeword statistics

	File Type	Number of Files
1.	Plain Text Files (TEXT)	122
2.	Image Files (JPEG)	71
3.	HTML Source Files (HTML)	71
4.	Binaries (BIN)	82
5.	Bitmap Images (BMP)	345

Note that for both approaches, the MAC layer protocol must provide the support for communicating to the receiver the information regarding the BpS mapping. For Scheme 1, the mapping needs to be transmitted once at the beginning of transmission of a set of data frames; for Scheme 2 the mapping needs to be transmitted with each frame.

Based on the above discussions, the transmitter and receiver side operations for an energy-efficient communication scheme are presented below as *CNSTransmit* and *CNSReceive*, respectively.

For the *CNSTransmit* procedure, $data\_frame$ refers to the binary string of length $frame\_len$ that needs to be transmitted. The transmitter first determines the BpS mapping of symbols, using either Scheme 1 or Scheme 2 described above. For Scheme 2, the BpS mapping needs to be transmitted for every frame, while for Scheme 1, transmitting just once for an entire session is sufficient. Using this mapping, the transmitter then interprets every bit pair, starting from the lsb and transmits the corresponding symbol sym.

On the other side, the receiver first receives the BpS mapping of the symbols from the transmitter, once for the entire session for Scheme 1 or with every frame for Scheme 2. Then, it uses this mapping for each received symbol from the transmitter to determine whether the transmitted symbol corresponds to the bit pair 00, 01, 10 or 11.

8.5.2.3 Synchronization Issues

For the above described scheme to work, the receiver and the transmitter must remain synchronized for the duration of the transmission. This synchronization can be achieved by one of the techniques described in Section 8.3.4.2.

```
procedure CNSTransmit(IN bit vector data_frame, IN integer frame_len)
    var trans_len : Integer;            ▷ trans_len is half of binary frame length

    trans_len = frame_len / 2;
    Transmit trans_len;                 ▷ length of CNS data
    Determine the BpSMapping using either Scheme 1 or Scheme 2; ▷ bit pair to symbol mapping
    Transmit BpSMapping;                ▷ obtained using either Scheme 1 or 2
    for i = 0 to frame_len step 2 do
        Determine symbol sym to be transmitted based on
        BpS Mapping of bits data_frame[i] and data_frame[i + 1];
        Transmit symbol sym;            ▷ either A, B, C or N
    end for
end procedure
```

```
procedure CNSReceive(OUT bit vector recv_data)
    /* time_slot[i] refers to the i^th time slot */
    var j : Integer;                    ▷ counter for output buffer recv_data

    j ← 0;
    Receive trans_len from transmitter; ▷ length of the transmitted CNS data
    Receive BpSMapping;                 ▷ transmitter first sends the BpS Mapping
    for i = 0 to trans_len do
```

$recv\_sym \leftarrow time\_slot[i];$ ▷ received symbol in time slot i, either A, B, C or N

Use received *BpSMapping* to determine output bit pair, $recv\_data[j]$

and $recv\_data[j+1]$ corresponding to symbol $recv\_sym;$ ▷ either 00, 01, 10 or 11

$j \leftarrow j + 2;$

end for

end procedure

8.5.2.4 Collision Avoidance Issues

For contention-based medium access protocols, an issue that arises with the use of N symbols in the transmitted data is how the neighboring nodes determine whether a channel is free for transmission or whether there is an ongoing transmission containing a run of silent N symbols. An erroneous interpretation of the absence of any signal (carrier and/or baseband) can lead to corruption of an ongoing transmission (thus leading to retransmission and an increase in energy consumption by the communicating node pair) or unnecessary increase in communication delay due to waiting, when the channel is actually free.

To avoid the above mentioned issue, we can use one of the three schemes described in Section 8.3.4.3. Following the third scheme described there, a node attempts a transmission if it finds the channel silent for, say, eight consecutive symbol periods. In CNS, eight consecutive N symbols correspond to sixteen consecutive zero bits in the original message. Thus, occurrence of eight successive N symbols in the message implies that there is a run of zeros of length 16 in the message. Depending on the symbols before and after these consecutive eight N symbols, the original message may actually have a run of zeros of length 16, 17 or 18 corresponding to this occurrence of eight consecutive N symbols. Using the results from [39] (which have also been discussed in the previous sections, we can see that for a binary string of length $n \geq 18$, the sum of the numbers of all possible runs of length less than or equal to 16 is given by

$$S_{16} = \sum_{i=1}^{16} \mu_i = (65535n + 65551)2^{n-18}, \tag{8.32}$$

which is derived from the following recurrence relation for μ_i, the number of runs of length i in all possible binary strings of length n:

$$\mu_{n-k} = \begin{cases} 1, & k = 0 \\ 2, & k = 1 \\ (k+3) \cdot 2^{k-2}, & \text{for } n > k \geq 2. \end{cases}$$

Thus, the fraction f of all possible numbers of runs of length greater than 16 is given by

$$f = \frac{S - S_{16}}{S} = \frac{n - 15}{2^{16}(n+1)}, \tag{8.33}$$

where the term S is defined as

$$S = \sum_{i=1}^{n} \mu_i = (n+1)2^{n-2}$$

.

We see from Equation (8.33) that the asymptotic value of f is 1.5259×10^{-5}, which implies that the probability of eight consecutive N symbols and hence the collision probability due to erroneously assuming the channel to be free is extremely low.

Similarly, for the second scheme discussed in Section 8.3.4.3, we can insert an additional symbol (A or B or C) after every run of, say, seven N symbols (which would correspond to a run of 14 zeros in the original binary string) to correctly detect the channel status during every CCA. Proceeding as above, we can show that the probability of occurrence of a run of fourteen zeros in a binary string of length $n \geq 16$ is $\frac{n-13}{2^{14}(n+1)}$, which tends to 6.1035×10^{-5} asymptotically. This implies that for the second scheme the increase in energy and transmission time would be nominal.

Thus, depending on the typical application environment, one may choose any one of the above three schemes to handle collisions while using CNS in contention-based medium access protocols.

8.5.2.5 Effect of Device Characteristics

For most commercially available radio devices, the power drawn in the *transmit* or *receive* state is considerably more than that drawn in the *idle* or *active* state, i.e., when the radio is in *low power* operation mode [22]. For example, in case of the CC1100 chip which supports O-QPSK modulation, only the voltage regulator and the local oscillator are on in the *idle* state. For the rest of our discussions, we will use the terms *idle* and *active* interchangeably to refer to this low power operation state.

The penalty paid for this low power operational state is the switching time from the idle to transmit state and vice versa. Thus, considering the CC1100 chip, the time required for moving from idle to transmit state is 88.4 μS (with no calibration). This means that when the symbol to be transmitted is changed from N to A, B or C, the minimum time required for the transmitter to start again is of the order of 88.4 μs and the symbol duration cannot be made smaller than this value. Typically, we may use symbol duration around 400 μS, corresponding to a data rate of 2.5 Kbps. However, this does not, in fact, put any limitation on the applicability of CNS protocol in a sensor network which is essentially a low data rate system. With chips having faster switching times the data rate can be increased.

We now investigate the effect of the device characteristics on the transmit energy saving aspect of the CNS protocol in comparison to the energy required for transmitting an n-bit binary data frame using conventional techniques that do not utilize the *silent symbol* strategy.

Let V_{cc} be the supply voltage and I_{high} and I_{low} denote currents drawn in the *transmit* (TX) and *idle* states, respectively. Let $I_0 = I_{high} - I_{low}$. We assume that the device moves from idle to TX state in time t_{rise} and during this transient period draws a current I_{tran}. Also, let t_p denote the duration of a symbol (A, B, C or N) which is sufficiently larger than the switching time t_{rise} . Typically, for the CC1100 chip, the value of I_{tran} is 8.7 mA. Similarly, when the device moves from the TX state to the idle state, it takes a finite time, say, t_{fall}. Usually t_{fall} is very small. For example, for the CC1100 or CC2420 chips, the device moves from TX state to idle state very quickly drawing insignificant extra energy over that in the idle state. We therefore neglect the energy consumed by the device during this t_{fall} time.

The energy expenditure for transmitting the equivalent radix 4 data frame using the CNS protocol can be represented as the sum of the following three terms:

1. A base energy value E_{base} that is needed throughout the duration of the transmission, and is equal to the energy required when the device is in the *idle* state. Considering that the original binary data is of length n bits, this energy required for transmitting all possible $4^{n/2}(= 2^n)$ strings of length $n/2$ symbols each of the CNS data is given by

$$E_{base} = \frac{n}{2} 2^n t_p I_{low} V_{cc}$$

.

2. Extra energy E_{tx} in addition to the base energy required to transmit an A, B or C symbol in TX state. For an n-digit binary data frame, considering all

possible strings, we have $\frac{3n}{8}$ A, B or C symbols in the TX state. Thus,

$$E_{tx} = \frac{3n}{8} 2^n t_p I_0 V_{cc}$$

.

3. Extra energy E_{rise} in addition to the base energy required during the edge transitions from an N symbol to A, B or C. Considering that the probability of occurrence of the N symbols is $\frac{1}{4}$ in an $\frac{n}{2}$-symbol data stream, we get

$$E_{rise} = \frac{n}{8} 2^n t_{rise} (I_{tran} - I_{low}) V_{cc}$$

.

Hence, the total energy required to transmit an $n/2$-digit CNS data frame is

$$\begin{aligned} E_{cns} &= E_{base} + E_{tx} + E_{rise} \\ &= V_{cc} 2^n \left[\frac{n}{2} t_p I_{low} + \frac{3n}{8} I_0 t_p + \frac{n}{8} t_{rise} (I_{tran} - I_{low}) \right]. \end{aligned} \tag{8.34}$$

In contrast, the total energy consumed by using the EbT scheme for transmitting an n-bit binary data frame, is $E' = n2^n t_p (I_{low} + I_0) V_{cc}$.

Considering the device characteristics and a noiseless channel, the fractional energy savings generated by the CNS protocol over conventional EbT transmission schemes is

$$\gamma' = \frac{E' - E_{cns}}{E'} = \frac{\frac{I_{low}}{2} + \frac{5}{8} I_0 - \frac{1}{8}(I_{tran} - I_{low}) \frac{t_{rise}}{t_p}}{I_{low} + I_0}., \tag{8.35}$$

TABLE 8.12: γ' values for CC1100

Transmit Power (dBm)	I_{high} (mA)	I_{low} (mA)	I_{tran} (mA)	t_{rise} (μs)	t_p (μs)	γ'
10	30.3	1.9	8.7	88.4	400	61.1%
5	19.7	1.9	8.7	88.4	400	60.34%
0	16.6	1.9	8.7	88.4	400	59.94%
−5	14.0	1.9	8.7	88.4	400	59.46%

TABLE 8.13: γ' values for CC2420

Transmit Power (dBm)	I_{high} (mA)	I_{low} (mA)	I_{tran} (mA)	t_{rise} (μs)	t_p (μs)	γ'
0	17.4	0.426	0.426	0.1	400	62.19%
−5	14.0	0.426	0.426	0.1	400	62.12%
−10	11.0	0.426	0.426	0.1	400	62.02%
−15	9.9	0.426	0.426	0.1	400	61.96%
−25	8.5	0.426	0.426	0.1	400	61.87%

To estimate the effect of device characteristics on energy savings, we consider the following examples using two commercially available popular chips:

Example 35. *For the CC1100 chip supporting O-QPSK modulation, we have* $I_{low} = 1.9mA$ *and* $I_{tran} = 8.7mA$. *At 868/915 MHz, values of* I_{high} *are 30.3 mA, 19.7 mA, 16.6 mA and 14.0 mA for 10 dBm, 5 dBm, 0 dBm and* -5 *dBm output power respectively. Also,* $t_{rise} = 88.4\mu s$ *and assuming a data rate of 2.5 Kbps, we have* $t_p = 400\mu s$. *Thus,* $\frac{t_{rise}}{t_p}$ *may be taken as* 0.221. *Hence, from Equation (8.35), we get* $\gamma' = 61.1\%, 60.34\%, 59.94\%$ *and* 59.46% *for 10 dBm, 5 dBm, 0 dBm and* -5 *dBm output power, respectively. Thus,* γ' *is only marginally reduced from the theoretical value of* 62.5% *as given by Observation 5. The results are summarized in Table 8.12.*

Example 36. *For the CC2420 chip,* $I_{low} = 0.426$ *mA,* $I_{tran} \approx I_{low}$ *and values of* I_{high} *are 17.4 mA, 14 mA, 11 mA, 9.9 mA and 8.5 mA at 0 dBm,* -5 *dBm,* -10 *dBm,* -15 *dBm and* -25 *dBm output power respectively.* $t_{rise} = 0.1\mu s$. *As in the case of CC1100, we assume that* $t_p = 400\mu s$. *Hence, from Equation (8.35), we get* $\gamma' = 62.19\%, 62.12\%, 62.02\%, 61.96\%$ *and* 61.87% *for 0 dBm,* -5 *dBm,* -10 *dBm,* -15 *dBm and* -25 *dBm output power, respectively. The results are summarized in Table 8.13.*

8.5.3 Performance Analysis

In this section, we compare the performance of CNS with existing techniques as discussed below.

8.5.3.1 Performance Comparison with CtS and VarBaTaC

We first compare the performance of CNS with those of CtS [9] and VarBaTaC [10] in terms of the transmission length duration. We assume ideal channel conditions for this comparison. Because both CtS and VarBaTaC utilize complete silence for information communication, their performance degradation in the presence of noise will be much more than that suffered by CNS.

The biggest drawback of the CtS scheme is that the transmission time is exponential in the number of bits to be sent. VarBaTaC uses a variable coding base and controls the communication time by suitably tuning the base value. Suppose we want to transmit a value M, where $0 \leq M < 2^k$, $k \geq 0$. The representation of M in base r will have $m = \left\lceil \frac{k}{\log r} \right\rceil$ digits. Each of these m digits is then transmitted using CtS with a pulse separating each digit. For m digits, the number of pulses transmitted would be $m+1$ and the total time taken in the worst case would be $(m+1)+m\cdot(r-1)$. For example, the decimal value $(245)_{10}$ in base 16 is $F5$. Assuming base 16 is used for transmission, the number of pulses required by the VarBaTaC scheme would be 3 (i.e., $3e_b$ units of energy expenditure) and the total duration of transmission would be $3+(15+5) = 23$ time slots. To compare VarBaTac with CNS, we see that $(245)_{10}$ in binary is 11110101. The frequency of both the bit pairs 01 and 11 is 2. We arbitrarily choose 01 to be the N symbol and map 11 to the symbol A. Thus the sequence to be transmitted in CNS is $AANN$ and the total duration of transmission would be four time slots, with only $2e_b$ units of energy expenditure by the transmitter for the two A symbols.

Let t_{avg} be the average number of time slots required in the VarBaTaC scheme.

TABLE 8.14: Number of slots for VarBaTaC with radix as power of 2 and $k = 64$

p	radix r	$t_{avg} = \lceil \frac{64}{p} \rceil \left(\frac{2^p+1}{2} \right) + 1$
2	4	81
3	8	100
4	16	137
5	32	215.5
6	64	357.5
7	128	646
8	256	1029

Whatever the value of the message to be transmitted in the range 0 to M (both inclusive), we need to transmit $m+1$ pulses for the m digits to be transmitted. In addition, we need on an average, $\frac{m(r-1)}{2}$ time slots for the digits (assuming equal likelihood for all possible values of the message). Hence, t_{avg} is given by

$$t_{avg} = (m+1) + \frac{m(r-1)}{2} = \frac{m(r+1)}{2} + 1. \tag{8.36}$$

Let $r = 2^p$. Hence, $m = \left\lceil \frac{k}{p} \right\rceil$, and from Equation (8.36) we have

$$\frac{k}{p}\left(\frac{2^p+1}{2}\right) + 1 \leq t_{avg} < \left(\frac{k}{p} + 1\right)\left(\frac{2^p+1}{2}\right) + 1.$$

Thus, for a given k, t_{avg} monotonically increases with p due to the function $\frac{2^p+1}{p}$.

Table 8.14 shows the values of t_{avg} for different values of $r = 2^p$, while Table 8.15 shows the variation of t_{avg} with radix r, even when r is not a power of 2.

As we can see from Tables 8.14 and 8.15, on average, CNS will require fewer time slots than VarBaTaC. In general, the transmission time of CNS will be less than that of CtS and VarBaTaC, while generating significant savings in energy at both the transmitter and the receiver. Note that neither CtS nor VarBaTaC generates energy savings at the receiver. Thus, in CtS and VarBaTac, the energy savings generated during transmissions would be substantially outweighed by the excessive receiver

TABLE 8.15: VarBaTaC time slot requirement for k = 64 with different radix values

r	$t_{avg} = \left\lceil \frac{64}{\log r} \right\rceil \left(\frac{r+1}{2} \right) + 1$
3	83
4	81
5	85
6	88.5
7	93
8	100
9	106
10	111
12	118
14	128.5
16	137

energy due to longer duration of communication than binary data transmission or CNS. For single hop networks this is not a problem but for multi-hop WSNs, this may become an issue.

8.5.3.2 Performance Comparison in a Noisy Channel

We now turn to the realistic situation where a channel is noisy and compare the performance of CNS with binary FSK and similar techniques that employ silent symbol-based strategies like the RBNSiZeComm [39], TSS [3] and RZE [5]. We first need to find the dependency of the bit error rate (BER) on the signal-to-noise ratio (SNR) for the CNS scheme. We assume that the channel noise is *additive white Gaussian noise* (AWGN). Let us also assume that the transmitter uses FSK modulation with frequencies f_c, $(f_c - \Delta f)$ and $(f_c + \Delta f)$ to represent A, B and C in any order. Effectively this will be a hybrid modulation scheme involving FSK and ASK. As a representative example for showing the energy savings, we use a non-coherent detection-based receiver similar to the one used for RBNSiZeComm or TSS protocol, with a structure as shown in Figure 8.19.

In Figure 8.19, r is the signal received at the input of the receiver which consists of the transmitted signal s as well as the channel noise η, i.e, $r = s + \eta$. The received

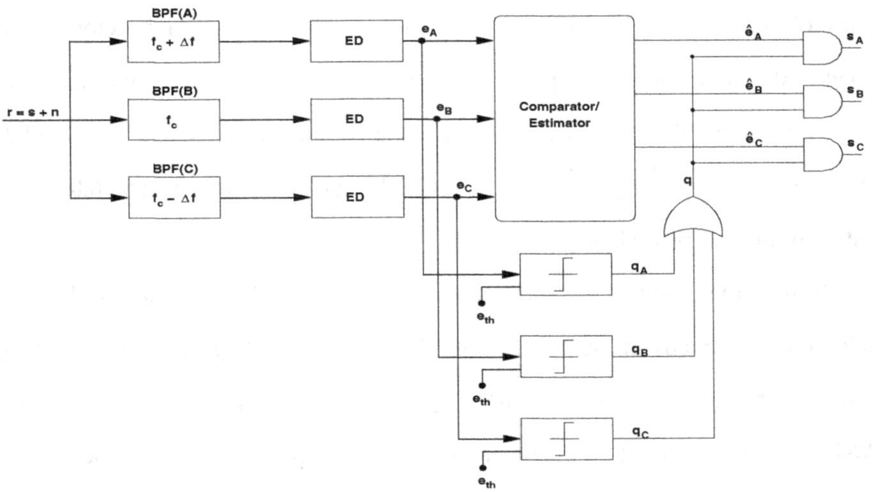

FIGURE 8.19: Representative FSK receiver for CNS scheme

signal r is first passed through three bandpass filters (BPFs) as shown in the figure, with center frequencies at $f_c + \Delta f$, f_c and $f_c - \Delta f$ (corresponding to A, B and C, respectively). The outputs of these bandpass filters are applied to the inputs of three envelope detectors (ED) to derive the values e_A, e_B and e_C at their outputs, corresponding to the symbols A, B and C respectively. In the ideal situation when there is no noise ($\eta = 0$), only one of these e_A, e_B and e_C will assume a non-zero value if the transmitted symbol is A or B or C, and all of e_A, e_B and e_C will be zero if the transmitted symbol is N. However, due to noise, even if the transmitted symbol is, for example, A, e_B and e_C may have non-zero values, although we may assume in this situation that $e_A > e_B$ and $e_A > e_C$. Similarly, if the transmitted symbol is B (or C), not only e_B (or e_C) will have a non-zero value, but e_A and e_C (or e_A and e_B) may also assume small non-zero values. If the transmitted symbol is N, all of e_A, e_B and e_C may assume small non-zero values.

To detect the transmitted symbol correctly in the presence of channel noise, we first pass e_A, e_B and e_C through a comparator to find which one is the largest, and also we use three threshold detectors to see whether e_A, e_B or e_C is above a certain threshold value e_{th}. The outputs q_A, q_B and q_C of these threshold detectors are applied to the inputs of an OR gate to get the value q at its output. If the transmitted symbol is N and if e_{th} is correctly chosen, the outputs of each of the threshold detectors will be 0, making q equal to 0. Thus, the symbol N will be generated at the output of the receiver when $q = 0$.

On the other hand, if the transmitted symbol is A, B or C, and the threshold value e_{th} is correctly chosen, then q will be 1. q and the largest output line of the comparator (only which will assume a logic level 1) are applied to the inputs of an AND gate whose output will generate the desired output A, B or C at the output $\hat{s}$ of the receiver. Thus, the desired output $\hat{s}$ of the receiver is:

i) If $q = 0$, $\hat{s} = N$, which corresponds to the case of $s_A = s_B = s_C = 0$.

ii) If $q = 1$,

$$\hat{s} = \begin{cases} A, & e_A > e_B, e_C \ (s_A = 1, s_B = s_C = 0) \\ B, & e_B > e_A, e_C \ (s_B = 1, s_A = s_C = 0) \\ C, & e_C > e_A, e_B \ (s_C = 1, s_A = s_B = 0). \end{cases} \tag{8.37}$$

8.5.3.3 Error Analysis

If k bits are blocked together to generate a symbol, the bit error rate (BER) is related to the symbol error rate (SER) by the relation

$$BER = \frac{2^{k-1}}{2^k - 1} \cdot P_S. \tag{8.38}$$

P_S is the SER. In the present case, $k = 2$ and thus, $BER = \frac{2}{3} P_S$.

Let P_N, P_A, P_B and P_C be the probabilities of occurrences of the symbols N, A, B and C, respectively, in the transmitted message. P_S can be written as

$$
\begin{aligned}
P_S = {} & P_N(1 - P\{q = 0|N\}) \\
& + P_A[1 - P\{q = 1 \text{ and } e_A > e_B, e_C|A\}] \\
& + P_B[1 - P\{q = 1 \text{ and } e_B > e_A, e_C|B\}] \\
& + P_C[1 - P\{q = 1 \text{ and } e_C > e_A, e_B|C\}].
\end{aligned} \tag{8.39}
$$

where $P\{q = 0|N\}$ is the probability that the symbol 0 is detected at the receiver, given that 0 was transmitted. It is possible to set the threshold value e_{th} so that the four constituent terms in the above expression for P_S are all equal. Under such a condition, we can show that the value of P_S will also be minimum.

$$
\begin{aligned}
P\{q = 0|N\} &= P\{q_A = 0|N\} \cdot P\{q_B = 0|N\} \\
&\quad \cdot P\{q_C = 0|N\} \\
&= [P\{q_A = 0|N\}]^3 \text{ (by symmetry).}
\end{aligned} \tag{8.40}
$$

Note that $P\{q_A = 0|N\}$ follows Rayleigh distribution. Assuming a zero mean and standard deviation of σ, we can write

$$
\begin{aligned}
P\{q_A = 0|N\} &= \int_0^{e_{th}} \frac{e_A}{\sigma^2} e^{-e_A^2/2\sigma^2}\, de_A \\
&= (1 - e^{-e_{th}^2/2\sigma^2}) \cdot u(e_A)
\end{aligned}
$$

Hence, we have $P\{q = 0|N\} = (1 - e^{-e_{th}^2/2\sigma^2})^3$. Now let $X = P\{q = 1 \text{ and } e_A > e_B, e_C|A\}, \forall e_A|A$,

$$
\begin{aligned}
X &= P\{q = 1|A\}P\{e_A > e_B, e_C|A\} && \text{(for all } e_A|A) \\
&= X_1 X_2 && \text{(say).}
\end{aligned}
$$

where $X_1 = P\{q = 1|A\}$ and $X_2 = P\{e_A > e_B, e_C|A\}, \forall e_A|A$. From the diagram of the receiver in Figure 8.19, it follows that X_1 can be written as

$$
\begin{aligned}
X_1 = \int_0^{\infty} \Big[& 1 - P\{q_A = 0|A\} \cdot P\{q_B = 0|A\} \\
& \times P\{q_C = 0|A\} \cdot p_A(e_A) \Big] \, de_A.
\end{aligned}
$$

$P\{q_B = 0|A\}$ and $P\{q_C = 0|A\}$ are governed by a Rayleigh distribution, while $P\{q_A = 0|A\}$ is governed by a Rician distribution. $p_A(e_A)$ is the probability density function for e_A which also follows a Rician distribution. Thus, we can write

$$
P\{q_A = 0|A\} = \int_0^{e_{th}} \frac{e_A}{\sigma^2} e^{-\frac{e_A^2 + s_A^2}{2\sigma^2}} I_0\left(\frac{s_A e_A}{\sigma^2}\right) de_A \text{ and,}
$$

$$
\begin{aligned}
P\{q_B = 0|A\} = P\{q_C = 0|A\} &= \int_0^{e_{th}} \frac{e_B}{\sigma^2} e^{-e_B^2/2\sigma^2} \, de_B \\
&= 1 - e^{-e_{th}^2/2\sigma^2}.
\end{aligned}
$$

We have for X_1 and X_2:

$$X_1 = \int_0^\infty \left[1 - \left(1 - e^{-e_{th}^2/2\sigma^2}\right)^2 \right.$$
$$\times \left(\int_0^{e_{th}} \frac{e_A}{\sigma^2} e^{-\frac{e_A^2+s_A^2}{2\sigma^2}} I_0\left(\frac{s_A e_A}{\sigma^2}\right) de_A\right) \frac{e_A}{\sigma^2} e^{-\frac{e_A^2+s_A^2}{2\sigma^2}}$$
$$\left. \times I_0\left(\frac{s_A e_A}{\sigma^2}\right)\right] de_A. \tag{8.41}$$

$$X_2 = \int_0^\infty P\{e_A > e_B|A\}P\{e_A > e_C|A\}p_A(e_A)\, de_A$$
$$= \int_0^\infty (P\{e_A > e_B|A\})^2 p_A(e_A) de_A. \tag{8.42}$$

In Equation (8.42), we have:

$$P\{e_A > e_B|A\} = P\{e_A > e_C|A\}$$
$$= \int_0^{e_A} \frac{e_B}{\sigma^2} e^{-e_B^2/2\sigma^2}\, de_B$$
$$= 1 - e^{-e_A^2/2\sigma^2}.$$

Thus, X_2 is given by

$$X_2 = \int_0^\infty \left(1 - e^{-e_A^2/2\sigma^2}\right)^2 \frac{e_A}{\sigma^2} e^{-\frac{e_A^2+s_A^2}{2\sigma^2}}$$
$$\times I_0\left(\frac{s_A e_A}{\sigma^2}\right) u(e_A)\, de_A. \tag{8.43}$$

Finally, P_S can thus be expressed as:

$$
\begin{aligned}
P_S = P_N\left[1-(1-e^{-e_{th}^2/2\sigma^2})^3\right] \\
+ P_A(1-X_1X_2) + P_B(1-X_1X_2) \\
+ P_C(1-X_1X_2).
\end{aligned} \tag{8.44}
$$

Assuming $P_A = P_B = P_C$ for equal likelihood of the symbols A, B and C, we get

$$
P_S = P_N\left[1-(1-e^{-e_{th}^2/2\sigma^2})^3\right] + 3P_A(1-X_1X_2).
$$

Assuming $P_N = 0.25$ and $P_A + P_B + P_C = 0.75$, we get the following expression for P_S:

$$
P_S = 0.25\left[1-(1-e^{-e_{th}^2/2\sigma^2})^3\right] + 0.75(1-X_1X_2). \tag{8.45}
$$

Thus, from Equations (8.38) and (8.45), using $k = 2$ for the CNS scheme, the bit error rate is given by

$$
BER = \frac{2}{3}[0.25\left[1-(1-e^{-e_{th}^2/2\sigma^2})^3\right] + 0.75(1-X_1X_2)]. \tag{8.46}
$$

8.5.3.4 BER for Different SNR Values

To compute the bit error rate for different SNR values from the above equation, we set $s_A = 1$, i.e., signal power = $s_A^2/2 = 1/2$ and SNR value = $1/2\sigma^2$. The computations to find the optimum value of threshold voltage e_{th} that gives the minimum BER for a given SNR value were implemented using MATLAB. The results of these computations are shown in Table 8.16.

TABLE 8.16: BER values as computed from Equation (8.46)

SNR in dB	BER value	Corresponding e_{th} value
14.00	5.07959×10^{-4}	0.55
14.50	2.3079×10^{-4}	0.55
15.00	9.49895×10^{-5}	0.54
15.50	3.52053×10^{-5}	0.54
16.00	1.18318×10^{-5}	0.54
16.50	3.36706×10^{-6}	0.53
17.00	8.37597×10^{-7}	0.53
17.50	1.80442×10^{-7}	0.53
18.00	3.1863×10^{-8}	0.52

8.5.3.5 Effect of Silent Symbols on Energy Savings

From Equation (8.46), we get the required peak transmitter power for a given BER value, which corresponds to the scenario where the transmitter is switched on for the whole of message transmission. However, in the CNS scheme, the transmitter will be put in *active* or *idle* state during the N symbols, and hence the required SNR value for a given BER will be less than the value demanded by Equation (8.46). We keep the CNS symbol duration same as that of the original bit duration in the binary message. Assuming equal probability of occurrence of the symbols A, B, C and N, 25% of the time the transmitter consumes negligible power (while transmitting the symbol N). The required CNS transmitter power will be three-fourths of the values as required by Equation (8.46). Noting that $10 \log_{10}(3/4)$ dB = $-1.25dB$ implies that the required SNR values for CNS will be reduced from the values given in Table 8.16 by 1.25 dB.

When using non-coherent detection, the required scaled average transmitter power values for CNS along with those for binary FSK for different BER values are shown in Table 8.17. Figure 8.20 provides a graphical representation of Table 8.17. The SNR values for binary FSK in Table 8.17 were determined from the relation

TABLE 8.17: Comparison of SNR values

BER value	Required SNR in dB	
	CNS	Binary FSK
5.08×10^{-4}	12.75	11.394
2.31×10^{-4}	13.25	11.864
0.95×10^{-4}	13.75	12.34
3.52×10^{-5}	14.25	12.815
1.183×10^{-5}	14.75	13.284
3.367×10^{-6}	15.25	13.769
8.376×10^{-7}	15.75	14.249
1.804×10^{-7}	16.25	14.72
3.186×10^{-8}	16.75	15.203

$$BER = \frac{1}{2}e^{-\gamma/2}, \tag{8.47}$$

where SNR in dB = $10\log_{10}\gamma$ and γ is the ratio of signal power to noise power. From Table 8.17 we see that for a given BER in the range 10^{-4} to 10^{-7}, CNS requires nearly 1.5 dB more power than binary FSK.

8.5.3.6 Effect of Data Compression on Energy Savings

We now consider the effect of data compression on the overall transmitter energy. Let P_b and P_C be the transmitter power required for binary FSK and CNS respectively. Also, let T_b, T_C be the transmission times for binary FSK and CNS, respectively. Since CNS needs half the time for transmission than binary FSK needs, $T_C = T_b/2$. Noting that $10^{1.5/10} = 1.4125$, we have $P_C/P_b = 1.4125$. Let E_b, E_C be the transmitter energy requirements for binary FSK and CNS respectively. Hence we have

$$E_C = P_C T_C = \frac{1.4125}{2} \cdot P_b T_b = 0.7063 E_b. \tag{8.48}$$

Theorem 28. *Assuming an AWGN channel, for equal probability of occurrence of*

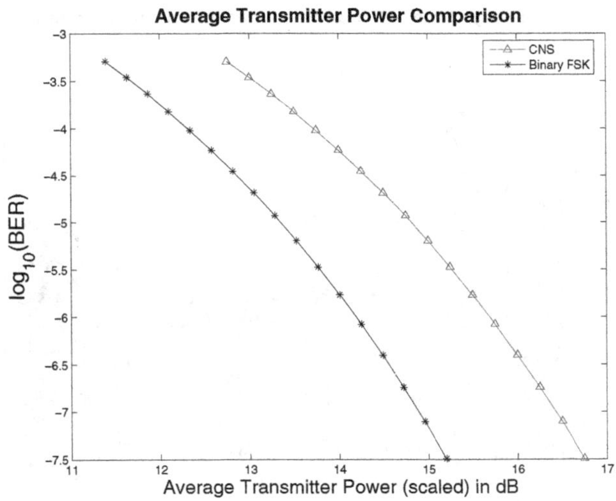

FIGURE 8.20: Comparison of average transmitter power (scaled) for given BER values

the symbols A, B, C and N, CNS requires approximately 70% *of the total energy as compared to binary FSK, resulting in a saving of* 30% *transmitter energy.*

In addition, CNS has the advantage of reducing receiver energy consumption by a factor of two, as the receiver receives all the symbol values in half the time required to transmit the original binary message. Figure 8.21 shows the comparison of the transmitter energy savings in energy generated by RBNSiZeComm [39], TSS [3], RZE [5] and CNS at the transmitter and receiver for equal likelihood of all possible binary strings of a given length, using non-coherent detection-based receiver over AWGN channels. Table 8.18 gives the values of the energy savings generated by the various schemes at the transmitter and the receiver, in comparison to a typical communication scheme utilizing binary FSK with non-coherent detection. We see that CNS outperforms TSS in saving energy at both the transmitter and the receiver sides. The results also show that while the transmitter energy savings by CNS is less than that of both RBNSiZeComm and RZE, it performs significantly better than these schemes at saving energy at the receiver. Most WSNs tend to be multi-hop in nature

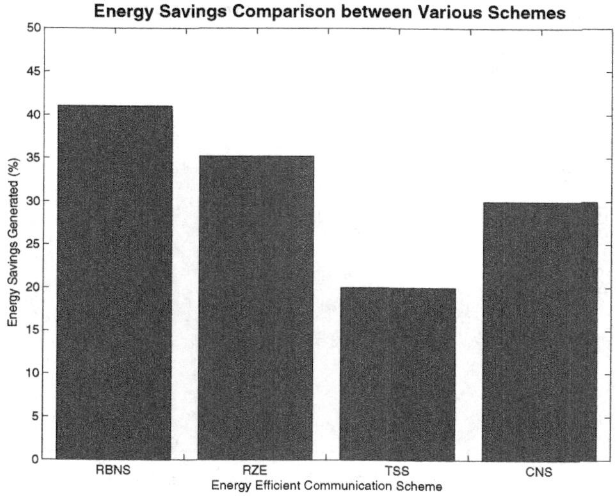

FIGURE 8.21: Comparison of transmitter energy savings among various energy efficient communication schemes

and this makes CNS more suitable than either RBNSiZeComm, RZE or TSS for use in low power WSNs.

However, in view of the results obtained from the discussions on the effect of device characteristics and considering the low encoding and decoding overhead, the above figures for transmitter and receiver energy savings obtained by using the CNS scheme may be only marginally reduced.

8.5.4 Simulation Results with Real-life Data

Let us consider the performance of CNS protocol on the data sets taken from some selected ECG signals, JPEG images of blood and tissue samples and ASCII text data to assess its effectiveness in potential application areas like i) remote healthcare through the use of body area networks and smartphones and ii) WSNs for agriculture.

Remote healthcare applications: With the emergence of 3G enabled cellular phones, remote healthcare systems involving body area networks and smartphones are becoming popular. As stated in [1], at low data rates of 60 bytes/s and transmis-

TABLE 8.18: Comparison of energy savings among various schemes

Communication Scheme	Energy Savings at Transmitter	Energy Savings at Receiver
RBNSiZeComm	53%	0%
RZE	35.2%	12.5%
TSS	20%	36.9%
CNS	30%	50%

sion frequency of about once every 0.5 to 1 sec (the usual required rate for transmitting physiological data like ECG signals and other vital body parameters) would drain a cell phone's battery completely within 10 to 12 hours, even while transmitting data over a strong connection [23]. In other words, this system needs a recharge cycle of every half day.

For evaluation of energy savings generated by CNS for healthcare applications, we assume the applications involve the transmission of ECG signals, JPEG images of blood samples and ASCII text data representing physiological parameters like blood pressure, blood sugar level, and temperature. [1, 23]. The performance of CNS noted below is based on simulation results, [1], on 50 real-life ECG signal samples taken from the QT database [27] and the JPEG images of blood, tissue and bone samples [28].

WSNs for agriculture: In WSNs deployed for monitoring crop health, information collected through sensor networks in the fields include climate data such as temperature, relative humidity, solar radiation and soil composition. Network architectures typically consist of a network of *acquisition stations* and a *gateway* node. Data acquisition is done through low-cost acquisition stations equipped with wireless transceivers and deployed in a mesh topology [24, 26]. The acquisition stations collect data such as temperature, relative humidity, solar radiation and soil composition, and also detect pest infestation and other data. The usual practice is to collect one sample per minute. Communications are based on low-power radio frequency links provided by inexpensive transceivers. A single gateway node collects the data from the acquisition stations for processing [25]. These data are assumed to be transmitted as short ASCII text messages [25, 26, 24], usually a few bytes long at the most. In a typical field deployment with EbT-based transmissions, considering a mesh network deployment for a field of size $60m \times 60m$, radio range of 0 to $80m$ with multi-hop communication, data rate of 300 packets/min = 0.5 packets/s and duty cycle of 15% at $20mA$ with a battery of $7.2Ah$ capacity, a radio battery would last for only 100 days [25, 26, 24].

From the simulation results, it follows that P_N is around 0.4751 for ECG signals, 0.2884 for JPEG data from blood samples and 0.388 for text data, as shown in Table 8.19. Using these values of P_N in Equation (8.44) and noting that the transmitter is in *active* or *idle* state during the transmission period of the N symbols, the required SNR values for given BERs were computed for each of the above types of data. From these computations, we find that for ECG signals at the same BER, the energy saving from using CNS for ECG signals as compared to binary FSK will be approximately 50%.

Similarly, the energy savings for transmission of JPEG image files of blood and tissue samples was 33% for CNS, when compared to binary FSK. For text data, the energy saving was 42.6%. All these results are summarized in Table 8.19.

The amount of transmitter energy savings is between 33% and 50% for CNS, depending on the nature of the data to be transmitted. For the same data set, the average transmitter energy savings generated by RBNSiZeComm varied from 33% to 61% [39]. Figure 8.22 shows the comparison of transmitter energy savings generated by CNS and RBNSiZeComm for the different data types. CNS also simultaneously generates a saving in energy of about 50% compared to binary FSK at the receiver, regardless of the type of data. However, no such saving was generated at the receiver by RBNSiZeComm [39].

TABLE 8.19: Energy savings for different application data

Signal Type	N Symbol Frequency	**TX Energy Savings Generated**	
		Noiseless Channel	CNS
ECG signals	47.51%	73.75%	50%
JPEG blood/tissue sample images	28.84%	64.02%	33%
ASCII text data	38.8%	69.44%	42.6%

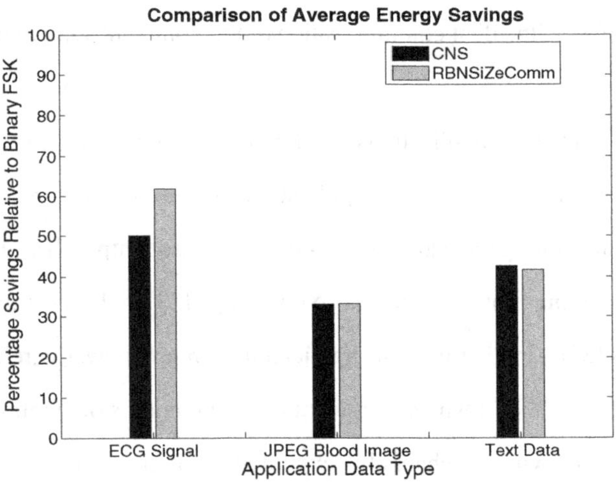

FIGURE 8.22: Comparison of energy savings generated by CNS and RBNSiZeComm at transmitter for various application data types to binary FSK

We can thus conclude that CNS provides energy savings at both the transmitter and the receiver. The sum of transmitter and receiver energy savings offered by CNS over conventional EbT schemes employing BFSK, is more than that with RBNSiZEComm and TSS. Thus, CNS will be very useful in real-life low-power applications, e.g., in wireless sensor networks and ad hoc networks where multi-hop wireless communication will be involved.

8.6 Tri-Digit Fibonaccian Number System (TFNS) Communication Protocol

We describe another energy-efficient communication protocol based on a variant of Fibonaccian number system, called the *Tri-digit Fibonaccian number system (TFNS)*, coupled with silent communication of the dominant symbol of the encoded message.

In many engineering applications, the Fibonacci number system (FNS) has been used with remarkable success. Such applications include VLSI chip design (in minimizing crosstalk among the data buses inside a processor chip [37]), synchronization control [35], compression of large sparse bit strings [33] and robust transmission of binary strings [32]. Another possible application is in low-power wireless communication systems [34, 39, 6] where asymmetry in the numbers of symbols to be transmitted through the wireless channel is exploited to minimize the transmitter and/or receiver energy required. If the Fibonaccian number system is used to give positional weights to binary digits for representation of decimal numbers, a marked asymmetry may be observed in the numbers of 0's and 1's in the resulting codewords. These asymmetric characteristics can be exploited for saving energy, particularly in low-power wireless sensor networks. However, if conventional FNS is used, the number of bits representing a given decimal number increases compared to the binary number system. As a result, the savings of energy for low-power wireless communication systems may not be substantial. A variant of FNS, called the *Tri-digit Fibonaccian number system* (TFNS) has been proposed by Sinha et al. [40] provides asymmetric distribution in the number of 0's, 1's and 2's used in representing all possible numbers, along with a substantial reduction in the number of digits. This brings about significant energy savings in wireless communication systems using the scheme [4] described below.

8.6.1 Preliminaries

Consider the Fibonacci sequence $< 0, 1, 1, 2, 3, 5, 8, 13, 21, \ldots >$ and denote the $n^{th}(n \geq 0)$ number of this sequence by F_n, where $F_0 = 0, F_1 = F_2 = 1, F_3 = 2, F_4 = 3, F_5 = 5$ and so on. In FNS, the ith bit position ($i \geq 0$, $i = 0$ being the least significant bit) has the positional weight of F_{i+2}. In the literature, there exist different versions of FNS which have different representations of the same decimal number [37]. However, the normal form of FNS [35] always provides a unique representation of a given decimal number N in FNS as shown in Table 8.20. Another interesting result regarding the Fibonacci numbers is that a positive integer can be expressed uniquely in terms of a sum of non-consecutive Fibonacci numbers as in Zeckendorf representation of numbers [38]. However, the positional weights in all such representations [35, 38] consist of all the Fibonacci numbers in order, leading to a representation of a number in terms of a string of 0's and 1's only.

One interesting property of the normal form of FNS is that if the $i^{th}(i > 0)$ bit is 1, the $(i-1)^{th}$ bit must be a 0. That is, two consecutive 1's do not appear in any of these normal codewords. The total numbers of 0's and 1's appearing at any bit position (starting from the most significant bit (msb)) of an n-bit FNS representation of all decimal numbers in the range $0 \leq N \leq F_{n+2} - 1$ can be visualized by an edge-weighted binary tree as shown in Figure 8.23.

The root node labeled * has the left child labeled as 0 with the corresponding edge weight of F_{n+1}, and a right child labeled as '1' with the corresponding edge weight of F_n; considering the representation of all possible numbers, there are F_{n+1} 0's and F_n 1's at the msb (i.e., $(n-1)^{th}$ bit position). This left child 0 of *, in turn, has a left child labeled 0 with the corresponding edge weight of F_n and a right child

TABLE 8.20: Comparison of FNS, TFNS and decimal number system

FNS						Corresponding decimal number	TFNS			
Weights $\Longrightarrow$	$F_6 = 8$	$F_5 = 5$	$F_4 = 3$	$F_3 = 2$	$F_2 = 1$		$F_8 = 21$	$F_6 = 8$	$F_4 = 3$	$F_2 = 1$
	0	0	0	0	0	0	0	0	0	0
	0	0	0	0	1	1	0	0	0	1
	0	0	0	1	0	2	0	0	0	2
	0	0	1	0	0	3	0	0	1	0
	0	0	1	0	1	4	0	0	1	1
	0	1	0	0	0	5	0	0	1	2
	0	1	0	0	1	6	0	0	2	0
	0	1	0	1	0	7	0	0	2	1
	1	0	0	0	0	8	0	1	0	0
	1	0	0	0	1	9	0	1	0	1
	1	0	0	1	0	10	0	1	0	2
	1	0	1	0	0	11	0	1	1	0
	1	0	1	0	1	12	0	1	1	1

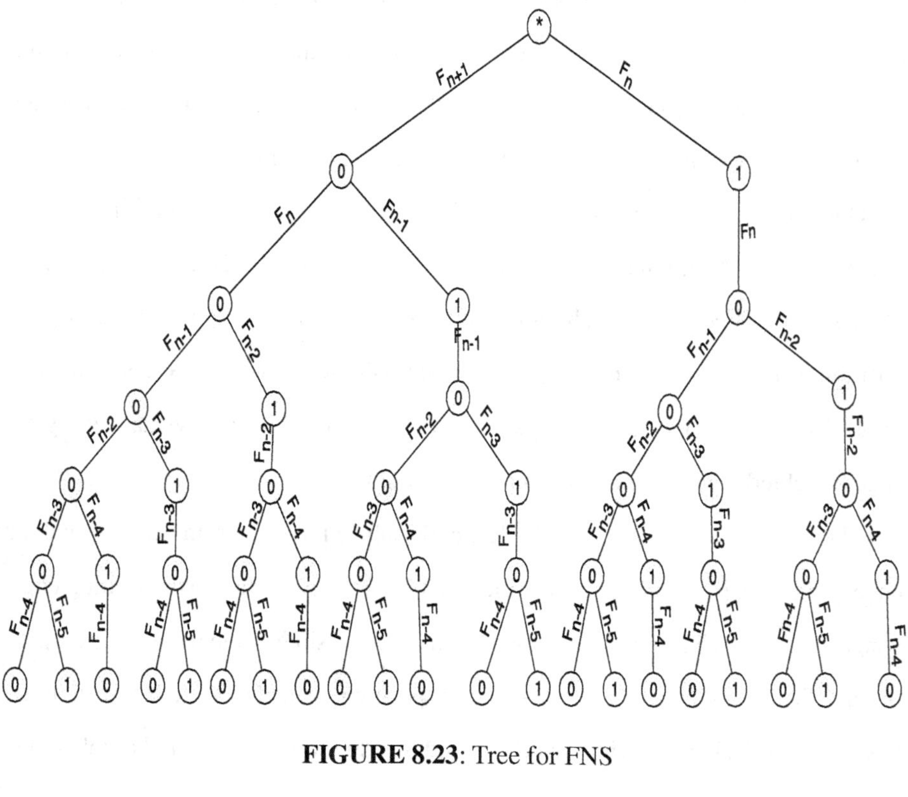

FIGURE 8.23: Tree for FNS

labeled 1 with the corresponding edge weight of F_{n-1}, meaning that the codes with 0's at the msb have a total of F_n 0's and F_{n-1} 1's at the $(n-2)^{th}$ bit position. The right child 1 of * has only one (left) child labeled as 0 (as two consecutive bits cannot be 1) with the corresponding edge weight of F_n, meaning that the codes with 1's at the msb have a total of only F_n 0's at the $(n-2)^{th}$ bit position.

It follows from the binary tree in Figure 8.23 that the total number of 0's in n-bit representation of all decimal numbers in the range 0 to $F_{n+2}-1$ (both inclusive) is given by:

$$\nu_0 = F_{n+1} + (F_n + F_n) + (F_{n-1} + F_{n-1} + F_{n-1}) + \ldots \quad (8.49)$$

$$= F_{n+1} + 2F_n + 3F_{n-1} + 5F_{n-2} + 8F_{n-3} + \ldots \quad (8.50)$$

$$= F_2F_{n+1} + F_3F_n + F_4F_{n-1} + F_5F_{n-2} + F_6F_{n-3} + \ldots + F_{n+1}F_2. \quad (8.51)$$

Similarly, the number of 1's denoted by ν_1 is given by:

$$\nu_1 = F_n + F_{n-1} + (F_{n-2} + F_{n-2}) + (F_{n-3} + F_{n-3} + F_{n-3}) + \ldots$$

$$= F_n + F_{n-1} + 2F_{n-2} + 3F_{n-3} + 5F_{n-4} + 8F_{n-5} + \ldots$$

$$= F_1F_n + F_2F_{n-1} + F_3F_{n-2} + F_4F_{n-3} + F_5F_{n-4} + F_6F_{n-5} + \ldots + F_nF_1.$$

Let $G(z) = F_0 + F_1z + F_2z^2 + F_3z^3 + F_4z^4 + \ldots$

Thus, $G^2(z) = (F_0 + F_1z + F_2z^2 + F_3z^3 + F_4z^4 + \ldots)^2$ and the coefficient of z^{n+3} in $G^2(z)$ is given by:

$$(F_{n+3}F_0 + F_{n+2}F_1 + F_{n+1}F_2 + \ldots + F_2F_{n+1} + F_1F_{n+2} + F_0F_{n+3})$$
$$= \nu_0 + 2F_{n+2}F_1.$$

Now, $G(z) = \frac{1}{\sqrt{5}}[(1-\phi z)^{-1} - (1-\bar{\phi}z)^{-1}]$ as in [36], where $\phi = \frac{\sqrt{5}+1}{2}$ and

$\bar{\phi} = \frac{\sqrt{5}-1}{2}$. So,

$$G^2(z) = \frac{1}{5}[(1-\phi z)^{-2} + (1-\bar{\phi} z)^{-2} - 2(1-\phi z)^{-1}(1-\bar{\phi} z)^{-1}].$$

Hence, the coefficient of z^{n+3} in $G^2(z)$ is:

$$\frac{1}{5}[(n+4).\phi^{n+3} + (n+4).\bar{\phi}^{n+3} - 2(\phi^{n+3}.\bar{\phi}^0 + \phi^{n+2}.\bar{\phi} + \ldots + \phi^1.\bar{\phi}^{n+2} + \phi^0.\bar{\phi}^{n+3}].$$

Let $S = \phi^{n+3} + \phi^{n+2}.\bar{\phi} + \ldots + \phi.\bar{\phi}^{n+2} + \bar{\phi}^{n+3}$. So, $S[1 - \frac{\bar{\phi}}{\phi}] = \phi^{n+3} - \frac{\bar{\phi}^{n+4}}{\phi} = \frac{\phi^{n+4} - \bar{\phi}^{n+4}}{\phi}$.

Hence, $S = \phi^{n+4} - \bar{\phi}^{n+4}$, as $\phi - \bar{\phi} = 1$. Therefore,

$$\nu_0 = \frac{1}{5}[(n+4)(\phi^{n+3} + \bar{\phi}^{n+3}) - 2(\phi^{n+4} - \bar{\phi}^{n+4})]. \tag{8.52}$$

Putting the values of ϕ and $\bar{\phi}$ in the above expression of ν_0, we get the number of 0's in all *n-bit* valid codewords. The number of 1's in all *n-bit* valid codewords is given by

$$\nu_1 = nF_{n+2} - \nu_0. \tag{8.53}$$

For large n, $\nu_0 \approx \frac{(n+4-2\phi)\phi^{n+3}}{5} = \frac{(n+0.764)\phi^{n+3}}{5}$. The ratio of ν_0 to the total number of bits nF_{n+2} asymptotically tends to $\frac{\phi}{\sqrt{5}} = 72.36\%$. Such a large asymmetry in the number of 0's and 1's can be exploited in applications involving low energy wireless communication by keeping the transmitter silent or off during the dominant symbols (0's in this case), thereby reducing the average power of signal transmission.

On the other hand, corresponding to m-bit binary representation of numbers, we need n bits in FNS where n is the least integer satisfying the inequality $F_{n+2} - 1 > 2^m - 1$. Noting that $F_{n+2} \approx \frac{\phi^{n+2}}{\sqrt{5}}$ for large n, $\frac{\phi^{n+2}}{\sqrt{5}} > 2^m$, i.e., $n \approx \lceil \frac{m + \log\sqrt{5}}{\log\phi} \rceil - 2$. For m = 8, 16, 32, 64, 128, 256, 512 and 1024, the corresponding values of n are 12,

23, 46, 92, 185, 369, 738 and 1475, respectively. The ratio $\frac{n}{m}$ asymptotically tends to 1.44. This shows that the number of bits in FNS representation is 1.44 times higher than that of the original binary representation. Such an increase in the length of the code may not be desirable in many engineering applications.

8.6.2 Tri-digit Fibonaccian Number System

To overcome the problem of increase in the length of a codeword in FNS, we describe a number system [40] in which i) the digits at each position will be taken from the set $\{0, 1, 2\}$, and ii) the positional weight at the ith $(i \geq 0)$ digit position ($i = 0$ corresponds to the least significant digit position) is equal to F_{2i+2}. That is, the positional weights at different digit positions are $F_2, F_4, F_6, \ldots$, with F_2 as the weight of the least significant digit. We term this number system as the *tri-digit Fibonaccian number system* (TFNS) [40]. The weights at different digit positions are alternate numbers in the Fibonacci sequence. Given a decimal number $N < F_{2k+2}$ for some k, we can represent N in its equivalent TFNS form using k digits $d_0, d_1, d_2, \ldots, d_{k-1}$ by the decimal-to-TFNS conversion algorithm $DTFNS$ as given below.

Input: Decimal number N, integer k such that $N < F_{2k+2}$

Output Equivalent TFNS form for N

for $i = k - 1$ downto 0 **do**
  if $N \geq 2F_{2i+2}$ **then**
    $d_i \leftarrow 2$
    $N \leftarrow N - 2F_{2i+2}$
  else if $N \geq F_{2i+2}$ **then**
    $d_i \leftarrow 1$
    $N \leftarrow N - F_{2i+2}$
  else
    $d_i \leftarrow 0$

```
10:     end if
11: end for
```

Following the $DTFNS$ conversion algorithm, the representations of all decimal numbers in the range 0 to 12 (both inclusive) using four digits in TFNS are shown in Table 8.20. Note that the decimal number 8 is *uniquely* represented as 0100 in TFNS using the above conversion algorithm. Although another possible representation of the decimal number 8 in TFNS with the above positional weights could be 0022; this representation cannot be obtained by the algorithm $DTFNS$ because of the *if* statement in line 3 which, in effect, forces the number to be represented by using non-zero digits at the highest possible digit positions whenever possible.

Lemma 31. *Let N be a decimal number whose representation in TFNS is* $2111\ldots1$ *having* $(k-1)$ *1's following a 2. The decimal number* $N+1$ *will be represented by* $1000\ldots0$, *with k 0's after the 1.*

Proof: From the TFNS representation of N, it follows that:

$$
\begin{aligned}
N =& 2F_{2k} + F_{2k-2} + F_{2k-4} + \ldots + F_4 + F_2 \\
\Rightarrow N+1 =& 2F_{2k} + F_{2k-2} + F_{2k-4} + \ldots + F_4 + F_2 + 1 \\
=& 2F_{2k} + F_{2k-2} + F_{2k-4} + \ldots + F_4 + F_2 + F_1 \\
=& 2F_{2k} + F_{2k-2} + F_{2k-4} + \ldots + F_6 + F_4 + F_3 \\
=& 2F_{2k} + F_{2k-2} + F_{2k-4} + \ldots + F_8 + F_6 + F_5 + \ldots \\
=& 2F_{2k} + F_{2k-1} \\
=& F_{2k} + F_{2k+1} \\
=& F_{2k+2}.
\end{aligned}
$$

By virtue of the conversion algorithm $DTFNS$, $N+1$ will thus be represented as $1000\ldots0$ in TFNS, with k 0's after the 1. □

Because of Lemma 31 it follows that if we consider n-digit representation of numbers in TFNS, not all possible 3^n code values will be generated by the $DTFNS$ algorithm. To be more precise, whenever we come across some number N with the *lowermost* k *digits* in its TFNS representation as $2111\ldots1$, the next consecutive number $N+1$ will have its lowermost $k+1$ digits in TFNS as $1000\ldots0$, thus eliminating many of the possible code values between them. The number system differs from the Zeckendorf number system in having the alternate Fibonacci numbers as positional weights. Such a choice of alternate Fibonacci numbers leads to a number representation using 0, 1 and 2 for any digit position. Moreover, Zeckendorf representation does not have two consecutive 1's as we will see below; specifically there will not be two consecutive 2's in the number representation stated in the following lemma.

Lemma 32. *TFNS representation of any given number may have consecutive 1's but no consecutive 2's.*

The proof follows as a corollary of Lemma 31.

Thus, unlike the Zeckendorf representation which does not allow consecutive 1's in the representation, this number system allows two consecutive 1's but no consecutive 2's. The properties given in Lemma 31 and Lemma 32 lead to unequal distribution in the total numbers of 0's, 1's and 2's in the TFNS coding of numbers for any given n-digit representation. A careful look at the digit generation process in the $DTFNS$ algorithm would reveal the exact distribution of the numbers of 0's, 1's and 2's in the TFNS code values. We follow this digit generation process (starting from the most significant digit position) for all possible codes with n digits (to represent numbers in the range 0 to $F_{2n+2}-1$) by constructing a weighted ternary tree T shown in Figure 8.24.

First, we start at the root of T and insert its three children nodes labeled 0, 1 and 2 (with three edges connected from the root to these three nodes), meaning that

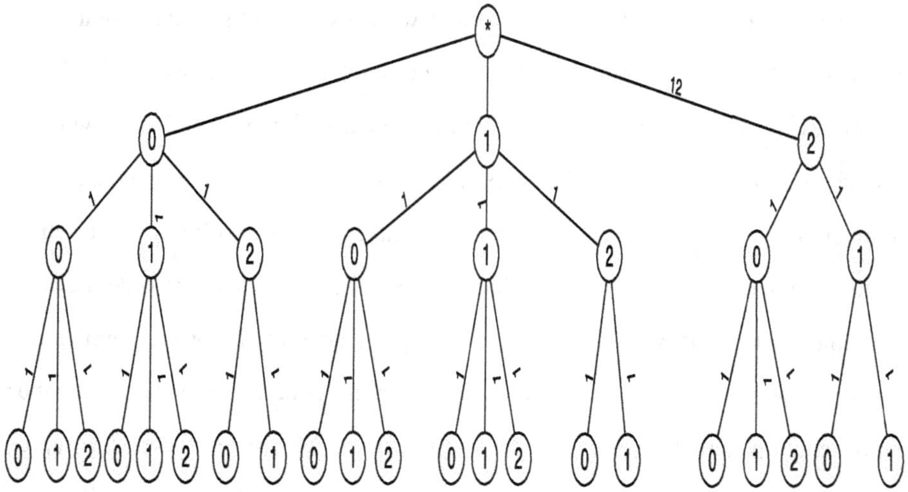

FIGURE 8.24: Tree for TFNS

the leftmost digit position of the code values can be filled by a 0, 1 or 2. Since a 0 at the leftmost digit position will appear for all the numbers having the decimal values ranging from 0 to F_{2n}, the weight of the edge connecting the root to its child node labeled 0, will be set as F_{2n}. This is also the case with the weight of the edge connecting the root to its child labeled 1. But the weight of the edge connecting the root to the child labeled 2 will be $F_{2n+2} - 2F_{2n} = F_{2n-1}$. We now derive expressions for the distribution of 0's, 1's and 2's in TFNS.

8.6.2.1 Deriving TFNS Digit Distributions

From Figure 8.24, we see that the total number of 0's in all the n-digit valid codewords representing all numbers in the range 0 to $F_{2n+2} - 1$ (both inclusive) is given by:

$$\nu_0' = F_{2n} + 3F_{2n-2} + 8F_{2n-4} + \ldots = F_2F_{2n} + F_4F_{2n-2} + F_6F_{2n-4} + \ldots + F_{2n}F_2.$$

Similarly, the total number of 1's in all valid codewords is:

$$\nu_1' = F_{2n} + (2F_{2n-2} + F_{2n-3}) + \ldots = F_1F_{2n} + (F_3F_{2n-2} + F_2F_{2n-3})$$
$$+ (F_5F_{2n-4} + F_4F_{2n-5}) + \ldots + (F_{2n-1}F_2 + F_{2n-2}F_1).$$

The total number of 2's in all the valid codewords is:

$$\nu_2' = F_1F_{2n-1} + 2F_{2n-3} + 5F_{2n-5} + \ldots$$
$$= F_1F_{2n-1} + F_3F_{2n-3} + F_5F_{2n-5} + \ldots + F_{2n-1}F_1.$$

To express ν_0', ν_1' and ν_2' in closed form we proceed as follows.

Let $G(z) = F_0 + F_1z + F_2z^2 + F_3z^3 + F_4z^4 + \ldots$

Thus, $G(z)$ is given by $G(z) = \frac{z}{1-z-z^2}$ as in [36]. Now, $G(-z) = F_0 - F_1z + F_2z^2 - F_3z^3 + F_4z^4 + \ldots$.

Hence, we have:

$$\frac{G(z) + G(-z)}{2} = F_0 + F_2z^2 + F_4z^4 + F_6z^6 + \ldots = H(z), \text{ (say)}.$$

.

That is,

$$H(z) = \frac{z^2}{(1-z^2)^2 - z^2} = \frac{z^2}{1 - 3z^2 + z^4}$$
$$= \frac{z^2}{(1-\alpha z^2)(1-\beta z^2)}, \text{ (say)}$$

where α and β are given by:

$$\left.\begin{aligned} \alpha + \beta &= 3 \\ \alpha\beta &= 1. \end{aligned}\right\} \tag{8.54}$$

Hence, $\alpha = \frac{3+\sqrt{5}}{2} = 1 + \phi$ and $\beta = \frac{3-\sqrt{5}}{2} = 1 - \bar{\phi}$. Also, $H^2(z) = (F_0 + F_2z^2 + F_4z^4 + \ldots)^2$, in which the coefficient of z^{2n+2} is $F_0F_{2n+2} + F_2F_{2n} + F_4F_{2n-2} +$

$\ldots + F_{2n+2}F_0 = \nu_0'$. Hence,

$$H(z) = \frac{z^2}{[1-(1+\phi)z^2][1-(1+\bar{\phi})z^2]}$$
$$\Rightarrow H^2(z) = \frac{z^4}{[1-(1+\phi)z^2]^2[1-(1+\bar{\phi})z^2]^2}.$$

Let $(1+\phi)z^2 = x$ and $(1-\bar{\phi})z^2 = y$. Then, $H^2(z) = \frac{z^4}{(1-x)^2(1-y)^2}$. The coefficient of z^{2n+2} in $H^2(z)$

$= $ the coefficient of z^{2n-2} in $(1-x)^{-2}(1-y)^{-2}$.

$=$ the sum of the coefficients of $x^i y^j$ in $(1-x)^{-2}(1-y)^{-2}$ with $i+j=n-1$.

$= (1+\phi)^0 \cdot 1 \cdot n(1-\bar{\phi})^{n-1} + 2(1+\phi)(n-1)(1-\bar{\phi})^{n-2}$

$+ 3(1+\phi)^2(n-2)(1-\bar{\phi})^{n-3} + \ldots + n(1+\phi)^{n-1} \cdot 1 \cdot (1-\bar{\phi})$.

8.6.2.2 Computation of ν_0'

Collecting the terms from the right side of the above expression and noting that $(1+\phi)(1-\bar{\phi}) = 1$, we get $\nu_0' = n(1+\phi)^{n-1} + 2(n-1)(1+\phi)^{n-3} + 3(n-2)(1+\phi)^{n-5} + \ldots + n(1+\phi)^{1-n}$. Replacing $(1+\phi)$ by α, we get

$$\nu_0' = n\alpha^{n-1} + 2(n-1)\alpha^{n-3} + 3(n-2)\alpha^{n-5} + \ldots + n\alpha^{1-n}. \quad (8.55)$$
$$\frac{\nu_0'}{\alpha^2} = n\alpha^{n-3} + 2(n-1)\alpha^{n-5} + \ldots + (n-1).2.\alpha^{1-n} + n\alpha^{-n-1}. \quad (8.56)$$

Subtracting Equation (8.56) from Equation (8.55) we get:

$$\nu_0'(1-\frac{1}{\alpha^2}) = n\alpha^{n-1} + (n-2)\alpha^{n-3} + (n-4)\alpha^{n-5}$$
$$+ (n-6)\alpha^{n-7} + \ldots + [n-2(n-1)]\alpha^{1-n} - n\alpha^{-n-1}$$
$$= n(\alpha^{n-1} + \alpha^{n-3} + \alpha^{n-5} + \ldots) - 2(\alpha^{n-3} + 2\alpha^{n-5} + 3\alpha^{n-7} + \ldots)$$

$$= \frac{n\alpha^{n-1}(1-\gamma^n)}{1-\gamma} - 2\alpha^{n-3}\left[\frac{1-\gamma^{n-1}}{(1-\gamma)^2} - \frac{(n-1)\gamma^{n-1}}{1-\gamma}\right] - n\alpha^{-1-n},$$

where $\gamma = \frac{1}{\alpha^2}$.

Thus,

$$\nu_0' = \frac{1}{1-\gamma}\left(\frac{n\alpha^{n-1}(1-\gamma^n)}{1-\gamma} - 2\alpha^{n-3}\left[\frac{1-\gamma^{n-1}}{(1-\gamma)^2} - \frac{(n-1)\gamma^{n-1}}{1-\gamma}\right] - n\alpha^{-1-n}\right). \tag{8.57}$$

That is,

$$\nu_0' = \frac{n\alpha^{n-1}(1-\gamma^n)}{(1-\gamma)^2} - 2\alpha^{n-3}\left[\frac{1-\gamma^{n-1}}{(1-\gamma)^3} - \frac{(n-1)\gamma^{n-1}}{(1-\gamma)^2}\right] - \frac{n\alpha^{-1-n}}{1-\gamma}. \tag{8.58}$$

8.6.2.3 Computation of ν_1' and ν_2'

We have, $\nu_2' = F_1F_{2n-1} + F_3F_{2n-3} + \ldots + F_{2n-1}F_1$. Let $H'(z) = \frac{1}{2}[G(z) - G(-z)] = F_1z + F_3z^3 + F_5z^5 + \ldots$.

ν_2' is the coefficient of z^{2n} in $H'^2(z)$ and $H'(z)$ can be written as:

$$\frac{1}{2}\left[\frac{z}{1-z-z^2} - \frac{-z}{1+z-z^2}\right] = z\left[\frac{1-z^2}{1+z^4-3z^2}\right].$$

Denoting z^2 by δ, we get

$$\frac{1-z^2}{1+z^4-3z^2} = \frac{1-\delta}{1-3\delta+\delta^2} = \frac{1-\delta}{(1-\alpha\delta)(1-\beta\delta)}$$

where α and β are given by Equation (8.54).

Now,

$$\frac{1}{1-\alpha\delta}+\frac{1}{1-\beta\delta} = \frac{2-(\alpha+\beta)\delta}{1-3\delta+\delta^2}$$
$$= \frac{2-3\delta}{1-3\delta+\delta^2}$$

and
$$\frac{1}{1-\alpha\delta}-\frac{1}{1-\beta\delta} = \frac{(\alpha-\beta)\delta}{1-3\delta+\delta^2}$$
$$= \frac{\sqrt{5}\delta}{1-3\delta+\delta^2}$$

$$\Rightarrow \frac{2(1-\delta)}{1-3\delta+\delta^2} = \frac{2-3\delta+\delta}{1-3\delta+\delta^2}$$
$$= \frac{1}{1-\alpha\delta}+\frac{1}{1-\beta\delta}+\frac{1}{\sqrt{5}}\left[\frac{1}{1-\alpha\delta}-\frac{1}{1-\beta\delta}\right].$$

We can therefore express $H'(z)$ as $\frac{z}{2}\left[\frac{C_1}{1-\alpha z^2}+\frac{C_2}{1-\beta z^2}\right]$, where $C_1 = 1+\frac{1}{\sqrt{5}}$ and $C_2 = 1-\frac{1}{\sqrt{5}}$. Hence,

$$H'^2(z) = \frac{z^2}{4}\left[\frac{C_1}{1-\alpha z^2}+\frac{C_2}{1-\beta z^2}\right]^2 = \frac{z^2}{4}\left[C_1(1-\alpha z^2)^{-1}+C_2(1-\beta z^2)^{-1}\right]^2$$
$$= \frac{z^2}{4}[C_1^2(1+2\alpha z^2+3\alpha^2 z^4+4\alpha^3 z^6+\ldots)$$
$$+ C_2^2((1+2\beta z^2+3\beta^2 z^4+4\beta^3 z^6+\ldots)$$
$$+ 2C_1C_2(1+\alpha z^2+\alpha^2 z^4+\alpha^3 z^6+\ldots)(1+\beta z^2+\beta^2 z^4+\beta^3 z^6+\ldots)].$$

Thus, $\nu_2' =$ coefficient of z^{2n} in $H'^2(z)$ and is given by the expression

$$\frac{1}{4}[C_1^2 n\alpha^{n-1}+C_2^2 n\beta^{n-1}+2C_1C_2(\beta^{n-1}+\beta^{n-2}\alpha+\beta^{n-3}\alpha^2+\alpha^{n-1})].$$

Let $S' = \beta^{n-1}+\beta^{n-2}\alpha+\beta^{n-3}\alpha^3+\ldots+\beta\alpha^{n-2}+\alpha^{n-1}$. Thus we get:

$$S'(1-\frac{\beta}{\alpha}) = \alpha^{n-1}-\frac{\beta^n}{\alpha}$$
$$\Rightarrow S' = \frac{\alpha^n-\beta^n}{\sqrt{5}}, \qquad \text{since} \quad \alpha-\beta=\sqrt{5}.$$

TABLE 8.21: Comparison of values of ν_0', ν_1' and ν_2' for different values of n

m	n	ν_0'	ν_1'	ν_2'
8	6	954	864	444
16	12	632916	556416	267384
32	23	4.87153×10^{10}	4.22352×10^{10}	1.96227×10^{10}
64	46	4.03077×10^{20}	3.46847×10^{20}	1.5813×10^{20}
128	93	3.6107×10^{40}	3.09529×10^{40}	1.39756×10^{40}
256	185	2.04551×10^{79}	1.75032×10^{79}	7.86544×10^{78}
512	369	3.29995×10^{156}	2.82114×10^{156}	1.2647×10^{156}
1024	738	1.1288×10^{311}	9.64587×10^{310}	4.31886×10^{310}

Hence,

$$\nu_2' = \frac{1}{4}\left[C_1^2 n\alpha^{n-1} + C_2^2 n\beta^{n-1} + 2C_1C_2\frac{(\alpha^n - \beta^n)}{\sqrt{5}}\right]. \tag{8.59}$$

Using the values of ν_0' and ν_2' from Equations (8.58) and (8.59), respectively, the total number of 1's in all n-digit valid codewords is thus given by:

$$\nu_1' = nF_{2n+2} - \nu_0' - \nu_2'. \tag{8.60}$$

The number of bits required for representing all m-bit binary numbers is given by the smallest integer n satisfying the inequality $F_{2n+2} - 1 > 2^m - 1$. Noting that $F_{2n+2} \approx \frac{\phi^{2n+2}}{\sqrt{5}}$ for large n, $\frac{\phi^{2n+2}}{\sqrt{5}} > 2^m$. That is,

$$n \approx \left\lceil \frac{m + \log\sqrt{5}}{2\log\phi} \right\rceil - 1. \tag{8.61}$$

For $m = 8$, 16, 32, 64, 128, 256, 512 and 1024, the corresponding values of n are 6, 12, 23, 46, 93, 185, 369 and 738, respectively. The ratio $\frac{n}{m}$ asymptotically tends to 0.72. In Table 8.21, we have shown the values of ν_0', ν_1' and ν_2' for some values of n with $m = 8$, 16, 32, 64, 128, 256, 512 and 1024. It appears that for large n, the frequencies of occurrences of 0, 1 and 2 in all the valid codewords of TFNS asymptotically tend to 42.2%, 38.2% and 19.6%, respectively. This fact, together with the reduction in lengths of codewords compared to binary numbers, may be exploited in many engineering applications, e.g., in reducing substantial amount of energy consumed by both the transmitters and receivers in wireless communication systems.

8.6.3 Energy-efficient Communication Using TFNS

We apply TFNS to communication systems to save energy during communication. Digital data in binary format are first converted to decimal form and the decimal number is encoded to the tri-digit Fibonaccian number using the appropriate conversion algorithm. If digital data are transmitted in this encoded form, i.e., TFNS-based format, the numbers of 1's and 2's in such data streams for large values of data are substantially reduced. The numbers of 0's, 1's and 2's in such data streams are found to be in the ratio 44 to 38.2 to 17.8, respectively as against 50 to 50 for 0's and 1's in normal binary coding.

8.6.3.1 Implementation of TFNS Protocol

The TFNS transmission scheme involves the recoding of an outgoing binary message to its equivalent decimal form and converting the decimal number into a tri-digit Fibonaccian number and transmitting it. The receiver then reconverts the received tri-digit Fibonaccian number to a decimal number and decodes into a binary number and passes it onto the higher layers of the communication stack for further processing. As in the case of TSS, we use a hybrid modulation scheme involving FSK and ASK. In the same way, the transmitter uses FSK modulation with two frequencies, such as f_c and $f_c + \Delta f$ corresponding to the symbols 1 and 2, respectively. The transmitter is switched to *sleep* state during the transmission of the silent symbol (0's). As in the case of TSS protocol, Figure 8.10 may be used as a receiver for implementation of the TFNS scheme based on non-coherent detection. We can use coherent detection at the receiver for the improvement of noise immunity. Figure 8.16 shows a coherent receiver.

We assume that the channel noise is *additive white Gaussian noise* (AWGN). Also, as in the RBNSiZeComm and TSS protocols, we would use FER instead of BER for the performance analysis of the TFNS scheme.

8.6.3.2 Performance Analysis

Referring to Figure 8.10, let r be the total signal voltage at the input of the receiver. The total signal voltage r is the sum of the desired received signal s and the channel noise n. Obviously, $r = s + n$. The channel noise is assumed to be AWGN. As in the case of TSS, we assume from Figure 8.10 that the frequencies f_c and $f_c + \Delta f$ will correspond in this case to 1 and 2, respectively. Also, q_A and q_B will now be the binary outputs of the threshold detectors corresponding to the symbols 1 and 2, respectively. Thus, in this case, the desired output $\hat{s}$ of the receiver can be expressed as follows:

i) If $q = 0$, $\hat{s} = 0$, corresponding to $s_A = s_B = 0$.

ii) If $q = 1$,

$$\hat{s} = \begin{cases} 1 & \text{if } e_A > e_B, \text{(when } s_A = 1,\ s_B = 0) \\ 2 & \text{if } e_B > e_A, \text{(when } s_B = 1,\ s_A = 0). \end{cases} \tag{8.62}$$

We are considering a message frame containing m binary bits. The binary message frame is converted to the TFNS form having a length of n digits. Hence, $n = 0.72m$. We know that for a binary frame BER $\approx$ FER/m.

8.6.3.3 Non-coherent Detection

Let α be the probability that a transmitted ternary symbol was a 0 and was also received correctly as a 0. Similarly, let β (or, δ) be the probabilities that a transmitted symbol was a 1 (or 2) and the received symbol is also a 1 (or 2). Proceeding in the same way as in the case of TSS, the *frame-correct rate (FCR)* is given by:

$$FCR = \alpha^{nP_0} \times \beta^{nP_1} \times \delta^{nP_2}. \tag{8.63}$$

where P_0, P_1 and P_2 are the probabilities of occurrences of 0, 1 and 2, respectively, in the transmitted message. The average number of 0's in the TFNS frame asymptotically (for large m) is 44% and those of all non-zero symbols, i.e., 1 and 2 taken together will be 56%. That is, $P_0 = 0.44$ and $P_1 + P_2 = 0.56$. From symmetry, $\beta = \delta$. Therefore, the probability that a frame in TFNS will be received correctly is given by:

$$FCR = \alpha^{0.44n} \times \beta^{0.56n}. \tag{8.64}$$

Hence, the FER for a TFNS frame is $FER_{tfns} = 1 - FCR$.

8.6.3.4 Performance Evaluation

Assume we transmit the original binary frame of m bits using non-coherent BFSK and the BER of the non-coherent BFSK is ρ. The probability that the whole frame will be received correctly is given by $(1 - \rho)^m$. Thus, the FER for BFSK will be $FER_{bfsk} = 1 - (1 - \rho)^m$. This may be approximated as $m\rho$, because ρ is usually very small. For a given frame size m, one can compare FER_{tfns} with $FER_{bfsk}(= m\rho)$. To know the performance of TFNS with respect to non-coherent BFSK we can compare FER_{tfns}/m with ρ. To compute FER_{tfns}/m, we denote $s_A = 1$, i.e., signal power = $s_A^2/2 = 1/2$. This corresponds to a signal-to-noise ratio (SNR) = $1/2\sigma^2$. We used MATLAB to implement the computations to find the optimum value of threshold voltage e_{th} that gives the minimum FER_{tfns}/m for a given SNR value. From Equation (8.64), we get the required peak transmitter power with TFNS for a given value of FER_{tfns}.

For non-coherent BFSK, the BER is $\rho = \frac{1}{2}e^{-\gamma/2}$, where γ is the signal-to-noise ratio (SNR). We use different values of frame size in the range $8 \leq m \leq 1024$ and the expressions for X_1 and X_2 to obtain SNR values for the TFNS scheme with a non-coherent receiver and the corresponding SNR values for non-coherent BFSK to meet the same FER values. The results are summarized in Table 8.22; on average, for a given FER/m, the SNR with the TFNS is nearly 2.68 dB higher than that with binary non-coherent FSK.

TABLE 8.22: Comparison of SNR of binary FSK with TFNS scheme for different values of m

SNR with TFNS (non-coherent)	Corresponding SNR with non-coherent BFSK for different frame sizes (m)						Average SNR with non-coherent BFSK
	16	64	128	256	512	1024	
14.0	11.4243	11.4316	11.4412	11.4602	11.4695	11.5686	11.4557
15.0	12.3623	12.3634	12.3649	12.3678	12.3736	12.3852	12.3678
16.0	13.3084	13.3086	13.3087	13.309	13.3096	13.3107	13.309
17.0	14.2566	14.2566	14.2567	14.2567	14.2567	14.2568	14.2566
18.0	15.2176	15.2176	15.2176	15.2176	15.2176	15.2176	15.2176

In the TFNS scheme, the transmitter will be on only during the non-zero symbols (1 and 2). The transmitter will be switched off during the zero symbols in a TFNS-encoded message. Thus, the required average transmitter power for a given BER will be less than the value demanded by Equation (8.64). We know that $P_0 = 0.44$ and $10\log_{10}(1 - 0.44)$ dB $= -2.518$ dB. Hence, the average transmitter power for TFNS will be reduced from the values given in Table 8.22 by 2.518 dB. It follows that the average transmitter power P_1 for TFNS communication will be greater than the corresponding average power P_b for non-coherent BFSK by $(2.68 - 2.518)$ dB = 0.162 dB. It makes $\frac{P_1}{P_b} = 1.038$. Figure 8.25 plots the scaled peak transmitter power for TFNS and BFSK.

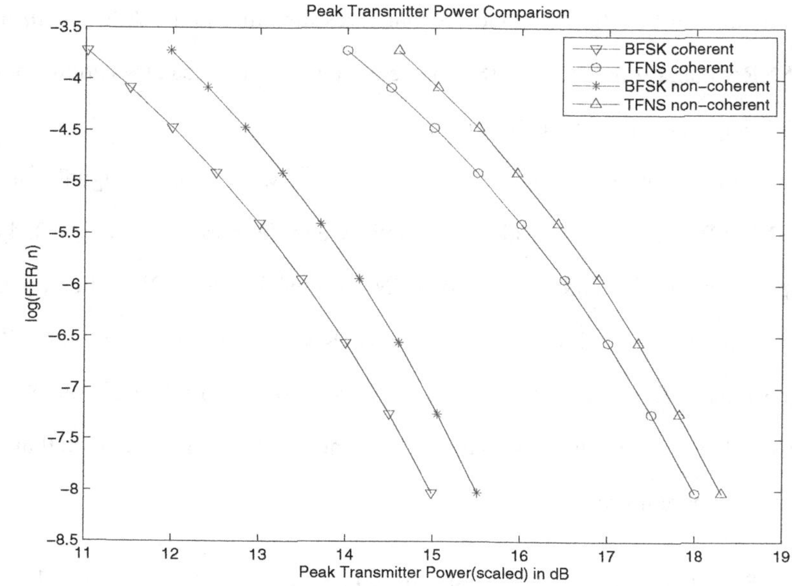

FIGURE 8.25: FER versus peak transmitter power

8.6.3.5 Energy Savings Analysis

Now we analyze the extent of energy savings in non-coherent detection using TFNS. We assume that the symbol duration of the TFNS-encoded message in TFNS is same as that in BFSK. It means that the transmission duration of the TFNS message will be smaller than that of the equivalent binary message due to encoding. Considering T_b and T_1 as the transmission duration of the binary message and its equivalent ternary message, respectively, we can write $\frac{T_1}{T_b} = \frac{n}{m}$. This is asymptotically (for large m) equal to 0.72.

If we denote the total (scaled) energy in TFNS with non-coherent receiver as E_1 and the energy in non-coherent BFSK as E_b,

$$\frac{E_1}{E_b} = \frac{P_1 \times T_1}{P_b \times T_b} = 1.038 \times 0.72 = 0.7474 \tag{8.65}$$

At the same time, because of the reduced value of n in the TFNS-encoded message, there is energy saving at the front-end amplifier section of the receiver. This energy saving in the receiver is given by $1 - 0.72 = 0.28$. Thus, we get:

Result 7. *The above implementation scheme of TFNS will result in energy savings of approximately* 25.26% *at the transmitter and* 28% *at the front-end amplifier section of the receiver simultaneously, compared to non-coherent BFSK.* □

8.6.3.6 Coherent Detection

We expect further improvement in energy savings if coherent detection is used in place of non-coherent detection in receivers. The receiver for coherent detection will be identical to that for coherent TSS as shown in Figure 8.16. The expression for the frame correct rate FCR in this case will be similar to the Equation (8.66) with appropriate values of the number of TFNS symbols in the message and the probabilities P_0, P_1 and P_2. Thus, for coherent TFNS, we get

$$FCR = \alpha_c^{0.44n} \times \beta_c^{0.56n} \tag{8.66}$$

Proceeding as in the TSS scheme, the SNR value required for a given FER/m can be computed. These SNR values and their corresponding FER/m values of TFNS with coherent detection are presented in Table 8.23 using $m = 128$.

TABLE 8.23: Comparison of peak and relative transmitter energy requirements in TFNS scheme

FER/m value	Required peak SNR in db		Relative energy units required for			
	TFNS (Coherent)	BFSK (Coherent)	TFNS (Coherent)	TFNS (Non-Coherent)	BFSK (Coherent)	BFSK (Non-Coherent)
1.88448×10^{-4}	14.0	11.02	9.502	10.14	11.02	11.9791
8.39127×10^{-5}	14.5	11.51	9.992	10.5643	11.51	12.4023
3.39341×10^{-5}	15.0	12.0051	10.4871	11.00	12.0051	12.8378
1.23487×10^{-5}	15.5	12.5	10.982	11.423	12.5	13.261
3.9986×10^{-6}	16.0	12.9977	11.4797	11.8681	12.9977	13.7061
1.13717×10^{-6}	16.5	13.4924	11.9744	12.3113	13.4924	14.1493
2.72429×10^{-7}	17.0	13.996	12.478	12.7625	13.996	14.6005
5.59824×10^{-8}	17.5	14.4955	12.9775	13.2153	14.4955	15.0533
9.5862×10^{-9}	18.0	14.9938	13.4758	13.6709	14.9938	15.5089

8.6.3.7 Energy Savings Analysis

Table 8.23 indicates that for different FER/m values in the range 10^{-4} to 10^{-9}, the peak transmitter power for TFNS with coherent detection is on average 3 dB higher than that with coherent BFSK. Now, $10\log_{10}(1 - 0.44)$ dB = -2.518 dB. Hence, the average transmitter power for TFNS with coherent detection will be $(3 - 2.518)$ dB = 0.482 dB higher than that with coherent BFSK scheme. We denote the average transmitter powers for coherent TFNS and coherent BFSK by P_t and P_b', respectively. Hence, we have $\frac{P_t}{P_b'} = 10^{0.482/10} = 1.1174$.

We denote E_t and E_b' as the transmitter energy requirements for the coherent TFNS and the coherent BFSK schemes, respectively. The reduction in transmission time is given by

$$\frac{E_t}{E_b'} = \frac{P_t \times T_t}{P_b' \times T_b} = 1.1174 \times 0.72 = 0.8045. \tag{8.67}$$

We consider FER/m in the range 10^{-4} to 10^{-8}. The comparative figures for relative energy requirements with non-coherent TFNS, coherent TFNS, coherent BFSK and non-coherent BFSK are shown in Table 8.23. From the values in this table, we see that the peak transmitter power for TFNS scheme with coherent detection is on average 2.28 dB higher than that with non-coherent BFSK and the average transmitter power for TFNS with coherent detection will be $(2.28 - 2.518)$ dB = - 0.238 dB (i.e., 0.9466 times) higher than that with non-coherent BFSK. Since $0.9466 \times 0.72 = 0.6815$, the TFNS scheme with coherent detection will require 0.6815 times the energy needed for non-coherent BFSK, indicating a savings of 31.85% of transmitter energy.

Result 8. *The above implementation scheme of TFNS with a coherent detection-based receiver will result in saving transmitter energy by approximately* 31.85% *and* 19.55% *relative to non-coherent BFSK and coherent BFSK schemes, respectively,*

TABLE 8.24: Comparison of energy savings among various schemes

Communication scheme	Energy savings at transmitter (%)	Energy savings at receiver (%)
RBNSZeComm	53	0
RZE	35.2	12.5
TSS (non-coherent)	23	36.9
TSS(coherent)	30	36.9
CNS	30	50
Non-coherent TFNS	25.26	28
Coherent TFNS	31.85	28

along with about 28% *energy saving at the front-end amplifier section of the receiver.*

□

We can compare transmitter and receiver energy savings (at the front-end amplifier section) for non-coherent TFNS, coherent TFNS, RBNsZeComm, RZE, non-coherent TSS, coherent TSS and CNS relative to non-coherent BFSK, as shown in Table 8.24.

We see that while the transmitter side saving for TFNS is less than that of RBNSiZeComm and RZE, it saves more energy at the front-end amplifier of the receiver section than they do.

We have presented several techniques for energy-efficient communication at the physical layer and it will be interesting to investigate further to find alternative means (possibly through source coding and/or channel coding) to achieve substantial savings in transmitter and receiver energy savings. Further investigations of energy-efficiency at the MAC layer, network layer and application layer are also very much needed.

8.7 Exercises

1. Show how the decimal number 222 will be represented in RBNS form after applying i) reduction rule 1 and ii) both reduction rules 1 and 2 as mentioned in the text. What will be the corresponding number of RBNS digits for which the transmitter will be in *transmit* state following the RBNSiZeComm protocol?

2. Describe the logic and draw the corresponding circuit diagram for converting a given binary string to its equivalent RBNS form suitable for use in the RBNSiZeComm protocol.

3. Describe the logic and draw the corresponding circuit diagram for deriving the binary string from the received RBNS digits (from the detector in the receiver circuit used by the RBNSiZeComm protocol).

4. Write a simple algorithm to convert a binary number into its equivalent ternary form that would not involve any division or multiplication operation in the hardware implementation.

5. Write a simple algorithm to convert a ternary number into its equivalent binary form that would not involve any division or multiplication operation in the hardware implementation.

6. Among the RBNSiZeComm, TSS and CNS protocols, which one do you think the most suitable in low-power sensor network applications and why?

7. Why is FER considered to be more appropriate than BER in performance analysis of receivers in the RBNSiZeComm protocol? How can FER be finally related to conventional BER for comparing the performance of RBNSiZeComm with that of conventional BFSK?

8. Why is FER considered to be more appropriate than BER in performance analysis of receivers in TSS protocol?

*9. Devise a suitable source coding technique and the associated communication protocol to obtain more energy savings for a transmitter and/or receiver than any of the techniques described in the text.

Bibliography

[1] K. Sinha, "An energy efficient communication scheme for applications based on low power wireless networks," *Proc. 6th IEEE Consumer Comm. and Networking Conf.*, Las Vegas, pp. 1–5, Jan. 2009.

[2] K. Sinha and B. P. Sinha, "On the distribution of runs of ones in binary strings," *Computers & Mathematics with Applications*, vol. 58(9), pp. 1816–1829, 2009.

[3] R. N. Ghosh, K. Sinha, B. P. Sinha and D. Datta, "TSS: an energy efficient communication scheme for low power wireless networks," *Proc. IEEE Intl. Perf. Comp. & Comm. Conf.*, pp. 85–92, Dec. 2008.

[4] R. N. Ghosh, K. Sinha, D. Datta and B. P. Sinha, "Energy-efficient communication in multihop wireless sensor networks using encoding and silent symbols," *International Journal of Communication Networks and Distributed Systems*, vol. 11, No. 4, pp. 376–389, 2013.

[5] K. Sinha and B. P. Sinha, "An energy efficient communication scheme for distributed computing applications in wireless sensor networks," *Proc. Intl. Conf. on Dist. Comp. & Internet Tech.*, pp. 139–144, Dec. 2008.

[6] K. Sinha, B. P. Sinha and D. Datta, "CNS: a new energy efficient transmission scheme for wireless sensor networks," *Wireless Networks Journal*, vol. 16(8), Nov. 2010, pp. 2087-2104.

[7] K. Sinha and P. K. Srimani, "Broadcast and gossiping algorithms for mobile ad hoc networks based on breadth-first traversal," *Proc. 6th International Workshop on Distributed Computing, Lecture Notes in Computer Science 3326, Springer*, pp. 459–470, Dec. 2004.

[8] R. Gandhi, S. Parthasarathy and A Mishra, "Minimizing broadcast latency and redundancy in ad hoc networks," *Proc. MobiHoc 2003*, pp. 222–232, 2003.

[9] Y. Zhu and R. Sivakumar, "Challenges: communication through silence in wireless sensor networks," *Proc. 11th Annual Intl.Conf. on Mobile Computing and Networking*, pp. 140–147, 2005.

[10] Y. P. Chen, D. Wang and J. Zhang, "Variable-base tacit communication: a new energy efficient communication scheme for sensor networks," *Proc. Intl. Conf. on Integrated Internet Ad Hoc and Sensor Networks*, 2006.

[11] I. Demirkol, C. Ersoy and F. Alagoz, "Mac protocols for wireless sensor networks, a survey," *IEEE Communications Magazine*, 2005.

[12] W. B. Heinzelman, A. Chandrakasan and H. Balakrishnan, "Energy efficient communication protocol for wireless microsensor networks," *Proc. 33rd Intl. Conf. on System Sciences*, Hawaii, 2000.

[13] W. Lou and J. Wu, "On reducing broadcast redundancy in ad hoc wireless networks," *Proc. 36th Hawaii Int. Conf. on System Sciences*, 2003.

[14] M. Cagalj, J.-P. Hubaux and C. C. Enz, "Energy-efficient broadcasting in all-wireless networks," *Wireless Networks*, vol. 11, nos. 1-2, pp. 177–188, 2005.

[15] S. Basagni, D. Bruschi and I. Chlamtac, "A mobility transparent deterministic broadcast mechanism for ad hoc networks," *IEEE Trans. Networking*, vol. 7, pp. 799–807, Dec. 1999.

[16] C. Intanagonwiwat, R. Govindan, D. Estrin, J. S. Heidemann and F. Silva, "Directed diffusion for wireless sensor networking," *IEEE/ACM Trans. on Networking*, vol. 11(1), pp. 2–16, Feb. 2003.

[17] W. Ye, J. Heidemann and D. Estrin, "An energy-efficient MAC protocol for wireless sensor networks," *Proc. IEEE Infocom*, pp. 1567–1576, 2002.

[18] Y. Li, W. Ye and J. Heidemann, "Energy and latency control in low duty cycle MAC protocols," *Proc. IEEE Wireless Communications and Networking Conference*, vol.2, pp. 676–682, 2005.

[19] C. Enz, A. El-Hoiydi, J.-D. Decotignie and V. Pereis, "WiseNET: an ultralow-power wireless sensor network solution," *IEEE Computer*, vol. 37, no. 8, pp. 62–70, 2004.

[20] Y. P. Chen A. L. Liestman and J. Liu, "Energy-efficient data aggregation hierarchy for wireless sensor networks," *Proc. QShine '05*, Orlando, 2005.

[21] N. Tagaki, H. Yassura and S. Yajima, "High-speed VLSI multiplication algorithm with a redundant binary addition tree," *IEEE Trans. Comp.*, vol. C-34, pp. 789–796, 1985.

[22] J. Polastre, R. Szewczyk and D. Culler, "Telos: enabling ultra-low power wireless research," *Proc. 4th Intl. Symp. on Information Processing in Sensor Networks*, pp. 364–369, 2005.

[23] D. Jea, J. Liu, T. Schmid and M. Srivastava, "Hassle free fitness monitoring," *2nd Intl. Workshop on Systems and Networking Support for Healthcare and Assisted Living Environments*, June, 2008.

[24] J. Burrell, T. Brooke and R. Beckwith, "Vineyard computing: sensor networks in agricultural production," *IEEE Pervasive Computing*, vol. 3(1), pp. 38–45, Jan.–Mar., 2004.

[25] R. Beckwith, D. Teibel and P. Bowen, "Report from the field: results from an agricultural wireless sensor network," *Proc. 29th Annual IEEE Intl. Conf. Local Comp. Networks*, 2004.

[26] P. Mestre, E. Peres and C. Serôdio, "Agricultural monitoring and control system using low-cost wireless sensors networks," *Proc. 5th EFITA/WCCA*, pp. 837–844, 2005.

[27] P. Laguna et. al., "A database for evaluation of algorithms for measurement of QT and other waveform intervals in the ECG," *Comp. in Cardiology*, (24), pp. 673–676, 1997.

[28] Systemic Pathology Laboratory, `http://www.tigerpath.com/unit_08.htm`, 2008.

[29] J. G. Proakis, *Digital Communication*. McGraw-Hill, 4th ed., 2000.

[30] O. Mirabella, M. Brischetto, A. Raucea, and P. Sindoni, "Dynamic continuous clock synchronization for IEEE 802.15.4 based sensor networks," *Proc. IEEE Annual Conf. on Industrial Electronics*, pp. 2438–2444, 2008.

[31] P. Song, X. Shan, K. Li and G. Qi, "Highly precise time synchronization protocol for ZigBee networks," *Proc. IEEE/ASME Intl. Conf. on Advanced Intelligent Mechatronics*, pp. 1254–1258, 2009.

[32] A. Apostolico and A. Fraenkel, "Robust transmission of unbounded strings using Fibonacci representations," *IEEE Transactions on Information Theory*, vol. 33, pp. 283–245, 1987.

[33] A. Fraenkel and S. Klein, *Combinatorial Algorithms on Words*. pp. 169–183, Springer, Heidelberg, 1985.

[34] R. N. Ghosh, K. Sinha, B. P. Sinha and D. Datta, "TSS: an energy efficient communication scheme for low power wireless networks," *Proc. 27th IEEE Intl. Performance Computing and Communications Conf.*, pp. 85–92, Dec. 2008.

[35] W. Kautz, "Fibonacci codes for synchronization control," *IEEE Transactions on Information Theory*, vol. 11, pp. 284–292, 1965.

[36] D. E. Knuth, *The Art of Computer Programming, 3rd ed., Volume 1*. Pearson Education, 1997.

[37] M. Mutyam, "Preventing crosstalk delay using Fibonacci representation," *IEEE Intl. Conf. on VLSI Design*, pp. 685–688, 2004.

[38] M.A.Sanders, "Counting bits with Fibonacci and Zeckendorf," *REU program in Mathematics at Oregon State University*, Summer 2006.

[39] K. Sinha, B. P. Sinha and D. Datta, "An energy-efficient communication scheme for wireless networks: A redundant radix-based approach," *IEEE Trans. Wireless Communications*, vol. 10, pp. 550 - 559, 2011.

[40] K. Sinha, R. Ghosh, B. P. Sinha, "A new number system using alternate Fibonacci numbers as the positional weights with Some engineering applications," *Proc. International Conference on Distribut. Comput. Internet Tech.*, Feb. 6-9, 2014, LNCS 8337, pp. 164–175, 2014.

[41] L. Wang and J. Manner, "Evaluation of data compression for energy-aware communication in mobile networks," *Proc. Intl. Conf. on Cyber-Enabled Distributed Computing and Knowledge Discovery*, pp. 69–76, 2009.

Chapter 9

Security in Wireless Communication

9.1 Introduction

As noted in Chapter 1, security is an important concern in all applications of wireless communication. There are different possible types of security threats in a wireless communication network, whether cellular, ad hoc, cognitive radio network, or wireless sensor network. Such threats threaten a network in multiple ways. An attacker may be a node of the network (called an *intruder*), or may be a node outside the network. For example, if we are concerned about message communication amongst the nodes of a MANET, an internal attacker may be a misbehaving or adversary node which is a member of the same network, while an external attacker may be a node which is not an integral part of the network.

9.1.1 Internal Attacks

Internal attacks may be of several types. For example, a node of the network may misbehave in such a way that it receives all messages coming from its neighbors, but never responds to those messages or never forwards them to their destinations. Thus, effectively, this *attacker node* behaves like an infinite sink for all messages.

Another kind of internal attack may be a reverse activity in the form of sending unnecessary or duplicate messages to other nodes of the network which would cause undue flooding of the network with useless messages, consuming a lot of communication bandwidth and node energy, thus disrupting or delaying the message communications. Some nodes may become non-operational if they are depleted of their stored battery energy due to undesirable flooding in the network.

In a third kind of more intelligent attack by an internal node, the attacker selectively suppresses received packets that are not forwarded to their destination nodes, The attacker node may also misdirect the packets and prevent them from reaching their true destinations by changing their destination addresses. Packets may be selected randomly or, more intelligently, by critical examination of their contents and limiting selection to packets containing important information.

A fourth kind of attack may be a little different. The attacker node misdirects a packet with changed route information in the message so that the packet moves around a local loop. Thus, the packet never reaches its destination, but the network resources of communication bandwidth, node energy, etc., are continuously wasted. This kind of attack is more serious than the third kind mentioned above.

In a fifth type, an internal node may be *jamming* by continually transmitting high intensity radio signals of the same frequency as that used by all the nodes in the network so that other nodes are unable to transmit their packets (for collision-free transmission) or nodes are unable to receive packets correctly due to the interference created by noise signals from the attacker node (bit-error rate of the receiver may be reduced due to reduced SNR). To measure the performance of the network under noisy situations, we usually define two parameters:

- **Packet Sending Ratio (PSR)** on the transmitter side, which is equal to s/n, where s is the number of packets actually sent by the transmitter (within a specified time-out period), given a total number of n generated packets (which are intended to be sent).

- **Packet Delivery Ratio (PDR)** on the receiver side, which is equal to r/s, where r is the number of correctly received packets out of the total of s packets sent by the transmitter.

Due to such jamming signals, both *PSR* and *PDR* are likely to be reduced. This type is called *constant jamming* to distinguish it from the other types of jamming which arise from the attack of the second type above. Note that the second type of attack has a flavour of jamming as transmission of regular packets is hampered by the introduction of unnecessary or superfluous or meaningless packets in the network by the attacking internal node. The difference between the two types of jamming is that the second type tries to 'jam' the network with regular packets (generated according to the format dictated by the protocol), instead of generating noise signals for jamming. The jamming effects from the second type of attack fall into three categories: i) *deceptive jamming*, in which regular packets are continually injected in the channel without any gap between transmissions, ii) *random jamming*, in which the jamming packets (of random lengths) are injected at random instants of time, and iii) *reactive jamming*, in which the attacker node continuously senses the channel and transmits a jamming packet as soon as it detects a transmission in the channel, so that a collision is experienced at the receiver node. Naturally, reactive jamming is the most intelligent type as it uses the minimum power from the attacker node to serve its purpose.

All these types of attacks effectively disrupt or degrade the expected services of the network in terms of successful packet transmission rate, energy expenditure of the constituent nodes and routing delays. These may be collectively called the **denial-of-service (DOS)** attack. There are also some other types of internal attacks as mentioned below.

In a sixth type of internal attack, an attacker node modifies the contents of a received packet according to its intention and forwards the tampered packet containing a different message to the destination. This may also be called a *man-in-the-middle* attack. Unless proper preventive measures are taken, such attacks can be disastrous as the receiving node can be completely misguided by this tampered message.

A seventh type of attack is simply pure eavesdropping, i.e., the attacker node silently deciphers a message in a secret packet (which is not intended for the node), and extracts the valuable information it was not supposed to learn and which may be used for adversary activity. Through such eavesdropping, an attacker may also gather data about the nodes which are very important in the network, e.g., which are often used as destinations of other nodes in the network. This information may be later used by the attacker for disrupting the communication to such important nodes.

Usually, the first five types of internal attacks causing denial of service are also called **active attacks**; the last two types are called **passive attacks** in the literature.

9.1.2 External Attacks

An external attack arises from an unauthorized node, i.e., a node which is not a part of the network. In the first type, the external attacker node injects a strong jamming signal in the communication band of a network, similar to *constant jamming* mentioned above. If an external node has enough information (from unauthorized sources) about the physical layer (communication band, modulation and demodulation mechanisms, etc.), MAC layer protocol and network layer protocol, it can generate false packets and insert those packets into the network continuously or at random intervals of time or as and when a packet is being transmitted over the channel, like similar attacks by an internal node. Packets injected by an external attacker node may eventually be ignored if the network has an appropriate authentication technique (which is often the case). The danger is that such an activity consumes a lot of network resources like communication band and node energy. However, considering that authentication will eventually be used by the network for allowing a node to take part in communication, an external attacker actually stands a very small chance of acceptance of its inserted packets by the nodes in the network. Based on this ultimate consequence, an external attacker may better choose to follow only constant jamming, generating random noise signals to jam the network, without taking the trouble of generating false or misguiding packets and trying to inject them in the network. It will suffice for an external attacker to create the desired *denial-of-service* through these jamming signals only.

The second type of attack by an external node is eavesdropping, i.e., the external attacker node is silently listening to the communication channel and deciphers the contents of the packets flowing in the network.

9.2 Measures against Attacks

Usually three types of measures can be taken against the internal and external attacks described above. An external attacker may be prevented from injecting any false or misguiding packets in a network by an appropriate **authentication** mechanism. Unless a node is properly authenticated, it would not be allowed to take part in the communication of the network. A misbehaving or malfunctioning internal node may be called an *intruder* and identifying such nodes by a suitable **intrusion detection** mechanism would allow the system to ignore the packets received from such nodes and not send packets through or to such nodes. Eavesdropping or man-in-the-middle attack may be prevented through a suitable message **encryption** and **decryption** mechanisms as described below.

9.2.1 Authentication

Authentication enables the recipient of a message to verify that the message was created by a node possessing certain key. Authentication depends on i) a set K of keys, ii) a set M of messages, iii) a set A of authenticators, iv) a secret function $S(k), k \in K$ which operates on a message $m \in M$ to generate an authenticator $a \in A$, i.e., $S(k)(m) = a$ and v) a verifying function $V(k)$ which operates on the tuple (m, a) to generate the authenticator a. The underlying idea is that if a node knows the function $S(k)$, it can generate the authenticator a on a message m generated by it, and if another node possesses the function $V(k)$, it can verify whether the tuple (m, a) generates a. If yes, then the node possessing $V(k)$ authenticates the node sending the (message, authenticator) tuple (m, a). A node not having the function $S(k)$ will not be able to correctly generate a from its message m and hence will not be authenticated by the node possessing the function $V(k)$. The authenticators are usually exposed in a network and everybody can see it. Hence, $S(k)$ should not only be secret, but also not be easily derivable from the authenticators.

There can be two types of authentication algorithms. In the first type, known as **message authentication code (MAC)** (not to be confused with the *media access control protocol*), knowledge of $S(k)$ and $V(k)$ are equivalent, i.e., one function can be derived easily from the other. Thus, anyone who knows $V(k)$, can also find $S(k)$ from it, and vice versa. An example is to have $S(k)(m) = f(k, m)$, where the function f cannot be derived easily by knowing k and $f(k, m_1) = f(k, m_2)$, when and only when $m_1 = m_2$. In other words, the function f is collision resistant on m. A corresponding verification algorithm will then be $V(k)(m, a) \equiv (f(k, m) = a)$, i.e., if $f(k, m) = a$, then and only then $V(k)(m, a)$ will produce a true value. Thus, anyone who knows $S(k)$ knows f and also knows $V(k)$ and vice versa. For this reason, it is essential to keep both $S(k)$ and $V(k)$ secret in this scheme.

The other type of authentication algorithm is through **digital signature** in which it is computationally infeasible (or very difficult) to derive $S(k)$ from a knowledge of $V(k)$ and also $V(k)$ is a one-way function, i.e., knowing $V(k)$ makes it infeasible (or very difficult) to compute k. Thus, it is not necessary to keep $V(k)$ secret. $V(k)$ can be thought as a public key and k can be treated as a private key in asymmetric key cryptography.

As an example of the digital signature technique for authentication, assume that we use RSA, one of the most popular encryption algorithms introduced by Rivest, Shamir and Adleman in 1978. We assume that a large number N is publicly known where N is a product of two large prime numbers p and q. It is very difficult for even the most powerful computer to compute p and q within a reasonable time if it knows only N. We form a secret key k as $k = (d, N)$, where d is a private key although N is publicly known. The signature algorithm to be executed by a node to be authenticated is $S(d, N)(m) = m^d \ mod \ N$ (assuming that $m^d \ mod \ N$ is a collision-resistant function of d). Thus, $a = m^d \ mod \ N$. The corresponding verification function will be $V(d, N)(m, a) \equiv (a^e \ mod \ N = m)$, where $ed \ mod \ (p-1)(q-1) = 1$. Even if a node knows e and N, i.e., the verification function, it is computationally infeasible or very difficult to compute d from the knowledge of e and N, as p and q are difficult to compute from N. That is, the function S cannot be computed from the knowledge of the function V.

9.2.2 Intrusion Detection

Intrusion detection techniques depend on detecting deviations of the characteristics of an internal node. If we gather statistical data about the contribution of a node to the various network function related parameters, an idea about the normal behavior of that node can be estimated. Any deviation from this estimated normal behavior might be considered an abnormal behavior or an anomaly that might be used for identifying this node as an intruder. The network function-related parameters that might be used for anomaly detection include packet drop ratio, delay in packet forwarding and successful forwarding of packets following the designated route (if route is computed at the source). In fact, this depends on the statistical behavior of nodes over a long period and appropriate modeling of the behavior of each constituent node. Inference from such analysis may require a lot of engineering. The potential problems with these techniques involve failure to detect a *true intruder* and *false* identification of a good node as an intruder.

9.2.3 Message Encryption

To guard against eavesdropping or man-in-the-middle attack, it is necessary that the source node properly encrypts its message m which can be decrypted only by its intended recipient. Thus, the message cannot be known by an external attacker or an intermediate internal node. As a consequence, this message cannot be changed by the attacker. There are two different types of encryption algorithms. The first is based on a **symmetric key encryption** in which every pair of sender and receiver nodes share a common key which is kept secret from all other nodes. The example is the **Data Encryption Standard (DES)** adopted by the National Institute of Standards and Technology (NIST) in the U.S. and largely used for civilian applications . However, in 2001, DES was considered insecure for many applications as it used key values only 56 bits long which can be exhaustively searched with available computing resources. DES was replaced by the **Advanced Encryption Standard (AES)** which works on blocks of 128 bits and key lengths of 128 bits to 256 bits.

The main problem with the symmetric key encryption is key management. If there are n nodes in the network, the number of possible communicating pairs will be $n(n-1)/2$, and thus we need $O(n^2)$ secret keys. Managing such a large number of secret keys, one for each pair of communicating nodes, is a problem in practice when the security of communication channel is questionable.

The other type of encryption algorithm is based on **asymmetric key encryption** which started with a ground-breaking idea of Diffie and Hellman in 1976. They proposed the notion of public key cryptography involving two different keys, one *public key* known to all and the other a *private key* known only to the recipient of a message. Anyone can use the public key to encrypt a message, but someone who has the private key can only decrypt the message. Thus, effectively, for all n nodes, we need to manage only n secret keys, far better than the $O(n^2)$ secret keys required for symmetric key encryption. The success of this idea is heavily dependent on the proposition that for any communicating pair, even if the public key is known, specifically to the attacker, it is not possible for any party to compute the secret private key from the public key. The **RSA algorithm** is the first successful encryption algorithm based on the above idea.

The RSA algorithm is applicable for encryption of a block of data or message bits. In many real-life applications, message bits are transmitted as streams, for example, in a wireless communication channel. For such applications, a different class of encryption algorithms, commonly known as **stream ciphers** are needed which differ from the so-called **block ciphers** like the RSA algorithm.

Stream ciphers are broadly classified into two categories depending on the platform most suited to their implementation: software stream ciphers and hardware stream ciphers. RC4 is one of the widely used stream ciphers implemented in software. This cipher is used in network protocols such as SSL, TLS, WEP and WPA. The cipher also finds applications in Microsoft Windows, Lotus Notes, Apple AOCE and Oracle Secure SQL. Though several other efficient and secure stream ciphers were devised after RC4, it is still the most popular stream cipher algorithm due to its simplicity, ease of implementation, and speed.

9.2.4 RC4 Stream Cipher

The RC4 stream cipher was designed by Ron Rivest for RSA Data Security in 1987. It uses S-box S, an array of length N, where each location of S stores one byte (typically, $N = 256$). A secret key k of size l bytes is used to scramble this permutation (typically, $5 \leq l \leq 16$). Array K of length N holds the main key, with secret key k repeated as $K[y] = k[y \bmod l]$, for $0 \leq y \leq N-1$.

RC4 has two components, namely the key scheduling algorithm (KSA) and the pseudo-random generation algorithm (PRGA). The KSA uses the key K to generate a pseudo-random permutation S of $\{0, 1, \ldots, N-1\}$ and PRGA uses this pseudo-random permutation to generate arbitrary number of pseudo-random keystream bytes. The procedures are in Algorithms 1 and 2, respectively.

Algorithm 1 Key Scheduling Algorithm

procedure KSA(Secret Key K)
  Initialize $S \leftarrow \{0, 1, \ldots, N-1\}$ and $j \leftarrow 0$
  for $i = 0, \ldots, N-1$ **do**
    Increment: $j \leftarrow j + S[i] + K[i]$
    Swap: $S[i] \leftrightarrow S[j]$
  end for
  return S-box S
end procedure

Algorithm 2 Pseudo-Random Generation Algorithm

procedure PRGA(S-box S)
  Initialize indices: $i \leftarrow 0, j \leftarrow 0$
  while TRUE **do**
    Increment: $i \leftarrow i + 1, j \leftarrow j + S[i]$
    Swap: $S[i] \leftrightarrow S[j]$
    output $Z \leftarrow S[S[i] + S[j]]$
  end while
end procedure

Any arithmetic addition used in context of RC4 is in general addition modulo N, unless specified otherwise. The output keystream Z is XOR-ed with the plaintext (byte per byte) to generate the ciphertext at the sender end ($C = M \oplus Z$), and is XOR-ed with the ciphertext to get back the plaintext at the receiver end ($M = C \oplus Z$).

Efficiency of keystream throughput is used as a benchmarking parameter for stream ciphers. The efficiency of the RC4 obviously depends on the efficiency of KSA and PRGA. While the KSA invokes a fixed cost for generating the initial pseudo-random state S, the PRGA incurs a variable cost based on the number of keystream bytes to be generated. An efficient implementation of RC4 would minimize the cost per round of KSA and PRGA to provide better throughput. The software implementation of RC4 is simple, and a detailed comparison of the software performance of eSTREAM and RC4 is given in [2]. It is rumored in the cryptographic community that an efficient RC4 software implementation can produce keystream bytes at a rate of 3 cycles per byte. However, there is no documented evidence to prove this claim. The best software performance of RC4 stated in the literature [2] provides an account of 7 to 15 cycles per byte of RC4 keystream, depending on the processor and the clock speed on which the cipher was tested. This rate is slow for typical wireless communication applications.

In contrast to the software implementation, a hardware implementation of the RC4 stream cipher may possibly be faster and find potential applications in encrypting and decrypting message streams in a wireless communication network. A number of attempts for efficient hardware implementation of RC4 have already been reported in the literature.

A 3-cycle-per-byte efficient implementation of RC4 on custom pipelined hardware was proposed by Kitsos et al. [11] in 2003. Later, a patent by Matthews [17] was disclosed and provided a similar 3-cycles-per-byte architecture using hardware pipelining. Five years later, Matthews disclosed another patent [18] that included a design for RC4 hardware based on pipeline architecture that could increase the efficiency of the cipher to obtain 1-byte-per-cycle in RC4 PRGA. In 2013, Sen Gupta et al. [1] introduced a hardware implementation of RC4 that also produced 1 byte per cycle as in the Mathews' patent [18]. We describe below a hardware implementation of RC4 [1] which is based on loop unfolding and instruction rescheduling with a three-stage pipelining, resulting in less hardware cost and complexity than in [18]. Note that wireless communication devices are often constrained by the computing resources and optimization of resources and speed are always beneficial.

The approach followed in [1] is based on twofold unrolling of the loop in the PRGA algorithm, rescheduling the instructions in the two consecutive iterations of the loop, and generating one cipher byte per cycle in a three-stage pipelined fashion. Section 9.3 describes the detailed design strategy and analysis of the circuit as presented in [1].

9.3 Hardware Implementation of RC4 for One Byte per Clock

We consider the generation of two consecutive values of Z together for the two consecutive plaintext bytes to be encrypted. Assume that the initial values of the variables i, j and S are i_0, j_0 and S_0, respectively. After the first execution of the PRGA loop, these values will be i_1, j_1 and S_1, respectively, and the output byte is Z_1, say. Similarly, after the second execution of the PRGA loop, these will be i_2, j_2, S_2 and Z_2, respectively. Thus, for the first two loops of execution to complete, we have to perform the operations shown in Table 9.1.

TABLE 9.1: Two consecutive loops of RC4 stream generation

First Loop	Second Loop
$i_1 = i_0 + 1$	$i_2 = i_1 + 1 = i_0 + 2$
$j_1 = j_0 + S_0[i_1]$	$j_2 = j_1 + S_1[i_2] = j_0 + S_0[i_1] + S_1[i_2]$
Swap $S_0[i_1] \leftrightarrow S_0[j_1]$	Swap $S_1[i_2] \leftrightarrow S_1[j_2]$
$Z_1 = S_1[S_0[i_1] + S_0[j_1]]$	$Z_2 = S_2[S_1[i_2] + S_1[j_2]]$

9.3.1 Design of Individual Components

To store S-array in hardware, a bank of 8-bit registers, 256 in total are used. The output lines of any one of these 256 registers can be accessed through a 256-to-1 multiplexer (MUX), with its control lines set to the required address i_1, j_1, i_2 or j_2. Thus, we need four 256-to-1 MUX units to simultaneously read $S[i_1]$, $S[i_2]$, $S[j_1]$ and $S[j_2]$. We now explain how to compute the increments of i and j at each level.

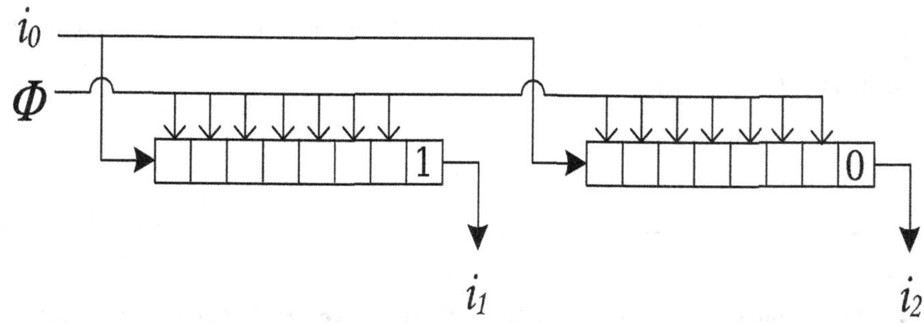

FIGURE 9.1: Circuit to compute i_1 and i_2

Step 1: Calculation of i_1 and i_2. Incrementing i_0 by 1 and 2 can be done by the same clock pulse applied to two synchronous 8-bit counters. The counter for i_1 is initially loaded with 00000001 and the counter for i_2 is loaded with 00000010, the initial states of these two indices. This serves the first two rounds of RC4, in both KSA and PRGA.

Thereafter, in every other cycle, the clock pulse is applied to all the flip-flops except the ones at the LSB position for both the counters, as shown in Figure 9.1. This will result in proper increments of i_1 that assume only the odd values $1, 3, 5, \ldots$, and for i_2 assuming only the even values $2, 4, 6, \ldots$, as required in RC4. This is assured as the LSB of i_1 will always be 1 and that of i_2 will always be 0 as shown in Figure 9.1.

Step 2: Calculation of j_1 and j_2. The values of j_1 and j_2 will be computed and stored in two 8-bit registers. To compute j_1, we need a 2-input parallel adder unit. It may use a carry lookahead adder or use *scan* operation as proposed by Sinha and Srimani [24], or use *carry-lookahead-tree* as proposed by Lynch and Swarzlander [13]. For computing j_2, there are two special cases:

$$j_2 = j_0 + S_0[i_1] + S_1[i_2] = \begin{cases} j_0 + S_0[i_1] + S_0[i_2] \text{ if } i_2 \neq j_1 \\ j_0 + S_0[i_1] + S_0[i_1] \text{ if } i_2 = j_1. \end{cases}$$

The only change from S_0 to S_1 is the swap $S_0[i_1] \leftrightarrow S_0[j_1]$, and hence we need to check whether i_2 is equal to either of i_1 or j_1. Now, i_2 cannot be equal to i_1 as they differ only by 1 modulo 256. Therefore, $S_1[i_2] = S_1[j_1] = S_0[i_1]$ if $i_2 = j_1$, and $S_1[i_2] = S_0[i_2]$ otherwise. In both cases, three binary numbers are to be added.

Let us denote the kth bit of j_0, $S_0[i_1]$ and $S_1[i_2]$ (either $S_0[i_2]$ or $S_0[i_1]$) by a_k, b_k and c_k, respectively, where $0 \le k \le 7$. We first construct two 9-bit vectors R and C, where the kth bits ($0 \le k \le 8$) of R and C are given by:

$$R_k = \text{XOR}(a_k, b_k, c_k) \text{ for } 0 \le k \le 7,\ R_8 = 0,\ C_0 = 0,$$

$$C_k = a_{k-1}b_{k-1} + b_{k-1}c_{k-1} + c_{k-1}a_{k-1} \text{ for } 1 \le k \le 8.$$

In RC4, all additions are done modulo 256. Hence, we can discard the ninth bit ($k = 8$) of the vectors R, C while adding them together, and carrying out normal 8-bit parallel addition considering $0 \le k \le 7$. Therefore, one may add R and C by a parallel full adder as used for j_1. The circuit to compute j_1 and j_2 is shown in Figure 9.2.

TABLE 9.2: Different cases for register-to-register transfers in swap operation

#	Condition	Register-to-Register Transfers
1 1	$i_2 \neq j_1$ & $j_2 \neq i_1$ & $j_2 \neq j_1$	$S_0[i_1] \to S_0[j_1]$, $S_0[j_1] \to S_0[i_1]$, $S_0[i_2] \to S_0[j_2]$, $S_0[j_2]$ $\to S_0[i_2]$
2	$i_2 \neq j_1$ & $j_2 \neq i_1$ & $j_2 = j_1$	$S_0[i_1] \to S_0[i_2]$, $S_0[i_2] \to S_0[j_1] = S_0[j_2]$, $S_0[j_1] \to S_0[i_1]$
3	$i_2 \neq j_1$ & $j_2 = i_1$ & $j_2 \neq j_1$	$S_0[i_1] \to S_0[j_1]$, $S_0[i_2] \to S_0[i_1] = S_0[j_2]$, $S_0[j_1] \to S_0[i_2]$
4	$i_2 \neq j_1$ & $j_2 = i_1$ & $j_2 = j_1$	$S_0[i_1] \to S_0[i_2]$, $S_0[i_2] \to S_0[i_1] = S_0[j_1] = S_0[j_2]$
5	$i_2 = j_1$ & $j_2 \neq i_1$ & $j_2 \neq j_1$	$S_0[i_1] \to S_0[j_2]$, $S_0[j_2] \to S_0[j_1] = S_0[i_2]$, $S_0[j_1] \to S_0[i_1]$
6	$i_2 = j_1$ & $j_2 \neq i_1$ & $j_2 = j_1$	$S_0[i_1] \to S_0[j_1] = S_0[i_2] = S_0[j_2]$, $S_0[j_1] \to S_0[i_1]$
7	$i_2 = j_1$ & $j_2 = i_1$ & $j_2 \neq j_1$	Identity permutation, no data transfer.
8	$i_2 = j_1$ & $j_2 = i_1$ & $j_2 = j_1$	Impossible, as it implies $i_1 = i_2 = i_1 + 1$.

Step 3: Swapping the S values. In Table 9.1, the two swap operations in the third row result in one of eight possible data transfer requirements of the S-register bank, depending on the possible values of i_1, j_1, i_2 and j_2. We have to check whether i_2 and j_2 can be equal to i_1 or j_1 (we only know that $i_2 \neq i_1$). All the cases in this direction can be listed as in Table 9.2. A more detailed explanation for each case is presented as follows.

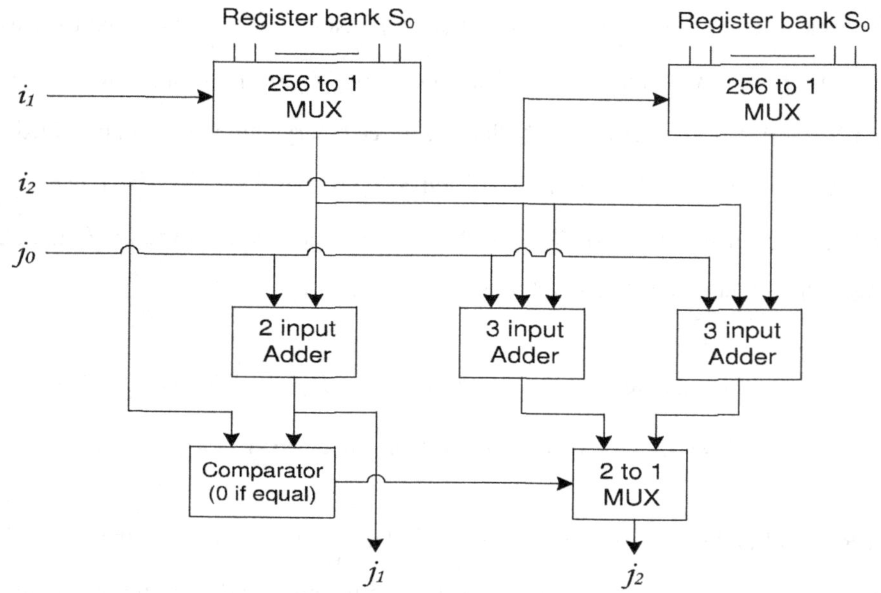

FIGURE 9.2: Circuit to compute j_1 and j_2

Case 1: $i_2 \neq j_1$ and $j_2 \neq i_1$ and $j_2 \neq j_1$

These data transfers are represented by the following permutation on data in S_0.

$$\begin{pmatrix} i_2 & j_2 \\ j_2 & i_2 \end{pmatrix} \circ \begin{pmatrix} i_1 & j_1 \\ j_1 & i_1 \end{pmatrix}.$$

This involves four simultaneous register-to-register data transfers as follows: $S_0[i_1] \rightarrow S_0[j_1]$, $S_0[j_1] \rightarrow S_0[i_1]$, $S_0[i_2] \rightarrow S_0[j_2]$ and $S_0[j_2] \rightarrow S_0[i_2]$.

Case 2: $i_2 \neq j_1$ and $j_2 \neq i_1$ and $j_2 = j_1$

In this case the data transfers are represented by

$$\begin{pmatrix} i_2 & j_1 \\ j_1 & i_2 \end{pmatrix} \circ \begin{pmatrix} i_1 & j_1 \\ j_1 & i_1 \end{pmatrix}.$$

This involves three data transfers: $S_0[i_1] \rightarrow S_0[i_2]$, $S_0[i_2] \rightarrow S_0[j_1] = S_0[j_2]$ and $S_0[j_1] \rightarrow S_0[i_1]$.

Case 3: $i_2 \neq j_1$ and $j_2 = i_1$ and $j_2 \neq j_1$

The data transfers are represented by

$$\begin{pmatrix} i_2 & i_1 \\ i_1 & i_2 \end{pmatrix} \circ \begin{pmatrix} i_1 & j_1 \\ j_1 & i_1 \end{pmatrix}.$$

This involves three data transfers: $S_0[i_1] \rightarrow S_0[j_1]$, $S_0[i_2] \rightarrow S_0[i_1] = S_0[j_2]$ and $S_0[j_1] \rightarrow S_0[i_2]$.

Case 4: $i_2 \neq j_1$ and $j_2 = i_1$ and $j_2 = j_1$

The data transfers are represented by

$$\begin{pmatrix} i_2 & i_1 \\ i_1 & i_2 \end{pmatrix} \circ \begin{pmatrix} i_1 & i_1 \\ i_1 & i_1 \end{pmatrix}.$$

This involves two data transfers: $S_0[i_1] \rightarrow S_0[i_2]$ and $S_0[i_2] \rightarrow S_0[i_1] = S_0[j_1] = S_0[j_2]$.

Case 5: $i_2 = j_1$ and $j_2 \neq i_1$ and $j_2 \neq j_1$

The data transfers are represented by

$$\begin{pmatrix} j_1 & j_2 \\ j_2 & j_1 \end{pmatrix} \circ \begin{pmatrix} i_1 & j_1 \\ j_1 & i_1 \end{pmatrix}.$$

This involves three data transfers: $S_0[i_1] \rightarrow S_0[j_2]$, $S_0[j_2] \rightarrow S_0[j_1] = S_0[i_2]$ and $S_0[j_1] \rightarrow S_0[i_1]$.

Case 6: $i_2 = j_1$ and $j_2 \neq i_1$ and $j_2 = j_1$

The data transfers are represented by

$$\begin{pmatrix} j_1 & j_1 \\ j_1 & j_1 \end{pmatrix} \circ \begin{pmatrix} i_1 & j_1 \\ j_1 & i_1 \end{pmatrix}.$$

This involves two data transfers: $S_0[i_1] \rightarrow S_0[j_1] = S_0[i_2] = S_0[j_2]$ and $S_0[j_1] \rightarrow S_0[i_1]$.

Case 7: $i_2 = j_1$ and $j_2 = i_1$ and $j_2 \neq j_1$

The data transfers are represented by

$$\begin{pmatrix} j_1 & i_1 \\ i_1 & j_1 \end{pmatrix} \circ \begin{pmatrix} i_1 & j_1 \\ j_1 & i_1 \end{pmatrix}.$$

This is the identity permutation and does not involve data transfer.

Case 8: $i_2 = j_1$ and $j_2 = i_1$ and $j_2 = j_1$

This case cannot occur as it implies $i_1 = i_2$, which is impossible because $i_2 = i_0 + 2 = i_1 + 1$.

After the swap operation is completed successfully, we obtain S_2 from S_0. From the view of the receiving registers (in the S-register bank) in case of register-to-register transfers, we can summarize the cases as follows.

- $S_2[i_1]$ receives data from $S_0[i_1], S_0[j_1]$ or $S_0[i_2]$.
- $S_2[j_1]$ receives from $S_0[i_1], S_0[j_1], S_0[i_2]$ or $S_0[j_2]$.
- $S_2[i_2]$ receives from $S_0[i_1], S_0[j_1], S_0[i_2]$ or $S_0[j_2]$.
- $S_2[j_2]$ receives from $S_0[i_1], S_0[i_2]$ or $S_0[j_2]$.

The input data (1 byte) for each of the 256 registers in the S-register bank will be taken from the output of an 8-to-1 MUX unit, whose data inputs are taken from $S_0[i_1], S_0[j_1], S_0[i_2], S_0[j_2]$, and the control inputs are taken from the outputs of three comparators comparing i) i_2 and j_1, ii) j_2 and i_1, iii) j_2 and j_1. The circuit to realize the swap appears in Figure 9.3.

For the simultaneous register-to-register data transfer during the swap operation, we propose the use of master-slave JK flip-flops to construct the registers in the S-register bank. The read and write operations will respect the required order of functioning and the synchronization can be performed at the end of each clock cycle to update the S-state.

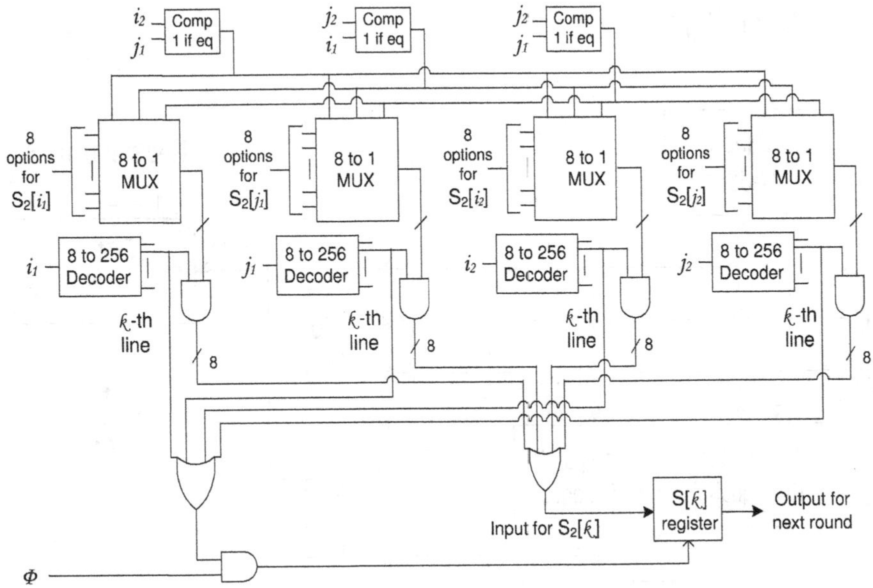

FIGURE 9.3: Circuit to swap S values (data lines shown only for fixed k)

Step 4: Calculation of Z_1 and Z_2. The main idea to get the most from loop unrolling in RC4 is to bypass the generation of S_1, and move directly from S_0 to S_2, as discussed. However, note that we require the state S_1 for computing the output byte $Z_1 = S_1[S_0[j_1] + S_0[i_1]]$. We apply the following method of cross-loop look-back to resolve this issue.

Using Step 4 of Algorithm 2, we can rewrite the output $Z_1 = S_1[S_1[i_1] + S_1[j_1]] = S_1[S_0[j_1] + S_0[i_1]]$ as

$$Z_1 = \begin{cases} S_2[i_2] & \text{if } S_0[j_1] + S_0[i_1] = j_2, \\ S_2[j_2] & \text{if } S_0[j_1] + S_0[i_1] = i_2, \\ S_2[S_0[j_1] + S_0[i_1]] & \text{otherwise.} \end{cases}$$

Computing Z_1 involves adding $S_0[i_1]$ and $S_0[j_1]$ first, which can be done using a two-input parallel adder. The 256-to-1 MUX used to extract appropriate data from S_2 will be controlled by another 4-to-1 MUX. This MUX is in turn controlled by the

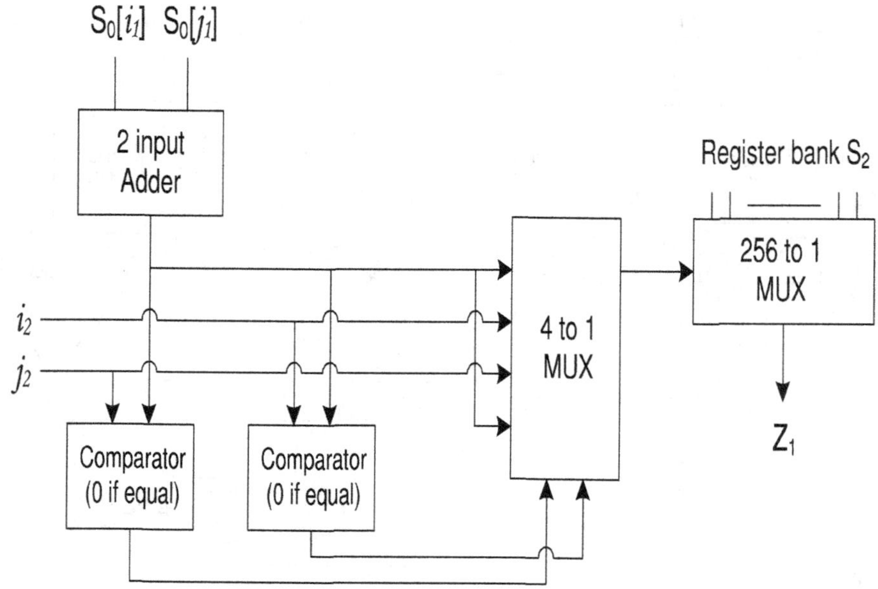

FIGURE 9.4: Circuit to compute Z_1

outputs of two comparators comparing i) $S_0[j_1]+S_0[i_1]$ and i_2, and ii) $S_0[j_1]+S_0[i_1]$ and j_2, as illustrated in Figure 9.4.

Computation of Z_2, however, involves adding $S_1[i_2], S_1[j_2]$, as in the following formula:

$$Z_2 = S_2\left[S_2[i_2] + S_2[j_2]\right] = S_2\left[S_1[j_2] + S_1[i_2]\right].$$

In this case, we unwrap one cycle of RC4 and gather the values of $S_1[i_2]$ and $S_1[j_2]$ from the S_0 state. $S_1[i_2]$ and $S_1[j_2]$ receive the values from the appropriate registers of S_0 as given below, depending on the conditions:

- $i_2 \neq j_1, j_2 \neq i_1, j_2 \neq j_1$: $S_1[i_2] = S_0[i_2]$, $S_1[j_2] = S_0[j_2]$
- $i_2 \neq j_1, j_2 \neq i_1, j_2 = j_1$: $S_1[i_2] = S_0[i_2]$, $S_1[j_2] = S_0[i_1]$
- $i_2 \neq j_1, j_2 = i_1, j_2 \neq j_1$: $S_1[i_2] = S_0[i_2]$, $S_1[j_2] = S_0[j_1]$
- $i_2 \neq j_1, j_2 = i_1, j_2 = j_1$: $S_1[i_2] = S_0[i_2]$, $S_1[j_2] = S_0[j_1]$

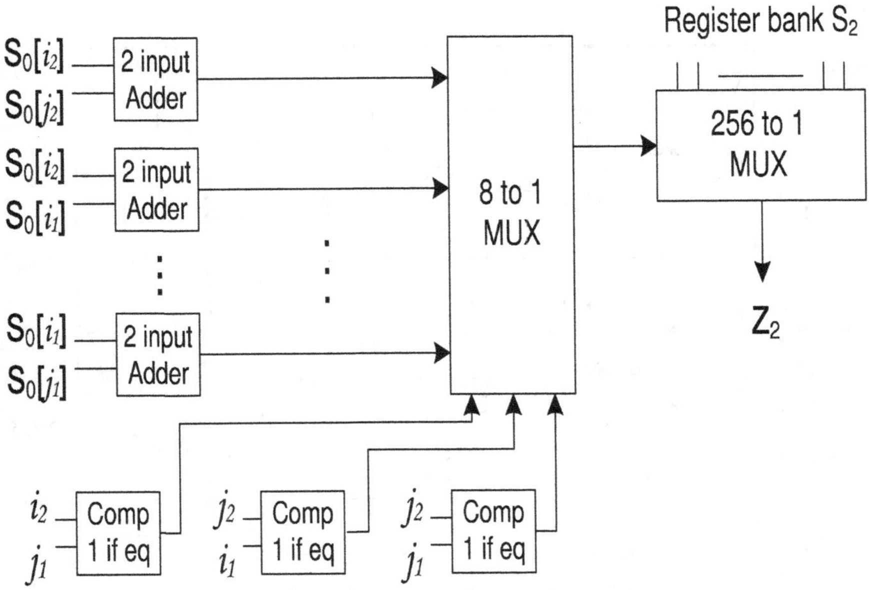

FIGURE 9.5: Circuit to compute Z_2

- $i_2 = j_1, j_2 \neq i_1, j_2 \neq j_1$: $S_1[i_2] = S_0[i_1]$, $S_1[j_2] = S_0[j_2]$
- $i_2 = j_1, j_2 \neq i_1, j_2 = j_1$: $S_1[i_2] = S_0[i_1]$, $S_1[j_2] = S_0[i_1]$
- $i_2 = j_1, j_2 = i_1, j_2 \neq j_1$: $S_1[i_2] = S_0[i_1]$, $S_1[j_2] = S_0[j_1]$

These conditions can be realized using an 8-to-1 MUX unit controlled by the outputs of three comparators comparing i) i_2 and j_1, ii) j_2 and i_1, iii) j_2 and j_1. We can use the same control lines as in case of the swapping operation. The circuit is shown in Figure 9.5.

9.3.2 Complete Circuit

The complete PRGA circuit is shown in Figure 9.6 in which the five blocks Circuit 1, Circuit 2, Circuit 3, Circuit 4 and Circuit 5 refer to the circuits shown in the Figures 9.1, 9.2, 9.3, 9.4 and 9.5, respectively. We denote the clock by ϕ and its cycles as ϕ_1, ϕ_2, etc., where ϕ_0 refers to the clock pulse that initiates PRGA.

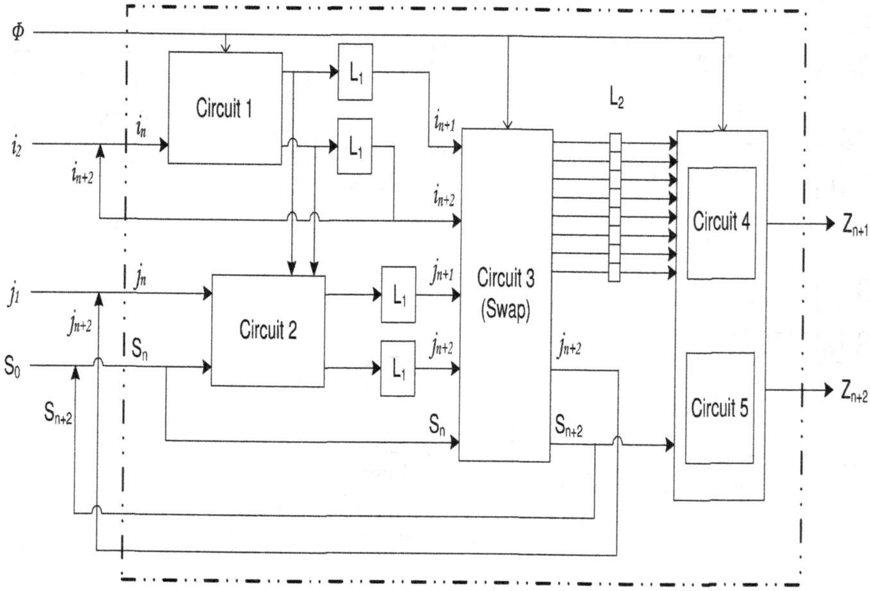

FIGURE 9.6: Circuit for PRGA stage of RC4 architecture

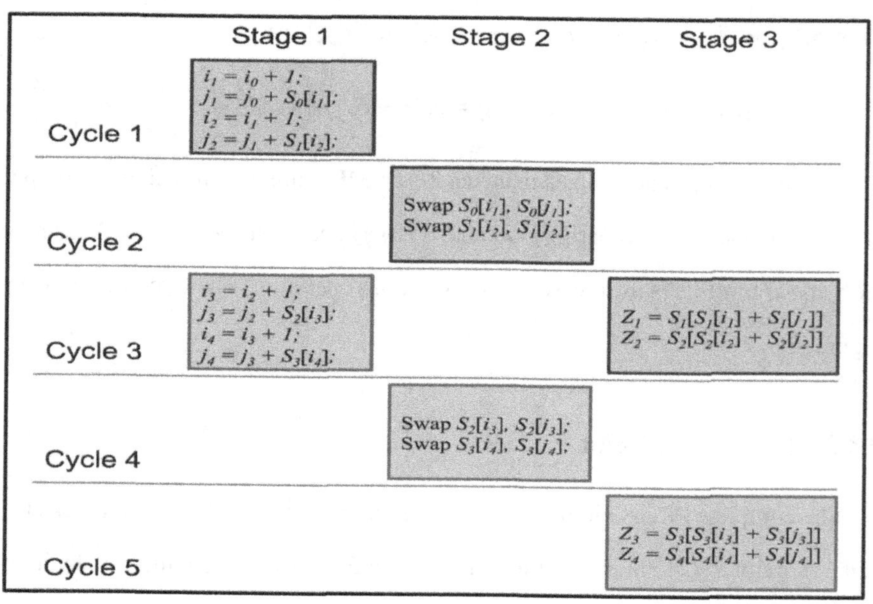

FIGURE 9.7: Pipeline structure for hardware design

In Figure 9.6, L_i denote the latches operated by the trailing edge of ϕ_{2n+i}, i.e., the $(2n + i)^{th}$ cycle of the master clock ϕ where $n \geq 0$. For example, the latches labeled L_1 (four of them) are released at the trailing edge of $\phi_1, \phi_3, \phi_5, \ldots$ and the latches labeled L_2 (eight of them) are released at the trailing edge of $\phi_2, \phi_4, \phi_6, \ldots$ etc. In the final implementation, these latches have been replaced by edge-triggered flip-flops which operate at the trailing edge of the clock.

The timing analysis for the circuit in Figure 9.6 is shown in the three-stage pipeline diagram of Figure 9.7. We illustrate the first two iterations, and the remainder fall along similar lines. The combinational logics operate between the clock pulses and all read, swap and increment operations are done at the trailing edges of the clock pulses. The first two bytes Z_1, Z_2 are obtained at the end of the third clock cycle and the next two bytes Z_3, Z_4 are obtained at the fifth clock cycle. We now generalize this observation.

Efficiency of the design for PRGA:

> *The hardware for the PRGA stage of RC4 as shown in Figure 9.6, produces one byte per clock after an initial delay of two clock cycles.*

Let us call the stage of the PRGA circuit shown in Figure 9.6 the nth stage. This denotes the nth iteration of the above model, which produces the output bytes Z_{n+1} and Z_{n+2}. The first block (Circuit 1) operates at the trailing edge of ϕ_n and increments i_n to i_{n+1}, i_{n+2}. During cycle ϕ_{n+1}, the combinational part of Circuit 2 operates to produce j_{n+1}, j_{n+2}. The trailing edge of ϕ_{n+1} releases the latches of type L_1, and activates the swap circuit (Circuit 3). The combinational logic of the swap circuit functions during cycle ϕ_{n+2} and the actual swap operation takes place at the trailing edge of ϕ_{n+2} to produce S_{n+2} from S_n. Simultaneously, the latch of type L_2 is released to activate Circuits 4 and 5. The combinational logic of these two circuits operates during ϕ_{n+3}, and we get Z_{n+1} and Z_{n+2} at the trailing edge of ϕ_{n+3}.

This complete block of architecture performs in a cascaded pipeline fashion, as the indices i_2, j_2 and the state S_{n+2} are fed back into the system at the end of ϕ_{n+2} (actually, i_{n+2} is fed back at the end of ϕ_{n+1} to allow for the increments at the trailing edge of ϕ_{n+2}). The operational gap between two iterations (e.g., n^{th} and $(n+2)^{th}$) of the system is thus two clock cycles (e.g., ϕ_n to ϕ_{n+2}), and we obtain two output bytes per iteration.

Hence, the circuit for PRGA shown in Figure 9.6, produces $2N$ bytes of output stream in N iterations, over $2N$ clock cycles. Note that the initial clock pulse ϕ_0 is an extra one, and the production of the output bytes lags the feedback cycle by one clock pulse in every iteration (e.g., ϕ_{n+3} in case of nth iteration). Therefore, this design produces $2N$ output bytes in $2N$ clock cycles or one byte per clock, after an initial lag of two clock cycles.

Issues with KSA. Note that the general KSA routine runs for 256 iterations to produce the initial permutation of the S-box. Moreover, the steps of KSA are similar to the steps of PRGA except that:

- Calculation of j involves key K along with S and i.
- Computing Z_1, Z_2 is not required or advised.

We use the *loop-unrolled* PRGA architecture (Figure 9.6) for the KSA with some minor modifications, as follows:

1. *K-register bank:* Introduce a new register bank for key K. It will contain l 8-bit registers, where $8 \leq l \leq 15$ in practice.
2. *K-register MUX:* To read key values $K[i_1 \bmod l]$ and $K[i_2 \bmod l]$ from the K-registers, we introduce two 16-to-1 multiplexer units. The first l input lines of this MUX will be fed data from registers $K[0]$ to $K[l-1]$, and the remaining $(16-l)$ inputs can be left floating (recall that $8 \leq l \leq 15$). The control lines

of these MUX units will be $i_1 \bmod l$ and $i_2 \bmod l$, respectively, and hence the floating inputs will never be selected.

3. *Modular Counters:* To obtain modular indices $i_1 \bmod l$ and $i_2 \bmod l$, we incorporate two modular counters (modulo l) for the indices. These are synchronous counters and the one for i_2 will have no clock input for the LSB position, similar to Figure 9.1.

4. *Extra 2-input Parallel Adders:* Two 2-input parallel adders are appended to Figure 9.2 for adding $K[i_1 \bmod l]$ and $K[i_2 \bmod l]$ to j_1 and j_2, respectively.

5. *No Outputs:* Circuits of Figure 9.4 and Figure 9.5 are removed from the structure so that no output byte is generated during KSA. If any such byte is generated, the key K may be compromised.

Using this modified hardware configuration, one can implement two rounds of KSA in two clock cycles or one round per clock after an initial lag of one cycle. Total time required for KSA is $256 + 1 = 257$ clock cycles.

9.3.3 Practical Implementation

The above design was implemented by Sen Gupta et al. [1] using synthesizable VHDL. The S-register box and K-register box were implemented as arrays of master-slave flip-flops. The VHDL code consists of approximately 1500 lines.

A major area impact of the circuit originates from the large number of accesses to the S- and K-boxes from the KSA and PRGA circuits. Since the PRGA and KSA will not run in parallel, the read and write ports of S- and K-boxes were shared by PRGA and KSA. From KSA, one read access to K-box, two read accesses to S-box and two write accesses to S-box are needed. From PRGA, six read accesses to S-box and four write accesses to S-box are needed. The two read accesses correspond to simultaneous generation of two Z values at the last step of PRGA. The four read and four write accesses correspond to the *double swap* operation. While sharing the mutually exclusive accesses, all the accesses from KSA can be merged amongst the PRGA accesses. Therefore, the total number of read ports to K-box is one, the total number of read ports to S-box is six and the total number of write ports to S-box is four. This sharing of storage access is shown in Figure 9.8.

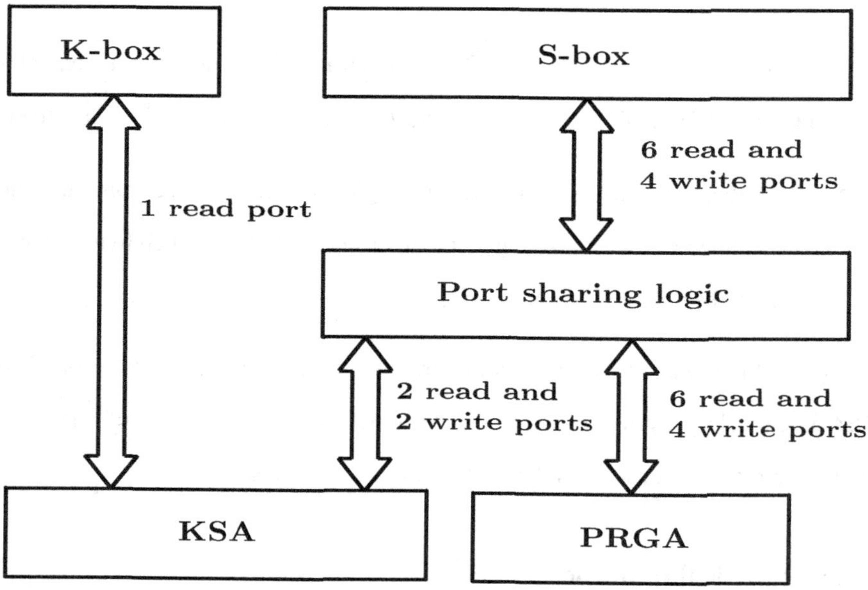

FIGURE 9.8: Access sharing of KSA and PRGA in Design 1

Sen Gupta et al. reported the detailed implementation of this design approach (Design 1 in [1]), based on the VHDL code synthesized with 90 nm and 65 nm fabrication technologies using Synopsys Design Compiler in topographical mode.

9.3.4 Comparison with Other Hardware Designs for RC4

Performance comparisons of the above design and other hardware implementations for RC4 in the literature are described below.

TABLE 9.3: Pipeline stages for design proposed in [18]

1st stage	$i_1 = i_0 + 1$	$i_2 = i_1 + 1$	$i_3 = i_2 + 1$			
2nd stage		Read $S[i_1]$	Read $S[i_2]$	Read $S[i_3]$		
		Store $S[i_1]$ into SI	Store $S[i_2]$ into SI	Store $S[i_3]$ into SI		
		$j_1 = j_0 + S[i_1]$	$j_2 = j_1 + S[i_2]$	$j_3 = j_2 + S[i_3]$		
3rd stage			Read $S[j_1]$	Read $S[j_2]$	Read $S[j_3]$	
			Store $S[j_1]$ in SJ	Store $S[j_2]$ in SJ	Store $S[j_3]$ in SJ	
			$t_1 = SI + S[j_1]$	$t_2 = SI + S[j_2]$	$t_3 = SI + S[j_3]$	
			Write SI into $S[j_1]$	Write SI into $S[j_2]$	Write SI into $S[j_3]$	
4th stage				Read $S[t_1]$	Read $S[t_2]$	Read $S[t_3]$
				Store $S[t_1]$ into K	Store $S[t_2]$ into K	Store $S[t_3]$ into K
				Write SJ into $S[i_1]$	Write SJ into $S[i_2]$	Write SJ into $S[i_3]$

Kitsos et al. [11] and Matthews [17]. Combining the KSA and PRGA architectures in the above design approach, we can obtain $2N$ output-stream bytes in $2N + 259$ clock cycles, counting the initial delay of 1 cycle for KSA and 2 cycles for PRGA. The hardware implementation of RC4 described in [11] and [17] provides an output of N bytes in $3N + 768$ clock cycles. A formal comparison of the timings is shown in Table 9.4. We can easily observe that for large N, the throughput of the design mentioned above [1] is three times that of the designs proposed in [11] and [17].

An exact comparison of area with [11] is not possible since we do not have access to the FPGA board for which the area of [11] is reported. Considering the design idea, both [11] and [17] modeled their storage using block RAMs. This implementation restricts the number of port accesses per cycle. To overcome that, three 256-byte dual-port RAM blocks are used in [11]. Even then, the design requires three cycles to produce 1 byte of data. An improved design is reported in [17] where only two 256-byte dual-port RAM blocks are used. Sen Gupta et al. [1] utilized register-based storage for the S and K arrays instead of RAM because RAM-based storage would incorporate port access restrictions and latency issues, resulting in a compromise of throughput. An alternative technique to maintain the high throughput with RAM based implementation may be partitioning the arrays according to the accesses and optimizing them accordingly.

Matthews [18]. This design proposes a 1-byte-per-cycle design of RC4 hardware using a technique different from that used by Sen Gupta et al. [1]. The claimed throughput is achieved in [18] by a hardware pipeline; instead of two iterations per cycle, one iteration per cycle in PRGA is performed. The pipeline design is as shown in Table 9.3.4 (as in [18, Table 1]). This design provides 1-cycle-per-byte output in PRGA and completes KSA in 256 cycles, with an initial lag of three cycles due to the four-stage pipeline structure (as in Table 9.3.4). Thus, N bytes of output are produced in approximately $N + 259$ clock cycles, which is comparable to the performance of the design by Sen Gupta et al. [1]. Detailed comparative results are presented in Table 9.4; Design 1 refers to the design approach described in Section 9.3.

TABLE 9.4: Timing comparison of Design 1 with those of [11], [17] and [18].

Operations	Number of clock cycles required for each operation		
	[11] and [17]	[18]	Design 1
Per KSA round	3	1	1
Complete KSA	$256 \times 3 = 768$	$256 + 3 = 259$	$256 + 1 = 257$
N bytes of PRGA	$3N$	$N + 3$	$N + 2$
N bytes of RC4	$3N + 768$	$259 + (N + 3) = N + 262$	$257 + (N + 2) = N + 259$
Cycles per byte	$3 + \frac{768}{N}$	$1 + \frac{262}{N}$	$1 + \frac{259}{N}$

The obvious advantage of the design proposed in [18] is its compactness. However, the documentation does not report area or timing figures corresponding to any technology node. Instead, a figurative summary of the logic structure is reported. Also in [18], neither the hardware implementation of KSA nor the port access sharing between PRGA and KSA is reported.

9.3.5 Yet Another Design Approach for RC4

This section presents a pipelined design approach for RC4 generation as proposed by Sen Gupta et al. [1].

Pipelined Design A: This design approach is based on an architecture pipelined in two stages, as in Figure 9.9. While [18] proposed a deep pipelining with bypass and data-forwarding, this scheme allows a simpler two-stage pipelining with the same throughput of 1 byte per cycle and similar memory port restrictions. The first pipeline stage is devoted to calculation of j and performing the swap and the second pipeline stage computes the value of Z. To minimize the read and write accesses to S-box, the index to be used for second pipeline is computed at the first stage. Note that the index computation at first stage does not alter the result as $S[i]$ and $S[j]$ are swapped; thus the addition result $S[i] + S[j]$ remains intact.

With the aforementioned structure, the pipeline in the PRGA circuit is consid-

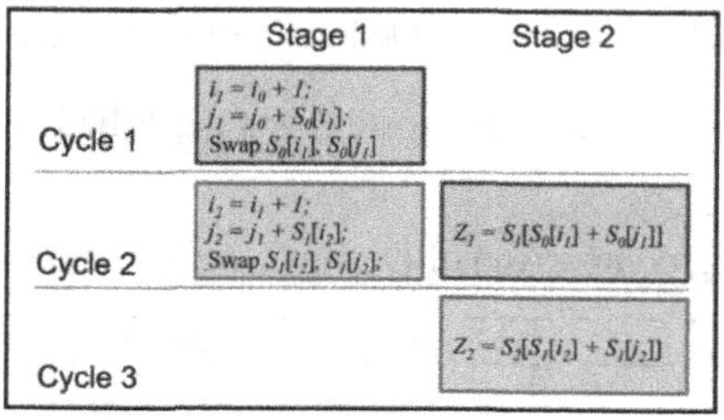

FIGURE 9.9: One byte/cycle by hardware pipelining

erably simplified compared to the design in Section 9.3. To improve the area and timing, the KSA circuit is reorganized to merge two iterations of key generation in one cycle. The benefit is to have the S-box created in 128 instead of 256 cycles.

TABLE 9.5: Two consecutive loops of RC4 key scheduling

First Loop	Second Loop
$i_1 = i_0 + 1$	$i_2 = i_1 + 1 = i_0 + 2$
$j_1 = j_0 + S_0[i_1] + K[i_1]$	$j_2 = j_1 + S_1[i_2] + K[i_2]$ $= j_0 + S_0[i_1] + S_1[i_2] + K[i_2]$
Swap $S_0[i_1] \leftrightarrow S_0[j_1]$	Swap $S_1[i_2] \leftrightarrow S_1[j_2]$

9.3.5.1 Area Optimization in Pipelined Design A

The idea of loop unrolling was also fused to merge two consecutive rounds of KSA. As a result, the number of read accesses to K-box from KSA grew to two. The numbers of read and write accesses from KSA to S-box are both four due to the double swap in one cycle. As discussed in [1], the following area optimization approaches can further be adopted.

Pipelined Design B: Modifications are performed on Design A to implement access sharing (read and write) for the S-box between KSA and PRGA. In Design A, KSA contains four read, four write accesses and PRGA contains three read and two write accesses to the S-box. Naturally, all the accesses from PRGA can be shared with accesses from KSA, resulting in a total of four read and four write accesses in Pipelined Design B. The synthesis indicated a compact circuit with the same throughput.

Pipelined Design C: Another idea for reducing circuit size further is to perform only one iteration of KSA per cycle. In this approach, KSA will require 256 cycles to initialize the S-box and the read and write accesses to S-box will become two per cycle. By applying access sharing on top of that, the total read and write accesses to S-box are reduced to three and two, respectively. Furthermore, the number of read accesses to K-box also drops to one. The synthesis of this circuit also indicated a more compact design.

A sharp reduction in S-box and KSA areas for Pipelined Design C in comparison with the previous designs was observed. The reduction in the area for S-box is most prominent as Pipelined Design C directly reduces the area requirements for the address decoders and multiplexers, due to fewer S-box access ports. However, the reduction in K-box access ports costs 256 cycles for KSA instead of 128 as in the previous two designs. A comparison of the area requirements for different circuit blocks in all these designs reported by Sen Gupta et al. [1], is presented in Table 9.6; Design 1 refers to the approach described in Section 9.3.

TABLE 9.6: Synthesis results for optimized designs with different schemes

Tech. (nm)	Design	Area (NAND gate equivalent)							Max. Clock Freq. (GHz)	KSA (cycles)	PRGA (bytes/cycle)	Speed (Gbps)
		KSA	PRGA	*S*-box	*K*-box	Sequential	Combinational	Total				
90	Design 1	310	1199	48669	1084	10942	40320	51262	1.22	256	1	9.76
	Pipelined Design A	560	428	52133	1084	10644	43561	54205	1.37	128	1	10.96
	Pipelined Design B	527	428	43454	1084	10611	34882	45493	1.25	128	1	10.00
	Pipelined Design C	484	612	39231	1084	10801	30610	41411	1.49	256	1	11.92
65	Design 1	310	1240	53638	1180	14109	42259	56368	1.85	256	1	14.80
	Pipelined C	550	690	48159	1180	13622	36957	50579	2.22	256	1	17.76

9.4 Exercises

1. List possible types of attacks on a wireless network.
2. Distinguish between passive and active attacks on wireless networks.
3. What is a denial-of-service attack and in how many ways may this be experienced by a wireless network?
4. What is a jamming attack?
5. Distinguish between deceptive jamming, random jamming and reactive jamming. How do these jamming attacks differ from constant jamming?
6. What is a man-in-the-middle attack? Is it different from pure eavesdropping? If yes, how?
7. Mention two ways of devising authentication algorithms.
8. In what ways does message authentication code differ from a digital signature for authentication purposes?
9. Does the data encryption standard use asymmetric key encryption?
10. What is the basic difference between symmetric key encryption and asymmetric key encryption? Which is better so far as secure data transfer over a wireless communication network is concerned?
11. What are the differences between a block cipher algorithm and a stream cipher algorithm? Which one is better suited for wireless communication networks?
12. Describe the RC4 stream cipher algorithm.

13. Describe briefly how the RC4 algorithm can be implemented in hardware to generate one byte of cipher data in every clock cycle.

14. Identify the weaknesses, if any, of the hardware implementation of the RC4 algorithm described in the text.

*15. Devise a suitable *light-weight* encryption algorithm appropriate for hardware implementation to be used in low cost low power wireless communication networks (e.g., MANETs or WSNs), in which the lifetime of an encryption key can be assumed to be short enough for providing the desired security and the encryption and decryption times and hardware cost requirement will also be very small.

Bibliography

[1] S. Sen Gupta, A. Chattopadhyay, K. Sinha, S. Maitra and B. P. Sinha, "High performance hardware implementation for RC4 stream cipher," *IEEE Trans. Computers*, vol. 62, No. 4, pp. 730–743, 2013.

[2] Software performance results from the eSTREAM Project. ECRYPT Stream Cipher Project. `http://www.ecrypt.eu.org/stream/perf/#results`

[3] The current eSTREAM portfolio. ECRYPT Stream Cipher Project. (`http://www.ecrypt.eu.org/stream/index.html`)

[4] S. R. Fluhrer and D. A. McGrew, "Statistical analysis of the alleged RC4 keystream generator," *LNCS*, vol. 1978, pp. 19–30, 2000.

[5] S. R. Fluhrer, I. Mantin and A. Shamir, "Weaknesses in the key scheduling algorithm of RC4," *Selected Areas in Cryptography 2001, LNCS*, vol. 2259, pp. 1–24, 2001.

[6] M. D. Galanis, P. Kitsos, G. Kostopoulos, N. Sklavos and C. E. Goutis, "Comparison of the hardware implementation of stream ciphers," *Int. Arab J. Inf. Tech.*, vol. 2, No. 4, pp. 267–274, 2005.

[7] J. Golic, "Linear statistical weakness of alleged RC4 keystream generator," *EUROCRYPT 1997, LNCS*, vol. 1233, pp. 226–238, 1997.

[8] T. Good and M. Benaissa, "Hardware results for selected stream cipher candidates," *ECRYPT Stream Cipher Project, SASC*, Report 2007/023, 2007.

[9] F. K. Gurkaynak, P. Luethi, N. Bernold, R. Blattmann, V. Goode, M. Marghitola, H. Kaeslin, N. Felber and W. Fichtner, "Hardware evaluation of eSTREAM candidates: Achterbahn, Grain, MICKEY, MOSQUITO, SFINKS, Trivium, VEST, ZK-Crypt," *ECRYPT Stream Cipher Project*, Report 2006/015, 2006.

[10] P. Hamalainen, M. Hannikainen, T. Hamalainen and J. Saarinen, "Hardware implementation of the improved WEP and RC4 encryption algorithms for wireless terminals," *Proc. Eur. Signal Processing Conf.*, pp. 2289–2292, 2000.

[11] P. Kitsos, G. Kostopoulos, N. Sklavos and O. Koufopavlou, "Hardware implementation of the RC4 stream cipher," *Proc. 46th IEEE Midwest Symposium on Circuits & Systems '03*, Cairo, 2003. (`http://dsmc.eap.gr/en/members/pkitsos/papers/Kitsos_c14.pdf`)

[12] J.-D. Lee and C.-P. Fan, "Efficient low-latency RC4 architecture designs for IEEE 802.11i WEP/TKIP," *Proc. Int. Symp. on Intelligent Signal Processing and Communication Systems (ISPACS) '07*, pp. 56–59, 2007.

[13] T. Lynch and E. E. Swartzlander Jr., "A spanning tree carry Lookahead adder," *IEEE Trans. Computers*, vol. 41, No. 8, pp. 931–939, 1992.

[14] I. Mantin, "Predicting and distinguishing attacks on RC4 keystream generator," *EUROCRYPT 2005, LNCS*, vol. 3494, pp. 491–506, 2005.

[15] I. Mantin, "A Practical attack on the fixed RC4 in the WEP mode," *ASIACRYPT 2005, LNCS*, vol. 3788, pp. 395–411, 2005.

[16] I. Mantin and A. Shamir, "A practical attack on broadcast RC4," *FSE 2001, LNCS*, vol. 2355, pp. 152–164, 2001.

[17] D. P. Matthews Jr., "System and method for a fast hardware implementation of RC4," US Patent 6549622, Campbell, CA, April 2003. (`http://www.freepatentsonline.com/6549622.html`)

[18] D. P. Matthews Jr., "Methods and apparatus for accelerating ARC4 processing," US Patent 7403615, Morgan Hill, CA, July 2008. (`http://www.freepatentsonline.com/7403615.html`)

[19] A. Maximov and D. Khovratovich, "New state recovering attack on RC4," *CRYPTO 2008, LNCS*, vol. 5157, pp. 297–316, 2008.

[20] I. Mironov, "(Not so) random shuffles of RC4," *CRYPTO 2002, LNCS*, vol. 2442, pp. 304–319, 2002.

[21] G. Paul and S. Maitra, "On biases of permutation and keystream bytes of RC4 towards the secret key," *Cryptography and Communications*, vol. 1, No. 2, pp. 225–268, 2009.

[22] A. Roos, "A class of weak keys in the RC4 stream cipher," Two posts in sci.crypt, 1995. (`http://marcel.wanda.ch/Archive/WeakKeys`)

[23] S. Sen Gupta, K. Sinha, S. Maitra and B. P. Sinha, "One byte per clock: A novel RC4 hardware," *Indocrypt 2010, LNCS*, vol. 6498, pp. 347–363, 2010.

[24] B. P. Sinha and P. K. Srimani, "Fast parallel algorithms for binary multiplication and their implementation on systolic architectures," *IEEE Trans. Computers*, vol. 38, No. 3, pp. 424–431, 1989.

[25] D. Wagner, "My RC4 weak keys," Post in sci.crypt, 1995. (`http://www.cs.berkeley.edu/~daw/my-posts/my-rc4-weak-keys`)

Index